GAUGE FIELD THEORIES: SPIN ONE AND SPIN TWO
100 Years after General Relativity

Günter Scharf
Physics Institute, University of Zürich

DOVER PUBLICATIONS, INC.
Mineola, New York

Copyright

Copyright © 2016 by Günter Scharf
All rights reserved.

Bibliographical Note

Gauge Field Theories: Spin One and Spin Two: 100 Years after General Relativity, first published by Dover Publications, Inc., in 2016, is a revised edition of *Quantum Gauge Theories: A True Ghost Story*, originally published in 2001 by John Wiley & Sons, Inc., New York.

International Standard Book Number
ISBN-13: 978-0-486-80524-5
ISBN-10: 0-486-80524-7

Manufactured in the United States by RR Donnelley
80524701   2016
www.doverpublications.com

Preface to the Dover Edition

In 1916, Albert Einstein published his general theory of relativity in the "Annalen der Physik 49." A hundred years later, this milestone event in the history of science will rightly be celebrated. However, the celebrations may be disturbed by physicists deep under the mountains of southern China who are trying to detect dark matter particles.[1] This hypothetical matter is necessary for the analysis of rotation curves in galaxies on the basis of standard general relativity. The Milky Way, too, must be full of dark matter. So far, however, the hunt for these new undiscovered particles has met with little success. If dark matter cannot be found, then the standard theory of general relativity is in serious trouble and the 100th anniversary celebrations are ruined. One should frankly admit that general relativity is not established on the scale of galaxies.

Standard general relativity is a geometric theory. Einstein postulated a fusion between geometry and gravity. In his theory, the metric tensor $g_{\mu\nu}$ describes both the gravitational field and the geometry of space and time. Here is the point where we deviate. Following H. Poincaré (in *Science and Hypothesis*) we consider geometry as a convention and the $g_{\mu\nu}(x)$ on the other hand describes the gravitational field only. The simplest geometry is the Minkowski space. So all the fields in this book will be defined on Minkowski space, $g_{\mu\nu}(x)$, as well. Otherwise, Einstein's equations remain unchanged. But now more solutions are physically possible, and there exist solutions with general form of the corresponding rotation curves. Consequently, in non-geometric general relativity no dark matter is needed for galactic dynamics and the same is true for cosmology. In this way the 2016 Einstein celebrations can be saved.

This is the third edition of my book *Quantum Gauge Theories: A True Ghost Story,* which was published in 2001 by John Wiley & Sons, Inc. The title of the second edition was changed to *Quantum Gauge Theories— Spin 1 and 2.* This already indicates the increasing importance of gravity (spin 2) in the framework of gauge theories. In this third edition, a further small variation of the title was necessary. By choosing the title *Gauge Field Theories* instead of quantum gauge theories, we take into consideration that gravity mainly acts as a classical field (although we shall derive it as a quantum gauge field). So the last chapter "Non-geometric

---

[1] Another dark matter search is being performed by the XENON Collaboration under the Gran Sasso mountain in Italy.

General Relativity" is 100% classical field theory. In fact, astrophysicists may read only this chapter to study the nonstandard approach to gravity and cosmology.

One may ask why I did not write one book for spin 1 concerning the microcosmos and a second one for spin 2 which is the macrocosmos. Maybe this has to be done in the future; at present it is my main aim to treat the electroweak and strong interactions and gravity on the same footing. In fact, it had quite often been said that the unification of gravity with quantum theory is the main open problem in theoretical physics. Our basic principle to achieve this unification is the appropriate formulation of quantum gauge invariance. In the case of the massless spin 2 theory, this gauge-theoretical foundation leads directly to the non-geometric interpretation of general relativity. So besides the dark matter problem, there is a strong theoretical reason to favor this approach above the standard geometrical theory. To see the microcosmos (i.e., particle physics) and the Universe in the large be governed by the same basic principle gives great intellectual satisfaction—or even more.

Zürich, April 2015

Günter Scharf

Preface (2011)

It has quite often been said that the unification of gravity with quantum theory is the main open problem in theoretical physics. The subtitle "spin one and two" means that the book is concerned with this problem, because spin-1 gauge theories include the successful standard theory of electroweak and strong interactions and spin-2 is gravitation. In fact we shall see that the notion of quantum gauge theory gives the natural framework for both spin-1 (non-abelian) gauge theories and gravity.

Quantum gauge theories are mostly treated with the functional method. For writing down the basic functional integral a classical Lagrangian must be given. This is a disadvantage because the correct choice of the classical Lagrangian (including ghost, Higgs and auxiliary fields) requires great skill. The resulting final theory has some artificial ad-hoc character so that it lacks strong evidence of being truly fundamental. Indeed, it is a common belief today that quantum gauge theory might only be the low energy limit of a more fundamental theory like string theory for example. A strong support for this belief comes from the necessity to include gravity.

In this monograph we use an alternative method which does not require any classical Lagrangian. This gives us the chance to discover new physics, and we shall see this in Chapter 5 when we consider massive gravity. In the spin-1 case we shall recover the results of the standard theory in all details. This now gives us stronger evidence that quantum gauge theories may be really fundamental. In the spin-2 case with mass zero we recover the coupling given by general relativity, of course, but we also find a very interesting modification of it when we consider the massless limit of massive gravity.

The reader now certainly asks: What is the basis of this alternative theory if we do not use a classical Lagrangian ? Answer: It is the proper definition of gauge invariance for the S-matrix. The S-matrix is defined on the space of free asymptotic fields which is the only thing that must be given. On these free quantum fields one introduces a gauge structure which involves ghost fields and further auxiliary fields if some gauge fields are massive. The gauge structure is given by the gauge variations of the asymptotic fields. These gauge variations are the quantum counterpart of classical gauge transformations and general coordinate transformations in general relativity. There is a formal similarity with the BRS transformation in the functional method. However, the BRS transformation operates on the interacting fields, and these can only be constructed if the coupling, i.e. the classical Lagrangian is given. We will not introduce interacting fields in this monograph.

The gauge variation will be denoted by d_Q where the suffix Q indicates that it is the (super)commutator with a nilpotent gauge charge Q ($Q^2 = 0$). The gauge charge Q serves for two purposes: (i) It defines the physical subspace of the big Fock space generated by the asymptotic fields. (ii) d_Q allows to calculate the gauge variation of the S-matrix. By S-matrix we always mean the scattering operator $S(g)$ where the coupling is smeared with a (Schwartz) test function g: $\int T(x)g(x)dx$. This $g(x)$ is a natural infrared regulator, so that $S(g)$ also exists if massless fields are present in the theory. The S-matrix is defined perturbatively by means of the time-ordered products $T_n(x_1,\ldots,x_n)$. The use of perturbation theory might look old-fashioned and restricts the possibility to calculate strong coupling phenomena. But it is good enough to gain the theory from scratch and to analyze its properties. Non-perturbative methods can be introduced at a later stage.

The time-ordered products T_n are constructed recursively from $T_1 = T$ by the causal method of Epstein and Glaser. In this way the nasty ultraviolet divergences which appear if one uses Feynman rules are avoided. No cut-off and no regularization is needed, everything is finite. This is of great advantage in the gravitational case. Gauge invariance of the S-matrix is now expressed by a condition on the time-ordered products. The basic condition called "causal gauge invariance" reads

$$d_Q T = \partial_\mu T^\mu. \tag{1}$$

Note that the gauge variation of T does not vanish, but it must be a divergence; T^μ is called Q-vertex in distinction to the ordinary vertex T. The solutions of equation (1) form the so-called relative cohomology group of d_Q. It is a crucial fact that this group contains very few non-trivial elements only. These non-trivial couplings T give the physical theories. It is a nice feature of this approach that T automatically contains all couplings of the quantum gauge theory including ghost, Higgs and auxiliary field couplings.

After this overview the plan of the book is clear. In Chapter 1 we introduce the free quantum fields and we prepare the ground for the gauge structure. In Chapter 2 we describe the inductive construction of the time-ordered products T_n starting from $T_1 = T$ by causal perturbation theory. The analysis of causal gauge invariance (1) begins with Chapter 3 where massless spin-1 gauge fields are considered. The solution is given by Yang-Mills theories (up to trivial modifications). In Chapter 4 the method is applied to massive gauge fields. Causal gauge invariance forces us to introduce unphysical and physical (Higgs) scalar fields and determines their couplings. Spontaneous symmetry breaking and the Higgs mechanism are not needed.

The construction of spin-2 gauge theories starts in Chapter 5 with the mass zero case. Needless to say that we use no input from general relativity. A stronger formulation of causal gauge invariance (in the form of descent equations) allows to derive the coupling T in an elegant way. The pure gravitational terms in T_1 and T_2 agree precisely with the expansion of the Einstein-Hilbert Lagrangian. We also consider the coupling to spin-1 gauge fields. This is a mixed spin-1 and spin-2 gauge theory, and we obtain some interesting new results in Section 5.11. There exists the technical problem of non-renormalizability of gravity. This problem gets simplified by using the cohomological definition of gauge invariance. But we do not discuss this issue because the complete solution is still not known.

We treat the massive spin-2 case parallel to ordinary massless gravity. The gauge structure, i.e. the nilpotency of d_Q, forces us to introduce a vector field v^μ with the same mass m as the graviton. To have a smooth limit for $m \to 0$ the physical modes of the massive graviton must be chosen as follows: two modes agree with the two transversal polarizations of the massless graviton, the remaining four modes are given by v^μ. (The physical subspace contains one mode more than a pure spin-2 field, 5+1=6.) It is a very interesting observation that the v-field does not decouple from the other degrees of freedom in the limit $m \to 0$. That means *the massless limit of massive gravity does not agree with massless gravity*, because the v-field survives.

Zürich, April 2011

Günter Scharf

Contents

1. **Free fields** 1
   - 1.1 Bosonic scalar fields 1
   - 1.2 Fermionic scalar (ghost) fields 8
   - 1.3 Massless vector fields 10
   - 1.4 Operator gauge transformations 18
   - 1.5 Massive vector fields 23
   - 1.6 Fermionic vector (ghost) fields 29
   - 1.7 Tensor fields 30
   - 1.8 Spinor fields 36
   - 1.9 Normally ordered products in free fields 39
   - 1.10 Problems 47

2. **Causal perturbation theory** 49
   - 2.1 The S-matrix in quantum mechanics 50
   - 2.2 The method of Epstein and Glaser 56
   - 2.3 Splitting of causal distributions in x-space 67
   - 2.4 Splitting in momentum space 74
   - 2.5 Calculation of tree graphs 80
   - 2.6 Calculation of loop graphs 87
   - 2.7 Normalizability 93
   - 2.8 Problems 96

3. **Spin-1 gauge theories: massless gauge fields** 98
   - 3.1 Causal gauge invariance 98
   - 3.2 Self-coupled gauge fields to first order 102
   - 3.3 Divergence- and co-boundary-couplings 107
   - 3.4 Yang-Mills theory to second order 112
   - 3.5 Reductive Lie algebras 117
   - 3.6 Coupling to matter fields 119
   - 3.7 Gauge invariance to all orders 122
   - 3.8 Unitarity 127
   - 3.9 Other gauges 131
   - 3.10 Gauge independence 137
   - 3.11 Appendix A: Cauchy problem for the iterated wave equation ... 141
   - 3.11 Problems 142

4. Spin-1 gauge theories: massive gauge fields 145
- 4.1 Massive QED and Abelian Higgs model 146
- 4.2 General massive gauge theory 150
- 4.3 First order gauge invariance 152
- 4.4 Second order gauge invariance 155
- 4.5 Third order gauge invariance 165
- 4.6 Derivation of the electroweak gauge theory 168
- 4.7 Coupling to leptons 172
- 4.8 More fermionic families 176
- 4.9 Gauge invariance to third order: axial anomalies .. 183
- 4.10 Problems 185

5. Spin-2 gauge theories 190
- 5.1 Causal gauge invariance with massless tensor fields ... 191
- 5.2 First order gauge invariance and descent equations ... 194
- 5.3 Massive tensor fields 197
- 5.4 Massive gravity 203
- 5.5 Expansion of the Einstein-Hilbert Lagrangian 205
- 5.6 Expansion in the massive case 210
- 5.7 Second order gauge invariance: graviton sector 213
- 5.8 Second order gauge invariance: ghost sector 221
- 5.9 Coupling to matter 224
- 5.10 Radiative corrections 231
- 5.11 Yang-Mills fields in interaction with gravity 235
- 5.12 Massive gravity: second order 239
- 5.13 Problems 245

6. Non-geometric general relativity 247
- 6.1 Geodesic equation 248
- 6.2 Einstein's equations and Maxwell's equations 250
- 6.3 Spherically symmetric fields and the circular velocity ... 252
- 6.4 Solutions of the vacuum equations 256
- 6.5 Cosmology in the cosmic rest frame 259
- 6.6 Failure of homogeneous cosmology 263
- 6.7 An inhomogeneous universe 265
- 6.8 Next to leading order 267
- 6.9 Calculation of the energy-momentum tensor 272
- 6.10 Null geodesics 276
- 6.11 The redshift 280
- 6.12 Area and luminosity distances 283
- 6.13 The Riemann and Weyl tensors 285

Bibliographical notes 287

Subject index 294

1. Free fields

Free fields are mathematical objects, they are not very physical. For example a free spin-1/2 Dirac field is a rather bad description of an electron because its charge and Coulomb field are ignored. In case of the photon the description by a free (transverse) vector field seems to be better, but still is not perfect. Elementary particles are complicated real objects, free fields are simple mathematical ones. Nevertheless, free fields are the basis of quantum field theory because the really interesting quantities like interacting fields, scattering matrix (S-matrix) etc. can be expanded in terms of free fields. We, therefore, first discuss all kinds of free fields which we will use later. Among them are some strange, but interesting guys called ghost fields. The German notion "spirit fields" (Geist-Felder instead of Gespenster-Felder) is more adequate. The reason is that these ghost fields define the infinitesimal gauge transformations of quantized gauge fields. That means they are at the heart of quantum gauge theory and so are never at any time negligible ghosts.

Our convention of the Minkowski metric is $g^{\mu\nu} = \mathrm{diag}(1, -1, -1, -1)$. If not explicitly written, we put the velocity of light and Planck's constant equal to 1, $c = \hbar = 1$. We sometimes refer for further discussion to the previous book G.Scharf "Finite quantum electrodynamics", Springer Verlag 1995, which will be abbreviated by FQED.

1.1 Bosonic scalar fields

First let us consider a neutral or real massive scalar field which is a solution of the Klein-Gordon equation

$$(\Box + m^2)\varphi(x) = 0, \quad \Box = \left(\frac{\partial}{\partial x^0}\right)^2 - \left(\frac{\partial}{\partial x^1}\right)^2 - \left(\frac{\partial}{\partial x^2}\right)^2 - \left(\frac{\partial}{\partial x^3}\right)^2. \quad (1.1.1)$$

A real *classical* solution of this equation is given by

$$\varphi(x) = (2\pi)^{-3/2} \int \frac{d^3p}{\sqrt{2E}} \left(a(\boldsymbol{p})e^{-ipx} + a^*(\boldsymbol{p})e^{ipx}\right), \quad (1.1.2)$$

where

$$px = p^0 x^0 - \boldsymbol{p} \cdot \boldsymbol{x} = p_\mu x^\mu, \quad E = +\sqrt{\boldsymbol{p}^2 + m^2} = p^0. \quad (1.1.3)$$

Free fields

In quantum field theory $a(\boldsymbol{p})$ and $a^*(\boldsymbol{p})$ become operator-valued distributions, that means

$$a(f) = \int d^3p\, f^*(\boldsymbol{p}) a(\boldsymbol{p}) \tag{1.1.4}$$

(f a test function) is an operator in some Hilbert space and

$$a^+(f) = \int d^3p\, f(\boldsymbol{p}) a^+(\boldsymbol{p}) \tag{1.1.5}$$

its adjoint. In the distribution $a(\boldsymbol{p})^+$ we make no difference about the place of the superscript $+$, before or behind the argument. The properties of the unsmeared objects $a(\boldsymbol{p}), a^+(\boldsymbol{p})$ are further analyzed in the problems 1.8-9 at the end of this chapter. In the following all equations between distributions mean that they become operator equations after smearing with test functions.

The crucial property of these operators is the fulfillment of the canonical commutation relations

$$[a(f), a^+(g)] = \int d^3p\, f^*(\boldsymbol{p}) g(\boldsymbol{p}) = (f, g), \tag{1.1.6}$$

the result is the L^2 scalar product of the test functions. The relation can be written in distributional form as follows

$$[a(\boldsymbol{p}), a^+(\boldsymbol{q})] = \delta^3(\boldsymbol{p} - \boldsymbol{q}). \tag{1.1.7}$$

all other commutators vanish. The quantized Bose field is now given by

$$\varphi(x) = (2\pi)^{-3/2} \int \frac{d^3p}{\sqrt{2E}} \left(a(\boldsymbol{p}) e^{-ipx} + a^+(\boldsymbol{p}) e^{ipx} \right). \tag{1.1.8}$$

It is obviously hermitian

$$\varphi^+(x) = \varphi(x). \tag{1.1.9}$$

Let us call the second term in (1.1.8) involving a^+ the creation part $\varphi^{(+)}$ and the first term with $a(\boldsymbol{p})$ the absorption part $\varphi^{(-)}$. Then by (1.1.7) their commutator is equal to

$$[\varphi^{(-)}(x), \varphi^{(+)}(y)] = (2\pi)^{-3/2} \int \frac{d^3p}{2E} e^{-ip(x-y)}. \tag{1.1.10}$$

To write this in Lorentz-covariant form we add the integration over p^0 and insert the one-dimensional δ-distribution

$$\delta(p^2 - m^2) = \frac{\delta(p^0 - E)}{2E} + \frac{\delta(p^0 + E)}{2E}, \tag{1.1.11}$$

note that E is positive (1.1.3). The commutator (1.1.10) is now equal to

$$[\varphi^{(-)}(x), \varphi^{(+)}(y)] = (2\pi)^{-3} \int d^4p\, \delta(p^2 - m^2) \Theta(p_0) e^{-ip(x-y)} =$$

$$\stackrel{\text{def}}{=} -iD_m^{(+)}(x-y). \tag{1.1.12}$$

In the same way we get

$$[\varphi^{(+)}(x), \varphi^{(-)}(y)] = -(2\pi)^{-3} \int d^4p\, \delta(p^2 - m^2)\Theta(-p_0)e^{-ip(x-y)} =$$

$$\stackrel{\text{def}}{=} -iD_m^{(-)}(x-y) = iD_m^{(+)}(y-x).$$

Then the commutation relation for the total scalar field reads

$$[\varphi(x), \varphi(y)] = (2\pi)^{-3} \int d^4p\, \delta(p^2 - m^2)\mathrm{sgn}\,(p_0)e^{-ip(x-y)} =$$

$$\stackrel{\text{def}}{=} -iD_m(x-y) = -i(D_m^{(+)} + D_m^{(-)})(x-y). \tag{1.1.13}$$

This is the so-called Jordan-Pauli distribution D_m. It has a causal support, that means its support lies in the forward and backward light cones (see problems 1.1-3 and FQED, Sect.2.3)

$$\mathrm{supp}\,\{D_m(x)\} \subseteq \{x \in \mathbb{R}^4 \mid x^2 \geq 0\}. \tag{1.1.14}$$

This property is crucial for the causal method in Sect.2. We already remark that D_m can be split into retarded and advanced functions

$$D(x) = D^{\mathrm{ret}}(x) - D^{\mathrm{av}}(x) \quad, \quad \text{with}$$

$$D^{\mathrm{ret}}(x) = \Theta(x^0)D(x) \quad, \quad D^{\mathrm{av}}(x) = \Theta(-x^0)D(x). \tag{1.1.15}$$

Our next task is to write the scalar field in Lorentz-invariant form, too. For this purpose we introduce the measure

$$d\mu_m(p) = \delta(p^2 - m^2)\Theta(p_0)d^4p = \frac{\delta(p_0 - E)}{2E}dp_0 d^3p = \frac{d^3p}{2p_0}\bigg|_{p_0=E}. \tag{1.1.16}$$

This is a Lorentz-invariant measure on the positive mass shell $\mathcal{M}^+ = \{p \in \mathbb{R}^4 \mid p^2 = m^2, p_0 > 0\}$. But the scalar field (1.1.8)

$$\varphi(x) = (2\pi)^{-3/2} \int_{\mathcal{M}^+} d\mu_m(p)\, \sqrt{2p_0}\left(a(\mathbf{p})e^{-ipx} + a^+(\mathbf{p})e^{ipx}\right) \tag{1.1.17}$$

still does not look covariant. Obviously, the operators

$$\tilde{a}(p) = \sqrt{2p_0}a(\mathbf{p}), \quad \tilde{a}^+(p) = \sqrt{2p_0}a^+(\mathbf{p}) \tag{1.1.18}$$

must be Lorentz scalars. According to (1.1.7) they obey the commutation relations

$$[\tilde{a}(p), \tilde{a}^+(q)] = 2E(\mathbf{p})\delta^3(\mathbf{p} - \mathbf{q}) \tag{1.1.19}$$

and all other commutators vanish. To get the corresponding operator equations, we smear (1.1.17) with a real test function $f(x) \in \mathcal{S}(\mathbb{R}^4)$ in 4-dimensional Schwartz space (see any book on distributions, for example

I.M.Gelfand et al., *Generalized functions*, Academic Press, New York 1964-68)

$$\int \varphi(x) f(x) \, d^4x = \sqrt{2\pi}(\tilde{a}(\hat{f}) + \tilde{a}^+(\hat{f})), \qquad (1.1.20)$$

where

$$\tilde{a}(\hat{f}) = \int d\mu_m(p) \, \tilde{a}(p) \hat{f}(-p) \qquad (1.1.21)$$

$$\tilde{a}^+(\hat{f}) = \int d\mu_m(p) \, \tilde{a}^+(p) \hat{f}(p) \qquad (1.1.22)$$

and

$$\hat{f}(p) = (2\pi)^{-2} \int f(x) e^{ipx} \, d^4x \qquad (1.1.23)$$

is the four-dimensional Fourier transform. Then

$$[\tilde{a}(\hat{f}), \tilde{a}^+(\hat{g})] = \int_{\mathcal{M}^+} \frac{d^3p}{2p_0} \, \hat{f}(-p) \hat{g}(p) =$$

$$= \int_{\mathcal{M}^+} d\mu_m(p) \, \hat{f}(p)^* \hat{g}(p) \stackrel{\text{def}}{=} (\hat{f}, \hat{g})_m \qquad (1.1.24)$$

is the Lorentz-invariant form of the commutation relation. The scalar product herein corresponds to the Hilbert space $L^2(\mathcal{M}^+, d\mu_m)$.

To show that the whole procedure is well defined and free of contradictions, we have to construct a concrete representation of the various operators in the so-called Fock-Hilbert space. To construct the latter we start from a normalized vector Ω, $|\Omega| = 1$ defined by

$$\tilde{a}(\hat{f})\Omega = 0, \quad \forall \hat{f} \in L^2. \qquad (1.1.25)$$

This vector is assumed to be unique and called the vacuum. Then the $\tilde{a}$ can be interpreted as absorption operators, because in Ω nothing can be absorbed according to (1.1.25). Next we consider the vectors $\tilde{a}^+(f)\Omega$ and calculate their scalar products

$$(\tilde{a}^+(\hat{f})\Omega, \tilde{a}^+(\hat{g})\Omega) = (\Omega, \tilde{a}(\hat{f})\tilde{a}^+(\hat{g})\Omega) = (\hat{f}, \hat{g})_m \qquad (1.1.26)$$

where the commutation relation (1.1.24) has been used to commute the absorption operator $\tilde{a}(\hat{f})$ to the right, giving zero on Ω by (1.1.25). We see that these vectors form a Hilbert space which is isomorphic to

$$\mathcal{H}_1 = L^2(\mathcal{M}^+, d\mu_m) \qquad (1.1.27)$$

and consists of complex functions $\hat{f}(p)$, $p_0 = +\sqrt{\mathbf{p}^2 + m^2}$, with

$$\int |\hat{f}|^2 d\mu_m < \infty. \qquad (1.1.28)$$

This is the one-particle space, so that $\tilde{a}^+$ can indeed be interpreted as an emission operator. It generates a one-particle state from the vacuum. As mentioned before the notion "particle" does not mean that this is a real physical particle. At best we have an approximate description of some real particle in terms of the free scalar field.

The n-particle space is defined as the symmetric tensor product

$$\mathcal{H}_n = S_n \mathcal{H}_1^{\otimes n} \tag{1.1.29}$$

where S_n is the symmetrization operator

$$S_n \varphi_n = \frac{1}{n!} \sum_P \varphi_n(p_{P1}, \ldots, p_{Pn}), \tag{1.1.30}$$

the sum goes over all permutations of the momenta of the n particles. This space is spanned by the vectors

$$\frac{1}{\sqrt{n!}} \prod_{j=1}^n \tilde{a}^+(f_j) \Omega \longleftrightarrow S_n f_1 \otimes \ldots \otimes f_n. \tag{1.1.31}$$

As in (1.1.26) one can verify that the mapping (1.1.31) is a unitary correspondence. The direct sum

$$\mathcal{F} = \oplus_{n=0}^\infty \mathcal{H}_n$$

gives the Fock-Hilbert space where the scalar field operates.

The representation of the field operators just constructed, the so-called Fock representation, realizes a unitary representation of the proper Poincaré group at the same time. By definition the vacuum is invariant

$$\mathbf{U}(a, \Lambda) \Omega = \Omega, \tag{1.1.32}$$

where Λ denotes proper Lorentz transformations (i.e. without reflections) and $a \in \mathbb{R}^4$ represents the translations. From (1.1.17) we then have

$$\varphi(\Lambda x + a) = (2\pi)^{-3/2} \int d\mu_m(p) \left[\tilde{a}(p) e^{-ip(\Lambda x + a)} + e^{ipa} \tilde{a}^+(p) e^{i(\Lambda^{-1} p)x} \right] \tag{1.1.33}$$

where we have used the Lorentz invariance of the Minkowski scalar product in the last term. The transformed field (1.1.33) must be equal to

$$= \mathbf{U}(a, \Lambda) \varphi(x) \mathbf{U}(a, \Lambda)^{-1}.$$

We smear the emission part $\varphi^{(+)}$ with $f(x)$ and apply it to the vacuum, using (1.1.32),

$$\mathbf{U}(a, \Lambda) \int d^4 x \, \varphi^{(+)}(x) f(x) \mathbf{U}(a, \Lambda)^{-1} \Omega = \mathbf{U}(a, \Lambda) \sqrt{2\pi} \int d\mu_m(p) \, \hat{f}(p) \tilde{a}^+(p) \Omega.$$

By (1.1.33) this is equal to

6   Free fields

$$= \sqrt{2\pi} \int d\mu_m(p)\, e^{ipa}\, \hat{f}(\Lambda^{-1}p)\tilde{a}^+(p)\Omega. \qquad (1.1.34)$$

This implies
$$(\mathbf{U}(a,\Lambda)\hat{f})(p) = e^{ipa}\hat{f}(\Lambda^{-1}p) \qquad (1.1.35)$$
which is an irreducible unitary representation of the Poincaré group in $\mathcal{H}_1$ (1.1.27). The representation in the n-particle sector $\mathcal{H}_n$ is the corresponding tensor representation

$$(\mathbf{U}(a,\Lambda)\varphi_n)(p_1,\ldots,p_n) = \exp\left(i\sum_{j=1}^n p_j a\right)\varphi(\Lambda^{-1}p_1,\ldots,\Lambda^{-1}p_n). \qquad (1.1.36)$$

It is no longer irreducible.

Next we want to find out how the emission and absorption operators operate in the Fock representation. From the correspondence (1.1.31) we immediately get

$$\tilde{a}^+(f)\frac{1}{\sqrt{(n-1)!}}\prod_{j=1}^{n-1}\tilde{a}^+(f_j)\Omega = \sqrt{n}\frac{1}{\sqrt{n!}}\tilde{a}^+(f)\prod_{j=1}^{n-1}\tilde{a}^+(f_j)\Omega$$

$$\longleftrightarrow \sqrt{n}S_n f \otimes f_1 \otimes \ldots \otimes f_{n-1}. \qquad (1.1.37)$$

By linearity this extends to
$$(\tilde{a}^+(f)\Phi)_n = \sqrt{n}S_n(f\otimes\varphi_{n-1}) \qquad (1.1.38)$$
where
$$\Phi = (\varphi_n)_{n=0}^N \in \mathcal{F}_N \qquad (1.1.39)$$
$\varphi_n = 0$ for $n > N$, Φ is a general vector containing not more than N particles. For arbitrary N this is a dense set in Fock space which is in the domain of $\tilde{a}^+(f)$.

In case of the absorption operator we use the commutation relation

$$\tilde{a}(f)\frac{1}{\sqrt{(n+1)!}}\prod_{j=1}^{n+1}\tilde{a}^+(f_j)\Omega = (f,f_1)_m\frac{1}{\sqrt{(n+1)!}}\tilde{a}^+(f_2)\ldots\tilde{a}^+(f_{n+1})\Omega$$

$$+\frac{1}{\sqrt{(n+1)!}}\tilde{a}^+(f_1)\tilde{a}(f)\tilde{a}^+(f_2)\ldots\tilde{a}^+(f_{n+1})\Omega.$$

In the next step we commute $\tilde{a}(f)$ with $\tilde{a}^+(f_2)$ and so on. This leads to

$$=\frac{1}{\sqrt{(n+1)!}}\sum_{j=1}^{n+1}(f,f_j)_m\prod_{k\neq j}\tilde{a}^+(f_k)\Omega \qquad (1.1.40)$$

$$\longleftrightarrow \frac{1}{\sqrt{n+1}}\sum_{j=1}^{n+1}(f,f_j)_m S_n f_1\otimes\ldots\otimes\overline{f}_j\otimes\ldots f_{n+1}, \qquad (1.1.41)$$

where the overlined f_j is lacking. Writing the scalar product as a p-integral and changing the symmetrization operator S_n into S_{n+1} we finally get

$$= \frac{n+1}{\sqrt{n+1}} \int d\mu_m(p)\, f(-p) S_{n+1} f_1(p_1) \otimes \ldots \otimes f_{n+1}(p).$$

By linearity this extends to

$$(\tilde{a}(f)\Phi)_n(p_1, \ldots, p_n) = \sqrt{n+1} \int d\mu_m(p)\, f(-p)\varphi_{n+1}(p_1, \ldots, p_n, p). \tag{1.1.42}$$

For completeness let us determine the adjoint operator of $\tilde{a}(f)$. Let $\Phi, \Psi \in \mathcal{F}_N$ (1.1.39) then the scalar product in Fock space is given by

$$(\Psi, \tilde{a}(f)\Phi) = \sum_n \sqrt{n+1} \int d\mu_m(p_1) \ldots d\mu_m(p_n) d\mu_m(p)$$

$$\times \psi_n(p_1, \ldots, p_n)^* f(-p)\varphi_n(p, p_1, \ldots, p_n) = (\tilde{a}^+(\overline{f})\Psi, \Phi), \tag{1.1.43}$$

where $\overline{f}(p) = f(-p)^*$ and the star denotes the complex conjugate. The Fourier transform of $\overline{f}(p)$ in x-space is just the complex conjugate function $f(x)^*$. From (1.1.43) we obtain the relation

$$\tilde{a}(f)^+ = \tilde{a}^+(\overline{f}).$$

For later use we write down the operation of the hermitian scalar field

$$\varphi(f) = \sqrt{2\pi}(\tilde{a}(f) + \tilde{a}^+(f)) = \varphi(\overline{f})^+ \tag{1.1.44}$$

in Fock space:

$$(\varphi(f)\Phi)_n(p_1, \ldots, p_n) = \sqrt{2\pi}\left[\sqrt{n+1} \int d\mu_m(p) f(-p)\varphi_{n+1}(p_1, \ldots, p_n, p)\right.$$

$$\left. + \frac{1}{\sqrt{n}} \sum_{j=1}^n f(p_j)\varphi_{n-1}(p_1, \ldots, \overline{p}_j, \ldots, p_n)\right]. \tag{1.1.45}$$

Here the symmetrization has explicitly been written out. In the Fock representation the commutation relation (1.1.12) can be written in terms of vacuum expectation values in the following form

$$[\varphi^{(-)}(x), \varphi^{(+)}(y)] = (\Omega, \varphi(x)\varphi(y)\Omega) = -iD_m^{(+)}(x-y), \tag{1.1.46}$$

because $\varphi^{(-)}(x)\varphi^{(+)}(y)$ is the only term which has a non-vanishing vacuum expectation value. This will later be generalized to more than two factors.

The charged or complex scalar field is a slight generalization of the neutral one:

$$\varphi(x) = (2\pi)^{-3/2} \int \frac{d^3p}{\sqrt{2E}} \left(a(\mathbf{p})e^{-ipx} + b^+(\mathbf{p})e^{ipx}\right). \tag{1.1.47}$$

It contains two different kinds of particles whose absorption and emission operators satisfy

$$[a(\boldsymbol{p}), a^+(\boldsymbol{q})] = \delta(\boldsymbol{p}-\boldsymbol{q}) = [b(\boldsymbol{p}), b^+(\boldsymbol{q})] \tag{1.1.48}$$

and all other commutators vanish. Then it follows

$$[\varphi(x), \varphi(y)^+] = -iD_m(x-y) \tag{1.1.49}$$

but

$$[\varphi(x), \varphi(y)] = 0. \tag{1.1.50}$$

The Lorentz-invariant form is given by

$$\varphi(x) = (2\pi)^{-3/2} \int_{\mathcal{M}^+} d\mu_m(p) \left(\tilde{a}(p) e^{-ipx} + \tilde{b}^+(p) e^{ipx} \right). \tag{1.1.51}$$

$\tilde{a}^+(\hat{f})\Omega$ spans the one-particle sector $\mathcal{H}_1^a$, but $\tilde{b}^+(\hat{g})\Omega$ spans the one-anti-particle sector $\mathcal{H}_1^b$ which is different. The many-particle sectors are again obtained by tensor products and the total Fock space is the direct sum

$$\mathcal{F} = \oplus_{n_1, n_2 = 0}^{\infty} \left(S_{n_1} \mathcal{H}_{n_1}^a \otimes S_{n_2} \mathcal{H}_{n_2}^b \right). \tag{1.1.52}$$

1.2 Fermionic scalar (ghost) fields

"Fermionic" means that we now quantize a scalar field with anticommutators. These fields occur as so-called ghost fields in gauge theory. This terminology is somewhat misleading because the ghost fields are genuine dynamical fields which interact with other fields in the theory. Their ghost character only expresses the fact that the ghost particles cannot occur as asymptotic scattering states. There seems to be a contradiction to the well-known theorem of spin and statistics. This theorem tells us that fields with integer spin should be quantized with commutators and those with half-integer spin with anticommutators. We will return to this point in detail below, for the moment we remark that the "wrong" commutation relation is possible here because the scalar field under consideration describes two different kinds of particles, similarly as the charged scalar field (1.1.46):

$$u(x) = (2\pi)^{-3/2} \int \frac{d^3p}{\sqrt{2E}} \left(c_2(p) e^{-ipx} + c_1(p)^+ e^{ipx} \right). \tag{1.2.1}$$

In addition, we introduce a second scalar field

$$\tilde{u}(x) = (2\pi)^{-3/2} \int \frac{d^3p}{\sqrt{2E}} \left(-c_1(p) e^{-ipx} + c_2(p)^+ e^{ipx} \right). \tag{1.2.2}$$

This is not the adjoint of $u(x)$. The absorption and emission operators c_j, c_k^+ obey the anticommutation relations

$$\{c_j(\boldsymbol{p}), c_k(\boldsymbol{q})^+\} = \delta_{jk}\delta^3(\boldsymbol{p}-\boldsymbol{q}). \tag{1.2.3}$$

The absorption and emission parts (with the adjoint operators) are again denoted by (-) and (+). They satisfy the following anticommutation relations

$$\{u^{(-)}(x), \tilde{u}^{(+)}(y)\} = (2\pi)^{-3} \int \frac{d^3p}{2E} e^{-ip(x-y)} = -iD_m^{(+)}(x-y) \tag{1.2.4}$$

$$\{u^{(+)}(x), \tilde{u}^{(-)}(y)\} = -(2\pi)^{-3} \int \frac{d^3p}{2E} e^{ip(x-y)} = -iD_m^{(-)}(x-y). \tag{1.2.5}$$

All other anticommutators vanish. This implies

$$\{u(x), \tilde{u}(y)\} = -iD_m(x-y) \tag{1.2.6}$$

and

$$\{u(x), u(y)\} = 0. \tag{1.2.7}$$

As before the fields can be written in Lorentz-covariant form by introducing

$$\tilde{c}_j(p) = \sqrt{2p_0} c_j(\boldsymbol{p}), \quad \tilde{c}_j(p)^+ = \sqrt{2p_0} c_j(\boldsymbol{p})^+. \tag{1.2.8}$$

Then we have

$$u(x) = (2\pi)^{-3/2} \int d\mu_m(p) \left(\tilde{c}_2(p) e^{-ipx} + \tilde{c}_1(\boldsymbol{p})^+ e^{ipx} \right) \tag{1.2.9}$$

$$\tilde{u}(x) = (2\pi)^{-3/2} \int d\mu_m(p) \left(-\tilde{c}_1(p) e^{-ipx} + \tilde{c}_2(\boldsymbol{p})^+ e^{ipx} \right). \tag{1.2.10}$$

The vectors $\tilde{c}_j(\hat{f})^+\Omega$, $j=1,2$ generate the one-particle sectors $\mathcal{H}_1^{(j)}$, and the n-particle sectors are obtained as antisymmetric tensor products

$$\mathcal{H}_n^{(j)} = S_n^- \mathcal{H}_1^{(j)\otimes n}, \tag{1.2.11}$$

where

$$S_n^- \varphi_n = \frac{1}{n!} \sum_P (-)^P \varphi_n(p_{P1}, \ldots, p_{Pn}) \tag{1.2.12}$$

is the antisymmetrization operator. The total Fock space is the direct sum

$$\mathcal{F} = \oplus_{n=0}^\infty \left(S_n^- \mathcal{H}_n^{(1)} \oplus S_n^- \mathcal{H}_n^{(2)} \right). \tag{1.2.13}$$

Let us now discuss the relation to the theorem of spin and statistics. This theorem can be expressed in the following form (see R.F. Streater, A.S. Wightman, *PCT, Spin and Statistics, and All That"*, *Benjamin 1964*):

Theorem 1.2.1.. In a quantum field theory with a Hilbert space with positive definite metric there cannot exist scalar fields different from zero which satisfy the *anti*-commutation relations

$$\{u(x), u(y)\} = 0 \qquad (1.2.14)$$

$$\{u(x), u^+(y)\} = 0 \qquad (1.2.15)$$

for all $(x-y)^2 < 0$.

The first condition is fulfilled (1.2.7), but the second one is not:

$$\{u(x), u(y)^+\} = -iD_m^{(+)}(x-y) + iD_m^{(-)}(x-y)$$
$$= -iD_m(x-y) + 2iD_m^{(-)}(x-y). \qquad (1.2.16)$$

The causal Jordan-Pauli distribution D_m vanishes for space-like arguments $(x-y)^2 < 0$, but the $D_m^{(-)}$-distribution does not. For example, in the massless case $m=0$ we have the simple expression

$$D_0^{(-)}(x) = \frac{1}{4\pi}\operatorname{sgn} x^0 \delta(x^2) + \mathrm{P}\frac{i}{4\pi^2 x^2} \qquad (1.2.17)$$

and the principal value contribution does not vanish for $x^2 < 0$. The situation in the massive case is similar (see FQED, Sect.2.3). Consequently, there is no contradiction to the spin-statistics theorem. The point is the minus sign in front of c_1 in (1.2.2) which implies $\tilde{u} \neq u^+$.

1.3 Massless vector fields

These fields obey the wave equation

$$\Box A^\mu(x) = 0. \qquad (1.3.1)$$

Examples of massless vector-particles are the photon and the gluons, so that these fields are the genuine gauge fields. The photon has only two physical transversal degrees of freedom. Therefore, two subsidiary conditions are necessary to eliminate the unphysical components. As one such condition we may choose the Lorentz condition

$$\partial_\mu A^\mu(x) = 0 \qquad (1.3.2)$$

which is Lorentz-invariant. But the second condition, for example the temporal gauge condition

$$A^0(x) = 0, \qquad (1.3.3)$$

cannot be chosen covariantly. This is the reason for the subtleties in the following. We recall that the free fields considered here are only the zeroth approximation to the real photon in the lab.

To start with we disregard the subsidiary conditions completely. We quantize $A^\mu(x)$ as four independent real scalar fields. Let

$$A^\mu(t, \boldsymbol{x}) = (2\pi)^{-3/2} \int \frac{d^3k}{\sqrt{2\omega}} \left(a^\mu(\boldsymbol{k})e^{-i(\omega t - \boldsymbol{k}\cdot\boldsymbol{x})} + a^\mu(\boldsymbol{k})^* e^{i(\omega t - \boldsymbol{k}\cdot\boldsymbol{x})} \right), \quad (1.3.4)$$

be a real classical solution of the wave equation with

$$\omega(\boldsymbol{k}) = |\boldsymbol{k}| \stackrel{\text{def}}{=} k^0, \quad (1.3.5)$$

the star denotes the complex conjugate. After quantization $a^\mu(\boldsymbol{k})$ become operator-valued distributions. Let us assume the usual commutation relations

$$[a^\mu(\boldsymbol{k}), a^\nu(\boldsymbol{k}')^+] = \begin{cases} \delta(\boldsymbol{k} - \boldsymbol{k}') & \text{for } \mu = \nu, \\ 0 & \text{for } \mu \neq \nu. \end{cases} \quad (1.3.6)$$

Then we know from Sect. 1.1 that $a^{\mu+}$ are emission operators and a^μ absorption operators in Fock space.

There is, however, a serious difficulty with Lorentz covariance in this approach: If we retain the classical expression (1.3.4) in the form

$$A^\mu(x) = (2\pi)^{-3/2} \int \frac{d^3k}{\sqrt{2\omega}} \left(a^\mu(\boldsymbol{k})e^{-ikx} + a^\mu(\boldsymbol{k})^+ e^{ikx} \right), \quad (1.3.7)$$

we obtain the following commutator

$$[A^\mu(x), A^\nu(y)] = (2\pi)^{-3} \int \frac{d^3k}{\sqrt{2\omega}} \int \frac{d^3k'}{\sqrt{2\omega'}} \times$$

$$\times \left\{ [a^\mu(\boldsymbol{k}), a^\nu(\boldsymbol{k}')^+] e^{-ikx + ik'y} + [a^\mu(\boldsymbol{k})^+, a^\nu(\boldsymbol{k}')] e^{ikx - ik'y} \right\}$$

$$= \delta^\mu_\nu (2\pi)^{-3} \int \frac{d^3k}{2\omega} \left(e^{-ik(x-y)} - e^{ik(x-y)} \right)$$

$$= \delta^\mu_\nu (2\pi)^{-3} \int d^4k\, \delta(k^2) e^{-ik(x-y)} \operatorname{sgn}(k^0)$$

$$= \delta^\mu_\nu \frac{1}{i} D_0(x-y). \quad (1.3.8)$$

The Lorentz invariant Jordan-Pauli distribution (1.1.13) for mass 0 appears here. However, the right-hand side is not a second rank Lorentz tensor of the same type as the left-hand side. We should have $g^{\mu\nu}$ instead of δ^μ_ν. The simplest way to remedy this defect is to change the sign in

$$[a^0(\boldsymbol{k}), a^0(\boldsymbol{k}')^+] = -\delta^3(\boldsymbol{k} - \boldsymbol{k}'). \quad (1.3.9)$$

After 3-dimensional smearing this implies

$$[a^0(f), a^0(f)^+] = -(f, f). \quad (1.3.10)$$

But this contradicts a positive definite metric in Hilbert space

$$(a^0(f)^+\Omega, a^0(f)^+\Omega) = (\Omega, a^0(f)a^0(f)^+\Omega) = -(f,f)\|\Omega\|^2. \quad (1.3.11)$$

We, therefore, will proceed differently, the "indefinite metric" will appear in a more satisfactory way.

Another possibility to solve the problem is to change the classical definition (1.3.4) of A^0 into

$$A^0(x) = (2\pi)^{-3/2} \int \frac{d^3k}{\sqrt{2\omega}} \left(a^0(\mathbf{k})e^{-ikx} - a^0(\mathbf{k})^+ e^{ikx}\right). \quad (1.3.12)$$

This makes A^0 a skew-adjoint operator instead of self-adjoint. As we shall discuss below, the physical Hilbert space will be defined in such a way that all expectation values of A^0 (and of any quantity derived from it) vanish. Then the non-self-adjointness of A^0 (1.3.12) causes no problems. But the spatial components remain hermitian

$$(A^j)^+ = A^j, \quad j = 1, 2, 3, \quad (1.3.13)$$

so that the adjoint operation is not Lorentz-invariant. We will introduce a second conjugation below, which is Lorentz-invariant.

With the new definition (1.3.12), the commutation relations for the vector field are

$$[A^\mu(x), A^\nu(y)] = g^{\mu\nu} i D_0(x-y). \quad (1.3.14)$$

We need also the commutators of the absorption and emission parts alone. Let

$$A^\mu_-(x) = (2\pi)^{-3/2} \int \frac{d^3k}{\sqrt{2\omega}} a^\mu(\mathbf{k}) e^{-ikx} \quad (1.3.15)$$

$$A^\mu_+(x) = (2\pi)^{-3/2} \int \frac{d^3k}{\sqrt{2\omega}} a^\mu(\mathbf{k})^+ e^{ikx} \cdot \begin{cases} -1 & \text{for } \mu = 0 \\ 1 & \text{for } \mu = 1,2,3, \end{cases} \quad (1.3.16)$$

then the only non-vanishing commutators are

$$[A^\mu_-(x), A^\nu_+(y)] = g^{\mu\nu} i D_0^{(+)}(x-y) \quad (1.3.17)$$

$$[A^\mu_+(x), A^\nu_-(y)] = g^{\mu\nu} i D_0^{(-)}(x-y). \quad (1.3.18)$$

We will briefly discuss the time evolution of the vector field. After construction $A^\mu(x)$ is a solution of the wave equation, therefore one may define

$$i\frac{\partial}{\partial t}A^n(t,\mathbf{x}) = (2\pi)^{-3/2} \int \frac{d^3k}{\sqrt{2\omega}} \left[\omega a^n(\mathbf{k}) e^{-i\omega t + i\mathbf{k}\cdot\mathbf{x}}\right.$$

$$\left. -\omega a^n(\mathbf{k})^+ e^{i\omega t - i\mathbf{k}\cdot\mathbf{x}}\right] \stackrel{\text{def}}{=} [A^n, \mathbf{H}_0] \quad (1.3.19)$$

and

$$i\frac{\partial}{\partial t}A^0(t,\mathbf{x}) = (2\pi)^{-3/2} \int \frac{d^3k}{\sqrt{2\omega}} \left[\omega a^0(\mathbf{k}) e^{-i\omega t + i\mathbf{k}\cdot\mathbf{x}}\right.$$

$$\left. +\omega a^0(\mathbf{k})^+ e^{i\omega t - i\mathbf{k}\cdot\mathbf{x}}\right] \stackrel{\text{def}}{=} [A^0(t,\mathbf{x}), \mathbf{H}_0]. \quad (1.3.20)$$

The operator $\mathbf{H}_0$ is uniquely determined by these two equations up to an additive constant. This is a consequence of the irreducibility of the Fock representation (see FQED, Sect.2.1). It is easy to verify that the positive definite operator

$$\mathbf{H}_0 = \int d^3k\, \omega(\boldsymbol{k}) \sum_{\mu=0}^{3} a^\mu(\boldsymbol{k})^+ a^\mu(\boldsymbol{k}) \tag{1.3.21}$$

satisfies (1.3.19) (1.3.20). As far as positive definiteness of the energy is concerned, our procedure of quantization of the massless vector field is satisfactory.

Now we want to define another conjugation K in Fock space which is Lorentz-invariant and such that $A^\mu(x)$ becomes self-conjugate. In order to achieve this, we introduce the bounded operator

$$\eta = (-1)^{\mathbf{N}_0}, \tag{1.3.22}$$

where $\mathbf{N}_0$ is the particle number operator for scalar "photons" $\mu = 0$. We obviously have

$$\eta^+ = \eta \quad , \quad \eta^2 = 1, \tag{1.3.23}$$

and η anticommutes with the emission and absorption operators for scalar "photons"

$$a^0(\varphi)\eta = -\eta a^0(\varphi) \quad , \quad a^0(\varphi)^+ \eta = -\eta a^0(\varphi)^+, \tag{1.3.24}$$

because these operators change the number of scalar "photons" by one. It commutes with all other emission and absorption operators. The conjugation K is now defined by

$$B^K = \eta B^+ \eta \tag{1.3.25}$$

for any (densely defined) operator in Fock space. It has all desired properties

$$(A+B)^K = A^K + B^K \quad , \quad (AB)^K = B^K A^K$$

$$A^{KK} = A \quad , \quad (\lambda A)^K = \lambda^* A^K, \tag{1.3.26}$$

for λ complex. Furthermore, if A can be inverted then

$$(A^{-1})^K = (A^K)^{-1}.$$

Since the skew-adjointness of A^0 is compensated by anticommuting with η, $A(\varphi)$ is indeed self-conjugate. It follows from (1.3.12) that

$$A^\mu(x) = (2\pi)^{-3/2} \int \frac{d^3k}{\sqrt{2\omega}} \left(a^\mu(\boldsymbol{k}) e^{-ikx} + a^\mu(\boldsymbol{k})^K e^{ikx} \right) \tag{1.3.27}$$

from which the self-conjugacy is evident. The so-called Krein operator η (1.3.22) can be used to define a bilinear form in Fock space

14   Free fields

$$\langle \Phi, \Psi \rangle = (\Phi, \eta \Psi). \tag{1.3.28}$$

This is the usual "indefinite metric" used in most textbooks, and the conjugation K then is denoted by $+$ or $*$. The correct distinction between a positive-definite scalar product which defines the topology in the Hilbert space and the indefinite bilinear form (1.3.28) was introduced by mathematicians (c.f. J.Bognar, *Indefinite inner product spaces*, Springer-Verlag, Berlin 1974). The conjugation K corresponds to the bilinear form (1.3.28):

$$\langle \Phi, B\Psi \rangle = (\Phi, \eta B \Psi) = (B^+ \eta \Phi, \Psi) =$$
$$= (\eta B^+ \eta \Phi, \eta \Psi) = \langle B^K \Phi, \Psi \rangle. \tag{1.3.29}$$

Hitherto we have not imposed any gauge condition. Now we shall investigate the Lorentz condition (1.3.2). As an operator equation, it cannot hold on the whole Fock space $\mathcal{F}$, because

$$\partial_\mu A^\mu(x) \Omega = \partial_\mu A^\mu_+(x) \Omega \neq 0. \tag{1.3.30}$$

However, the vacuum expectation value of (1.3.2) vanishes

$$(\Omega, \partial_\mu A^\mu \Omega) = (\Omega, \partial_\mu A^\mu_+ \Omega) = (\mp \partial_\mu A^\mu_- \Omega, \Omega) = 0,$$

where the upper minus sign corresponds to $\mu = 0$. The same is true for a large class of states. The expression

$$\partial_\mu A^\mu(x) = (2\pi)^{-3/2} \int \frac{d^3k}{\sqrt{2\omega}} \left[-i \Big(\omega a^0(\boldsymbol{k}) + k_j a^j(\boldsymbol{k}) \Big) e^{-ikx} \right.$$
$$\left. + i \Big(-\omega a^0(\boldsymbol{k})^+ + k_j a^j(\boldsymbol{k})^+ \Big) e^{ikx} \right] \tag{1.3.31}$$

may be written in a more transparent form by introducing the absorption and emission operators for longitudinal "photons"

$$a_\parallel(\boldsymbol{k}) \stackrel{\text{def}}{=} \frac{k_j}{\omega} a^j(\boldsymbol{k}),$$

$$\partial_\mu A^\mu(x) = (2\pi)^{-3/2} \int d^3k \sqrt{\frac{\omega}{2}} \left[-i \Big(a^0(\boldsymbol{k}) + a_\parallel(\boldsymbol{k}) \Big) e^{-ikx} \right.$$
$$\left. + i \Big(-a^0(\boldsymbol{k})^+ + a_\parallel(\boldsymbol{k})^+ \Big) e^{ikx} \right]. \tag{1.3.32}$$

If Φ, Φ' are states without scalar and longitudinal "photons", that means,

$$a^0 \Phi = 0 \quad , \quad a_\parallel \Phi = 0, \tag{1.3.33}$$

then we obviously have

$$(\Phi, \partial_\mu A^\mu(x) \Phi') = 0.$$

The subspace of states Φ (1.3.33) without scalar and longitudinal modes is called the physical subspace $\mathcal{F}_{\text{phys}}$ of the vector field. Only such states will be taken as incoming and outgoing photon states in scattering theory. In fact, as will be shown later on, the S-matrix of a gauge theory is only unitary if it is restricted to $\mathcal{F}_{\text{phys}}$.

We now turn to Poincaré covariance. For this purpose it is convenient to use 4-dimensional smearing and consider the field operators as distributions over real test functions $\varphi_\mu(x) \in S(\mathbb{R}^4)$ where φ is now a classical four-vector field in Minkowski space. We introduce the operator

$$A(\varphi) = \int d^4x \, \varphi_\mu(x) A^\mu(x) \qquad (1.3.34)$$

in Fock space. The Poincaré transformation of the test functions

$$\varphi'_\mu(x') = \Lambda_\mu{}^\nu \varphi_\nu(x) = \Lambda_\mu{}^\nu \varphi_\nu(\Lambda^{-1}(x'-a)) \qquad (1.3.35)$$

can be lifted into Fock space by the definition

$$\mathbf{U}(a,\Lambda) A(\varphi) \mathbf{U}(a,\Lambda)^{-1} \stackrel{\text{def}}{=} A(\varphi') \qquad (1.3.36)$$

$$= \int d^4x' \, \Lambda_\mu{}^\nu \varphi_\nu(\Lambda^{-1}(x'-a)) A^\mu(x')$$

$$= \int d^4x \, \varphi_\nu(x) \left(\Lambda^{-1}\right)^\nu{}_\mu A^\mu(\Lambda x + a).$$

This leads to the following transformation law

$$\mathbf{U}(a,\Lambda) A^\nu(x) \mathbf{U}(a,\Lambda)^{-1} = \left(\Lambda^{-1}\right)^\nu{}_\mu A^\mu(\Lambda x + a). \qquad (1.3.37)$$

Although we have used the suggestive letter $\mathbf{U}$ in these equations, the representation $\mathbf{U}(a,\Lambda)$ defined by (1.3.37) is not unitary. This is a consequence of the non-selfadjointness of $A^0(x)$. But it is pseudo-unitary in the following sense

$$\mathbf{U}(a,\Lambda)^K = \mathbf{U}(a,\Lambda)^{-1}. \qquad (1.3.38)$$

This follows by taking the conjugate of (1.3.36). However the pseudo-unitarity (1.3.38) cannot be the whole story. The physical content of relativistic invariance is the fact that two observers in uniform motion relative to each other observe the same physics. Accordingly, proper Poincaré transformations must give rise to a *unitary* mapping between physical states in $\mathcal{F}_{\text{phys}}$. We are now going to show how this comes about.

In the discussion so far, it was not necessary to use a concrete realization of the Fock space $\mathcal{F}$. To introduce such a representation, we consider a general time-dependent one-"photon" state

$$\Phi = \int d^3k \, \hat{f}_\mu(\mathbf{k}) e^{i\omega t} a_\mu(\mathbf{k})^+ \Omega \qquad (1.3.39)$$

and represent it by the four-vector potential f

Free fields

$$f^\mu(t, \boldsymbol{x}) = (2\pi)^{-3/2} \int \frac{d^3k}{\sqrt{2\omega}} \, \hat{f}^\mu(\boldsymbol{k}) e^{i(\omega t - \boldsymbol{k}\cdot\boldsymbol{x})}. \tag{1.3.40}$$

This is a complex solution of the wave equation with "positive frequencies", according to the usual terminology. The denominator $\sqrt{2\omega}$ has been introduced for reasons of covariance. It is convenient to define the Fock space scalar product by means of time derivatives in such a way that the factors 2ω drop out:

$$(f, g) \stackrel{\text{def}}{=} \sum_{\mu=0}^{3} i \int d^3x \, [f_\mu^* \partial_t g_\mu - (\partial_t f_\mu^*) g_\mu]. \tag{1.3.41}$$

In fact, from (1.3.40) we then obtain

$$(f, g) = \sum_{\mu=0}^{3} \int d^3k \, \hat{f}_\mu(\boldsymbol{k})^* \hat{g}_\mu(\boldsymbol{k}). \tag{1.3.42}$$

This is the usual positive definite L^2 scalar product, in agreement with (Φ, Ψ) computed from (1.3.39) by means of the commutation relations (1.3.6). Furthermore, since (1.3.42) is constant in time, the time evolution is unitary. But the sum over μ in (1.3.41) is not a Minkowski product. We therefore expect troubles with Lorentz invariance. Nevertheless, we can rewrite (1.3.41) in covariant form as a surface integral

$$(f, g) = \sum_{\mu=0}^{3} i \int_{t=\text{const}} d\sigma^\nu(x) \, [f_\mu^* \partial_\nu g_\mu - (\partial_\nu f_\mu^*) g_\mu], \tag{1.3.43}$$

which can be taken over an arbitrary smooth space-like surface S. This is a consequence of Gauss' theorem

$$\left(\int_{t=\text{const}} - \int_S \right) = i \int d^4x \, \partial^\nu \left[f_\mu^* \partial_\nu g_\mu - (\partial_\nu f_\mu^*) g_\mu \right] = 0, \tag{1.3.44}$$

because f_μ^* and g_μ are solutions of the wave equation. The generalization of the construction to many-particle states is straightforward.

The vectors $\Phi \in \mathcal{F}_\text{phys}$, defined by (1.3.33), obviously obey the radiation gauge, that means

$$\Phi_0 = 0 \quad, \quad \partial_j \Phi_j(x) = 0, \tag{1.3.45}$$

in the case of a one-"photon" state, because the first condition is the absence of scalar photons and then the absence of longitudinal photons implies the second condition. We are a little sloppy with the notation here, because we use the same symbol Φ for the element in Fock space and for its one-particle component, but this causes no confusion. Although the condition (1.3.45) depends on the frame of reference, the whole space $\mathcal{F}_\text{phys}$ is the same for all observers, as we will shortly see. The states which satisfy the Lorentz condition (1.3.2) form a subspace $\mathcal{F}_L$ in $\mathcal{F}$ (L stands for Lorentz). Any one-particle state Ψ in $\mathcal{F}_L$ is obtained from some physical state $\Phi \in \mathcal{F}_{\text{phys}}$ by a gauge transformation

$$\Psi_\mu(x) = \Phi_\mu(x) + \partial_\mu \Lambda(x). \tag{1.3.46}$$

Φ is uniquely determined by Ψ, because the radiation gauge is unique. This means, geometrically speaking, that there is a *fibration* in $\mathcal{F}_L$: $\mathcal{F}_{\text{phys}}$ is the base space, a general $\Psi \in \mathcal{F}_L$ is connected with a unique $\Phi \in \mathcal{F}_{\text{phys}}$ by a fibre which consists of gauge equivalent states. The projection on $\mathcal{F}_{\text{phys}}$ along the fibres, that means by gauge transformations (1.3.46), is different from the orthogonal projection in the Hilbert space sense. The latter is not used here.

Let us now consider two one-particle states Ψ, $\Phi \in \mathcal{F}_{\text{phys}}$. Since the zero components Φ_0, Ψ_0 vanish due to the radiation gauge, the scalar product (1.3.43) can be written as

$$(\Phi, \Psi) = -(\Phi_\mu, \Psi^\mu), \tag{1.3.47}$$

where the usual summation convention is applied if the Greek indices are written up and down. Under a Poincaré transformation (1.3.35) the states are rotated out of $\mathcal{F}_{\text{phys}}$, in general, because the transformed states Φ'_μ, Ψ'^μ have non-vanishing zero components. Nevertheless, (1.3.47) remains invariant

$$(\Phi, \Psi) = -(\Phi'_\mu, \Psi'^\mu), \tag{1.3.48}$$

because this is now a Minkowski product which renders the bilinear form (1.3.43) invariant. Furthermore, the transformed states are in $\mathcal{F}_L$ because they satisfy the Lorentz gauge condition

$$0 = \partial^\mu \Phi_\mu = \partial'^\mu \Phi'_\mu, \tag{1.3.49}$$

and the same for Ψ'. Next we transform back into $\mathcal{F}_{\text{phys}}$ by projecting along the fibres, that means by gauge transformations (1.3.46)

$$\Phi'_\mu = \tilde{\Phi}_\mu + \partial_\mu \chi \quad , \quad \Psi'_\mu = \tilde{\Psi}_\mu + \partial_\mu \Lambda. \tag{1.3.50}$$

The new states are in the radiation gauge

$$\tilde{\Phi}_0 = 0 \quad , \quad \partial_j \tilde{\Phi}_j = 0 \tag{1.3.51}$$

and the same for $\tilde{\Psi}$. Let us now consider

$$-(\Phi'_\mu, \Psi'^\mu) = -(\tilde{\Phi}_\mu, \tilde{\Psi}^\mu) - (\partial_\mu \chi, \tilde{\Psi}^\mu) - (\tilde{\Phi}_\mu, \partial^\mu \Lambda) - (\partial_\mu \chi, \partial^\mu \Lambda). \tag{1.3.52}$$

Rewriting the scalar products in the non-covariant form (1.3.41), we see that the second and third term vanish by (3-dimensional) partial integration. The gauge functions χ, Λ satisfy the wave equation because the gauge transformations are within the Lorentz class. Then the last term in (1.3.52) can be transformed as follows

$$(\partial_\mu \chi, \partial^\mu \Lambda) = (\partial_0 \chi, \partial_0 \Lambda) - (\partial_j \chi, \partial_j \Lambda)$$
$$= (\partial_0 \chi, \partial_0 \Lambda) + \tfrac{1}{2}(\triangle \chi, \Lambda) + \tfrac{1}{2}(\chi, \triangle \Lambda)$$
$$= (\partial_0 \chi, \partial_0 \Lambda) + \tfrac{1}{2}(\partial_0^2 \chi, \Lambda) + \tfrac{1}{2}(\chi, \partial_0^2 \Lambda)$$
$$= \tfrac{1}{2}\partial_0^2(\chi, \Lambda) = 0, \qquad (1.3.53)$$

because the scalar product between two solutions of the wave equation is constant in time (1.3.42). Hence we arrive at

$$(\Phi, \Psi) = -(\tilde\Phi_\mu, \tilde\Psi^\mu) = (\tilde\Phi, \tilde\Psi), \qquad (1.3.54)$$

which is the desired unitary mapping in $\mathcal{F}_\text{phys}$. In the whole process the pseudo-unitary Lorentz transformation $\mathbf{U}(\Lambda)$ cooperates with gauge transformations to give a unitary transformation $\tilde{\mathbf{U}}(\Lambda)$ in $\mathcal{F}_\text{phys}$. This establishes the important fact that $\mathcal{F}_\text{phys}$ is independent of the reference frame.

1.4 Operator gauge transformations

In the last section we have considered gauge transformations on states in Fock space. Those are classical gauge transformations in the sense of adding the gradient of a *function*. Now we consider gauge transformations of the field operators themselves of the form

$$A'^\mu(x) = A^\mu(x) + \lambda \partial^\mu u(x) + O(\lambda^2), \qquad (1.4.1)$$

where $u(x)$ will be a free quantum field. We require the wave equation

$$\Box u(x) = 0 \qquad (1.4.2)$$

because we want the transformed field $A'^\mu(x)$ to satisfy the wave equation also. In addition, $A'^\mu(x)$ should fulfill the same commutation relations (1.3.14) as $A^\mu(x)$. This is true if the gauge transformation (1.4.1) is of the following form

$$A'^\mu(x) = e^{-i\lambda Q} A^\mu(x) e^{i\lambda Q}, \qquad (1.4.3)$$

which can be expanded by means of the Lie series

$$= A^\mu(x) - i\lambda [Q, A^\mu(x)] + O(\lambda^2). \qquad (1.4.4)$$

Comparing this with (1.4.1) we conclude

$$[Q, A^\mu(x)] = i\partial^\mu u(x), \qquad (1.4.5)$$

which determines Q essentially uniquely, that means up to a C-number. This is again a consequence of the irreducibility of the Fock representation (see FQED, Sect.2.1). Therefore, to make life simple it is sufficient to write down an expression for Q and then verify (1.4.5).

Proposition 1.4.1. Q is given by

$$Q = \int d^3x \, [\partial_\nu A^\nu \partial_0 u - (\partial_0 \partial_\nu A^\nu) u]$$

$$\stackrel{\text{def}}{=} \int d^3x \, \partial_\nu A^\nu \overleftrightarrow{\partial}_0 u \qquad (1.4.6)$$

where the integrals are taken over any plane $x^0 = \text{const}$.

Proof.

$$[Q, A^\mu(x)] = \int_{y^0=t} d^3y \, i[\partial_y^\mu D_0(y-x)\partial_0 u(y) - \partial_0^y \partial_y^\mu D_0(y-x) u(y)]$$

Using $\partial_y^\mu = -\partial_x^\mu$ and $D_0(y-x) = -D_0(x-y)$ this is equal to

$$= i\partial_x^\mu \int_{y^0=t} d^3y \, [D_0(x-y)\partial_0 u(y) + \partial_0^x D_0(x-y) u(y)]. \qquad (1.4.7)$$

The causal distribution $D_0 = D_0^{\text{ret}} - D_0^{\text{av}}$ can be split into a retarded and advanced part (1.1.15). Then the integral with D_0^{ret} is just the solution of the Cauchy problem for the wave equation for $x^0 > t$ with initial data at time t. Similarly, the integral with $-D_0^{\text{av}}$ gives the solution for times $x^0 < t$, so that (1.4.7) is equal to $i\partial_x^\mu u(x)$. This proves (1.4.5). By a similar calculation we obtain the commutators of the positive and negative frequency parts

$$[Q, A_\pm^\mu(x)] = \int d^3y \, [\partial_\nu A_\mp^\nu \partial_0 u - \partial_0 \partial_\nu A_\mp^\nu u, A_\pm^\mu(x)]$$

$$= \int d^3y \, i \Big(\partial_y^\mu D^{(\pm)}(y-x) \partial_0 u(y) - \partial_0^y \partial_y^\mu D^{(\pm)}(y-x) u(y) \Big) = i\partial^\mu u_\pm(x). \qquad (1.4.8)$$

That all these expressions are actually well defined operators follows from the corresponding expressions in momentum space given below (1.4.16). □

The operator Q (1.4.6) will be called gauge charge because it is the infinitesimal generator of the gauge transformation (1.4.1). For the following it is important to have Q nilpotent

$$Q^2 = 0. \qquad (1.4.9)$$

Such a nilpotency is characteristic for Fermi operators. Therefore, we assume $u(x)$ to be a fermionic scalar field (1.2.1) with mass zero, a so-called ghost field. In fact, calculating Q^2 as the anticommutator

$$Q^2 = \tfrac{1}{2}\{Q, Q\}$$

and using the general identity

$$\{AB, C\} = A\{B, C\} - [A, C]B \qquad (1.4.10)$$

we find

$$\tfrac{1}{2}\{Q, Q\} = \tfrac{1}{2}\int d^3x\, \partial_\nu A^\nu \{\overleftrightarrow{\partial}_0 u, Q\} - \tfrac{1}{2}\int d^3x\, [\partial_\nu A^\nu, Q]\overleftrightarrow{\partial}_0 u. \qquad (1.4.11)$$

The first term is zero because u anticommutes with Q due to (1.2.7). The commutator in the second term gives $-i\partial_\nu \partial^\nu u = -i\Box u = 0$. This shows the nilpotency (1.4.9).

Using previous representations of $A^\mu(x)$ (1.3.27) and $u(x)$ (1.2.1) in momentum space we get the gauge charge in momentum space

$$Q = \int d^3k\, \omega(\boldsymbol{k})[(a_\|(\boldsymbol{k})^+ - a_0(\boldsymbol{k})^+)c_2(\boldsymbol{k})$$
$$+ c_1(\boldsymbol{k})^+(a_\|(\boldsymbol{k}) + a_0(\boldsymbol{k}))] \qquad (1.4.12)$$

where

$$a_\|(\boldsymbol{k}) = \frac{k_j}{\omega} a^j(\boldsymbol{k}) \qquad (1.4.13)$$

is the absorption operator for longitudinal vector bosons. It is convenient to introduce new bosonic operators

$$b_1(\boldsymbol{k}) = \frac{1}{\sqrt{2}}(a_\|(\boldsymbol{k}) + a_0(\boldsymbol{k}))$$

$$b_2(\boldsymbol{k}) = \frac{1}{\sqrt{2}}(a_\|(\boldsymbol{k}) - a_0(\boldsymbol{k})). \qquad (1.4.14)$$

They satisfy the ordinary commutation relations

$$[b_i(\boldsymbol{k}), b_j^+(\boldsymbol{k}')] = \delta_{ij}\delta^3(\boldsymbol{k} - \boldsymbol{k}'). \qquad (1.4.15)$$

Then Q and its adjoint assume the following form

$$Q = \sqrt{2}\int d^3k\, \omega(\boldsymbol{k})[b_2^+(\boldsymbol{k})c_2(\boldsymbol{k}) + c_1^+(\boldsymbol{k})b_1(\boldsymbol{k})] \qquad (1.4.16)$$

$$Q^+ = \sqrt{2}\int d^3k\, \omega(\boldsymbol{k})[b_1^+(\boldsymbol{k})c_1(\boldsymbol{k}) + c_2^+(\boldsymbol{k})b_2(\boldsymbol{k})]. \qquad (1.4.17)$$

These formulas show that Q is actually a well defined unbounded operator in Fock space. As one generates a sector of Fock space from the vacuum by applying the creation operators b_i^+, c_j^+ as in (1.1.31), one can easily specify the domains of these unbounded operators. The considered sector of Fock space contains unphysical particles (longitudinal and scalar modes of the gauge field and ghosts), only. This allows to give another equivalent definition of the physical subspace $\mathcal{F}_{\text{phys}}$.

Let us calculate the anticommutator

$$\{Q^+, Q\} = 2 \int d^3k\, \omega^2(\boldsymbol{k})[b_1^+(\boldsymbol{k})b_1(\boldsymbol{k}) + b_2^+ b_2 + c_1^+ c_1 + c_2^+ c_2]. \qquad (1.4.18)$$

This follows from the mixed Bose- Fermi-calculation

$$\{b_1^+(\boldsymbol{k})c_1(\boldsymbol{k}), c_1^+(\boldsymbol{k}')b_1(\boldsymbol{k}')\} = b_1^+(\boldsymbol{k})\{c_1(\boldsymbol{k}), c_1^+(\boldsymbol{k}')\}b_1(\boldsymbol{k}')$$
$$- [b_1^+(\boldsymbol{k}), c_1^+(\boldsymbol{k}')b_1(\boldsymbol{k}')]c_1(\boldsymbol{k})$$
$$= (b_1(\boldsymbol{k})b_1(\boldsymbol{k}) + c_1^+(\boldsymbol{k})c_1(\boldsymbol{k}))\delta(\boldsymbol{k} - \boldsymbol{k}'), \qquad (1.4.19)$$

where (1.4.10) has again been used. Apart from the factor $\omega^2(\boldsymbol{k})$, the anticommutator (1.4.18) is the sum of particle number operators of the four kinds of unphysical particles. Consequently, if a state Φ in Fock space satisfies

$$\{Q^+, Q\}\Phi = 0, \qquad (1.4.20)$$

then it contains physical particles, only, that means the physical Hilbert space is the kernel

$$\mathcal{F}_{\text{phys}} = \text{Ker}\,\{Q^+, Q\}. \qquad (1.4.21)$$

We remark that

$$\{Q^+, Q\} = Q^+ Q + Q Q^+$$

is always selfadjoint and positive:

$$(f, (Q^+Q + QQ^+)f) = \|Qf\|^2 + \|Q^+ f\|^2 \geq 0. \qquad (1.4.22)$$

This only vanishes if and only if $Qf = 0 = Q^+ f$ so that we have another characterization of the physical Fock space:

$$\mathcal{F}_{\text{phys}} = \text{Ker}\, Q \cap \text{Ker}\, Q^+. \qquad (1.4.23)$$

The structure of the total Fock space $\mathcal{F}$ can now be described as follows. $\text{Ker}\, Q$ is always a subspace and it is orthogonal to $\overline{\text{Ran}\, Q^+}$ where Ran is the range and the over-line means the closure. In fact, for $f \in \text{Ker}\, Q$ we have

$$(Qf, g) = 0 = (f, Q^+ g).$$

We claim that $\mathcal{F}$ has the direct decompositions

$$\mathcal{F} = \text{Ker}\, Q \oplus \overline{\text{Ran}\, Q^+} = \text{Ker}\, Q^+ \oplus \overline{\text{Ran}\, Q}. \qquad (1.4.24)$$

To prove this we notice that the domain $D(Q^+)$ is dense in $\mathcal{F}$, and if $(h, Q^+g) = 0$ for all $g \in D(Q^+)$ then $(Qh, g) = 0$ which implies $Qh = 0$, that means $h \in \text{Ker}\, Q$. Now we use the nilpotency $Q^2 = 0$. From $(Q^+g, Qf) = (g, Q^2 f) = 0$ we see that $\overline{\text{Ran}\, Q^+}$ is orthogonal to $\overline{\text{Ran}\, Q}$. Then $\mathcal{F}$ has the following direct decomposition

$$\mathcal{F} = \overline{\text{Ran}\, Q^+} \oplus \overline{\text{Ran}\, Q} \oplus \mathcal{F}_{\text{phys}}. \qquad (1.4.25)$$

Indeed, let P_1 and P_2 be the projection operators on the first two subspaces in (1.4.25). Since they are orthogonal, we have $P_1 P_2 = 0 = P_2 P_1$. Then the projection operator on their orthogonal complement is equal to

$$1 - (P_1 + P_2) = (1 - P_1)(1 - P_2),$$

and this is the projection onto $\operatorname{Ker} Q \cap \operatorname{Ker} Q^+$ which is the physical subspace (1.4.23). This proves (1.4.25). Comparing (1.4.24) and (1.4.25) we obviously have

$$\operatorname{Ker} Q = \mathcal{F}_{\text{phys}} \oplus \overline{\operatorname{Ran} Q}.$$

This implies the third expression for the physical subspace

$$\mathcal{F}_{\text{phys}} = \operatorname{Ker} Q / \overline{\operatorname{Ran} Q}. \tag{1.4.26}$$

Note that $\operatorname{Ran} Q = D(Q^{-1})$ is not closed because Q^{-1} is unbounded, if the gauge field A^μ is massless.

Now we return to the defining property of Q as being the infinitesimal generator of gauge transformations (1.4.3) (1.4.5). We introduce the notation

$$d_Q F = [Q, F],$$

if F contains only Bose fields and an even number of ghost fields, and

$$d_Q F = \{Q, f\} = QF + FQ, \tag{1.4.27}$$

if F contain an odd number of ghost fields. Then d_Q has all properties of an anti-derivation, in particular (1.4.10) implies the product rule

$$d_Q(F(x)G(y)) = (d_Q F(x))G(y) + (-1)^{n_F} F(x) d_Q G(y), \tag{1.4.28}$$

where n_F is the ghost number of F, i.e. the number of u's minus the number of $\tilde{u}$'s. The fundamental gauge variations are

$$d_Q A^\mu = i\partial^\mu u, \quad d_Q A^\mu_\pm = i\partial^\mu u_\pm \tag{1.4.29}$$

$$d_Q u = 0, \quad d_Q \tilde{u} = \{Q, \tilde{u}\} = -i\partial_\mu A^\mu, \quad d_Q \tilde{u}_\pm = -i\partial_\mu A^\mu_\pm. \tag{1.4.30}$$

The latter follows from the anticommutation relation (1.2.6) in the same way as (1.4.7). d_Q changes the ghost number by one, i.e. a Bose field goes over into a Fermi field and vice versa. Then the nilpotency $Q^2 = 0$ implies for a Bose field F_B

$$d_Q^2 F_B = \{Q, [Q, F_B]\} = Q(QF_B - F_B Q) + (QF_B - F_B Q)Q = 0,$$

and for a Fermi field F

$$d_Q^2 F = [Q, \{Q, F\}] = Q(QF_B + F_B Q) - (QF_B - F_B Q)Q = 0,$$

hence

$$d_Q^2 = 0 \tag{1.4.31}$$

is also nilpotent. In this situation we can use notions from homological algebra, for example if
$$F = d_Q G \tag{1.4.32}$$
then F is called a co-boundary (see W.S.Massey, *Homology and cohomology theory*, Dekker 1978, for example).

Finally let us consider finite operator gauge transformations (1.4.3). It is easy to see that the Lie series
$$A'^\mu(x) = A^\mu(x) - \frac{i\lambda}{1!}[Q, A^\mu] - \frac{\lambda^2}{2!}\Big[Q, [Q, A^\mu]\Big] + \ldots \tag{1.4.33}$$
terminates after the second order term. In fact, since
$$[Q, u(x)] = Qu(x) - u(x)Q = -2u(x)Q,$$
the next commutator with Q
$$[Q, u(x)Q] = [Q, u(x)]Q + u(x)[Q, Q] = 0$$
vanishes. Consequently, the finite gauge transformation of the vector field is given by
$$A'^\mu(x) = A^\mu(x) + \lambda \partial^\mu u(x) + i\lambda^2 \partial^\mu u(x) Q. \tag{1.4.34}$$
In a similar way one finds the gauge transformations of the ghost fields
$$u'(x) = u(x) + 2i\lambda u(x) Q \tag{1.4.35}$$
$$\tilde{u}'(x) = \tilde{u}(x) + \lambda(2i\tilde{u}(x)Q - \partial_\mu A^\mu(x)) - i\lambda^2 \partial_\mu A^\mu(x) Q. \tag{1.4.36}$$
The gauge variation d_Q is also called "infinitesimal gauge transformation".

1.5 Massive vector fields

These fields will be used to represent the $W^\pm$- and Z-bosons of the electroweak theory, for example. They obey the Klein-Gordon equation
$$(\Box + m^2) A^\mu(x) = 0. \tag{1.5.1}$$
Since a spin-1 field has three physical degrees of freedom, we only need one subsidiary condition, instead of two in the massless case. As this we choose the Lorentz condition
$$\partial_\mu A^\mu(x) = 0. \tag{1.5.2}$$
Since it is Lorentz-invariant, we do not expect problems with Poincaré covariance in the quantization procedure.

It is convenient to introduce real polarization vectors, first
$$\varepsilon_0^\mu(p) = \frac{p^\mu}{m}, \tag{1.5.3}$$

which is normalized to 1 because $p^2 = m^2$, and three transversal vectors

$$p_\mu \varepsilon_j^\mu(p) = 0, \quad j = 1, 2, 3. \tag{1.5.4}$$

The latter are also normalized such that

$$\varepsilon_\lambda^\mu \varepsilon_{\lambda'\mu} = g_{\mu\nu}\varepsilon_\lambda^\mu(p)\varepsilon_{\lambda'}^\nu(p) = g_{\lambda\lambda'}. \tag{1.5.5}$$

Since the four vectors form an orthonormal basis in Minkowski space we have

$$\sum_{\lambda=0}^{3} g_{\lambda\lambda}\varepsilon_\lambda^\mu(p)\varepsilon_\lambda^\nu(p) = g^{\mu\nu}. \tag{1.5.6}$$

For later use we write this equation in the form

$$\sum_{\lambda=1}^{3} \varepsilon_\lambda^\mu(p)\varepsilon_\lambda^\nu(p) = -g^{\mu\nu} + \frac{p^\mu p^\nu}{m^2}. \tag{1.5.7}$$

Similarly as in the massless case (1.3.27) a solution of the Klein-Gordon equation (1.5.1) is given by

$$A^\mu(x) = (2\pi)^{-3/2} \sum_{\lambda=0}^{3} \int \frac{d^3p}{\sqrt{2E}} \left(\varepsilon_\lambda^\mu(p) a_\lambda(\boldsymbol{p}) e^{-ipx} + \varepsilon_\lambda^\mu(p) a_\lambda^K(\boldsymbol{p}) e^{ipx} \right), \tag{1.5.8}$$

where the conjugation K is defined by

$$a_0(\boldsymbol{p})^K = -a_0(\boldsymbol{p})^+, \quad a_j(\boldsymbol{p})^K = a_j(\boldsymbol{p})^+, \quad j = 1, 2, 3. \tag{1.5.9}$$

Then the absorption and emission operators satisfy the usual commutation relations

$$[a_\lambda(\boldsymbol{p}), a_{\lambda'}(\boldsymbol{q})^+] = \delta_{\lambda\lambda'}\delta^3(\boldsymbol{p} - \boldsymbol{q}) \tag{1.5.10}$$

and can be represented on a Fock space with positive-definite metric as in sect.1.1. The vector field now obeys the same commutation relation as in the massless case (1.3.14)

$$[A^\mu(x), A^\nu(y)] = g^{\mu\nu}iD_m(x - y). \tag{1.5.11}$$

To get a Lorentz-covariant representation we again introduce the modified operators

$$\tilde{a}_\lambda(p) = \sqrt{2p_0}a_\lambda(\boldsymbol{p}), \quad \tilde{a}_\lambda^+(p) = \sqrt{2p_0}a_\lambda(\boldsymbol{p})^+$$

which obey the commutation relations

$$[\tilde{a}_\lambda(p), \tilde{a}_{\lambda'}(q)^+] = \delta_{\lambda\lambda'}2E(\boldsymbol{p})\delta^3(\boldsymbol{p} - \boldsymbol{q}) \tag{1.5.12}$$

and all other commutators vanish. We smear (1.5.8) with four-dimensional real test functions $f(x) \in \mathcal{S}(\mathbb{R}^4)$

$$A^\mu(f) = \sqrt{2\pi} \sum_{\lambda=0}^{3} \int d\mu_m(p)\, \varepsilon_\lambda^\mu(p) \left(\tilde{a}_\lambda(p) \hat{f}(-p) + \tilde{a}_\lambda(p)^K \hat{f}(p) \right)$$

$$\stackrel{\text{def}}{=} \sqrt{2\pi}(\tilde{a}^\mu(\hat{f}) + \tilde{a}^\mu(\hat{f})^K), \tag{1.5.13}$$

where the measure $d\mu_m(p)$ on the positive mass shell $\mathcal{M}^+$ is given by (1.1.16). The commutation relations for the smeared operators follow from (1.5.12) using (1.5.6):

$$[\tilde{a}^\mu(\hat{f}), \tilde{a}^\nu(\hat{g})^K] = \int_{\mathcal{M}^+} \frac{d^3p}{2p^0} \hat{f}(-p) g(p) \sum_\lambda \varepsilon_\lambda^\mu(p) \varepsilon_\lambda^\nu(p)(-g_{\lambda\lambda})$$

$$= -g^{\mu\nu}(\hat{f}, \hat{g})_m. \tag{1.5.14}$$

The factor $-g_{\lambda\lambda}$ is due to the minus sign in the definition of a_0^K (1.5.9). We see that the commutation relation with the conjugation K instead of the adjoint is Lorentz covariant. On the other hand, the form

$$[\tilde{a}^j(\hat{f}), \tilde{a}^k(\hat{g})^+] = \delta_{jk}(\hat{f}, \hat{g})_m \tag{1.5.15}$$

$$[\tilde{a}^0(\hat{f}), -\tilde{a}^0(\hat{g})^+] = -(\hat{f}, \hat{g})_m \tag{1.5.16}$$

shows that the four components are quantized as four scalar fields in a Hilbert space with positive definite metric. The one-particle sector is represented by

$$\tilde{a}^\mu(\hat{f})^+ \Omega \longleftrightarrow f^\mu(p) \in L^2(\mathcal{M}^+, d\mu_m) \tag{1.5.17}$$

and the n-particle sector $\mathcal{H}_n$ by

$$\frac{1}{\sqrt{n!}} \prod_{j=1}^{n} \tilde{a}^{\mu_j}(\hat{f}_j)^+ \Omega \longleftrightarrow S_n f_1^{\mu_1} \otimes \ldots \otimes f_n^{\mu_n}, \tag{1.5.18}$$

and the whole Fock space is the direct sum

$$\mathcal{F} = \oplus_{n=0}^{\infty} \mathcal{H}_n. \tag{1.5.19}$$

For arbitrary operators the conjugation K is defined as

$$B^K = \eta B^+ \eta, \tag{1.5.20}$$

where

$$\eta = (-1)^{\mathbf{N}_0} \tag{1.5.21}$$

is the Krein operator and

$$\mathbf{N}_0 = \int d^3p\, a_0(\mathbf{p})^+ a_0(\mathbf{p}) = \int d\mu_m(p)\, \tilde{a}_0(p)^+ \tilde{a}_0(p) \tag{1.5.22}$$

the particle number operator for the longitudinal mode (1.5.3). We now consider the representation of the proper Poincaré group in Fock space which is defined by

$$\mathbf{U}(a,\Lambda)A^\mu(x)\mathbf{U}(a,\Lambda)^{-1} = \Lambda^\mu{}_\nu A^\nu(\Lambda x + a). \tag{1.5.23}$$

We take the conjugate and use the self-conjugacy of A^ν

$$= \mathbf{U}(a,\Lambda)^{-1K} A^\mu(x) \mathbf{U}(a,\Lambda)^K.$$

It follows that $\mathbf{U}(a,\Lambda)^K \mathbf{U}(a,\Lambda)$ commutes with all $A^\mu(x)$. By the irreducibility of the Fock representation this implies

$$\mathbf{U}(a,\Lambda)^K = \mathbf{U}(a,\Lambda)^{-1}, \tag{1.5.24}$$

if an irrelevant phase is set equal to zero. That means, the representation is pseudo-unitary. But it is even unitary, in contrast to the massless case.

To see this we perform in

$$A^\nu(x) = (2\pi)^{-3/2} \sum_{\lambda=0}^{3} \int d\mu_m(p)\, \varepsilon_\lambda^\nu(p) \left(\tilde{a}_\lambda(p) e^{-ipx} + \tilde{a}_\lambda(p)^K e^{ipx} \right) \tag{1.5.25}$$

a Poincaré transformation

$$\Lambda^\mu{}_\nu A^\nu(\Lambda x + a) = (2\pi)^{-3/2} \sum_{\lambda=0}^{3} \int d\mu_m(p)\, \Lambda^\mu{}_\nu \varepsilon_\lambda^\nu(p) \left(\tilde{a}_\lambda(p) e^{-ip(\Lambda x + a)} \right.$$
$$\left. + \tilde{a}_\lambda(p)^K e^{ip(\Lambda x + a)} \right). \tag{1.5.26}$$

We smear this with $f(x)$ and consider the emission part

$$e^{ipa} \Lambda^\mu{}_\nu \tilde{a}^\nu(\hat{f}(\Lambda^{-1}p))^K = \mathbf{U}(a,\Lambda) \tilde{a}^\mu(\hat{f})^K \mathbf{U}(a,\Lambda)^{-1}. \tag{1.5.27}$$

Applying it to the invariant vacuum we find

$$e^{ipa} \Lambda^\mu{}_\nu f^\nu(\Lambda^{-1}p) = (\mathbf{U}(a,\Lambda) f^\mu)(p). \tag{1.5.28}$$

Under Lorentz transformations the $\lambda = 0$ polarization is transformed into itself

$$\Lambda^\mu{}_\nu \varepsilon_0^\nu = (\Lambda \varepsilon_0)^\mu = \frac{(\Lambda p)^\mu}{m},$$

therefore.

$$(\mathbf{U}(a,\Lambda) f_0^\mu)(p) = e^{ipa} f_0^\mu(\Lambda^{-1}p) \tag{1.5.29}$$

and similarly in the n-particle sector

$$(\mathbf{U}(a,\Lambda) \varphi_{n0}^{\mu_1,\ldots,\mu_n})(p_1,\ldots,p_n) = \left(\exp\left(i \sum_{j=1}^n p_j a \right) \varphi_{n0}^{\mu_1,\ldots,\mu_n} \right)(\Lambda^{-1}p_1,\ldots,\Lambda^{-1}p_n). \tag{1.5.30}$$

Then the particle number operator

$$(\mathbf{N}_0 \varphi_{n0}^{\mu_1,\ldots,\mu_n})(p_1,\ldots,p_n) = n \varphi_{n0}^{\mu_1,\ldots,\mu_n}(p_1,\ldots,p_n)$$
$$= 0 \quad \text{if} \quad \lambda = 1,2,3$$

commutes with $\mathbf{U}(a,\Lambda)$. Consequently, the Krein operator η also commutes with $\mathbf{U}$ and $\mathbf{U}(a,\Lambda)$ is unitary.

The longitudinal mode $\lambda = 0$ is unphysical. As in the massless case we would like to characterize the physical subspace with help of a nilpotent gauge charge Q. The old definition (1.4.6) does not work because in (1.4.11) we now get

$$Q^2 = \tfrac{1}{2}\{Q,Q\} = \tfrac{1}{2}i \int d^3x\, (\Box u)\overleftrightarrow{\partial}_0 u = -\tfrac{1}{2}im^2 \int d^3x\, u\overleftrightarrow{\partial}_0 u \neq 0.$$

To restore the nilpotency we modify the expression for Q by introducing a scalar field $\Phi(x)$ with the same mass m as the gauge field $A^\nu(x)$

$$Q = \int d^3x\, (\partial_\nu A^\nu + m\Phi)\overleftrightarrow{\partial}_0 u. \tag{1.5.31}$$

All fields satisfy the Klein-Gordon equation

$$(\Box + m^2)\Phi = 0 \tag{1.5.32}$$

$$(\Box + m^2)u = 0, \tag{1.5.33}$$

but, while $u(x)$ is a Fermi field, $\Phi(x)$ is quantized with commutation relations

$$[\Phi(x), \Phi(y)] = -iD_m(x-y), \tag{1.5.34}$$

and all other commutators are the same as in Sect.1.1. Now we can check the nilpotency:

$$Q^2 = -\tfrac{1}{2} \int d^3x\, [\partial_\nu A^\nu + m\Phi, Q] = 0, \tag{1.5.35}$$

because the first term in the commutator gives $-i\Box u$ and the second one $-im^2 u$, so that the sum is zero by (1.5.33). The infinitesimal gauge transformations or gauge variations are given by

$$d_Q A^\mu(x) = [Q, A^\mu(x)] = i\partial^\mu u(x) \tag{1.5.36}$$

$$d_Q \Phi(x) = [Q, \Phi(x)] = imu(x) \tag{1.5.37}$$

$$d_Q u(x) = \{Q, u(x)\} = 0 \tag{1.5.38}$$

$$d_Q \tilde{u}(x) = \{Q, \tilde{u}(x)\} = -i(\partial_\mu A^\mu + m\Phi(x)). \tag{1.5.39}$$

The last equation follows from (1.5.31); using $Q^2 = 0$ (1.5.39) implies (1.5.37).

Let us stress the difference between our approach to massive gauge fields and the conventional one (see the textbooks in the notes). In the usual approach one starts with massless gauge fields and the scalar field Φ is the so-called Goldstone boson. The fields become massive after "spontaneous breaking" of the gauge symmetry. We start directly with massive vector fields. To define a gauge variation with a nilpotent Q, we are forced to introduce the scalar field Φ, spontaneous symmetry breaking plays no role.

28   Free fields

Next we write Q in momentum space. For this purpose we use the ordinary representation (1.1.8) for the scalar field

$$\Phi(x) = (2\pi)^{-3/2} \int \frac{d^3p}{\sqrt{2E}} \left(b(\boldsymbol{p})e^{-ipx} + b^+(\boldsymbol{p})e^{ipx}\right). \tag{1.5.40}$$

and from (1.5.8) we obtain

$$\partial_\nu A^\nu = -i(2\pi)^{-3/2} \int \frac{d^3p}{\sqrt{2E}} \frac{p^2}{m} \left(a_0(\boldsymbol{p})e^{-ipx} + a_0^+(\boldsymbol{p})e^{ipx}\right). \tag{1.5.41}$$

Here we have used (1.5.3) and (1.5.4) which show that only the unphysical mode $\lambda = 0$ contributes to $\partial_\nu A^\nu$. Then from (1.5.31) we find

$$Q = \int d^3p \, \frac{p^2}{m} \Big[(c_1^+(\boldsymbol{p})(a_0(\boldsymbol{p}) + ib(\boldsymbol{p})) - (a_0^+(\boldsymbol{p}) + ib^+(\boldsymbol{p}))c_2(\boldsymbol{p}) \Big]. \tag{1.5.42}$$

We can now compute the anticommutator

$$\{Q, Q^+\} = 2 \int d^3p \, \frac{p^4}{m^2} \Big[a_0^+(\boldsymbol{p})a_0(\boldsymbol{p}) + b^+(\boldsymbol{p})b(\boldsymbol{p}) +$$

$$+ c_1^+(\boldsymbol{p})c_1(\boldsymbol{p}) + c_2^+(\boldsymbol{p})c_2(\boldsymbol{p}) \Big]. \tag{1.5.43}$$

The square bracket contains the particle number operators of all unphysical particles, in particular, the scalar field $\Phi(x)$ is also unphysical (it is the "eaten" Goldstone boson in the conventional theory). Then, in agreement with the massless case (1.4.21) the physical Fock space is given by the kernel

$$\mathcal{F}_{\text{phys}} = \text{Ker}\,\{Q^+, Q\}. \tag{1.5.44}$$

The physical part of the vector field contains the three transversal polarizations, only,

$$A^\mu_{\text{phys}}(x) = (2\pi)^{-3/2} \sum_{\lambda=1}^{3} \int d\mu_m(p)\, \varepsilon^\mu_\lambda(p) \left(\tilde{a}_\lambda(p)e^{-ipx} + \tilde{a}_\lambda(p)^+ e^{ipx}\right) \tag{1.5.45}$$

It is gauge invariant

$$d_Q A^\mu_{\text{phys}} = 0 \tag{1.5.46}$$

and satisfies the covariant commutation relation

$$[A^\mu_{\text{phys}}(x), A^\nu_{\text{phys}}(y)] = \left(g^{\mu\nu} + \frac{\partial^\mu \partial^\nu}{m^2}\right) iD_m(x-y). \tag{1.5.47}$$

This is a consequence of (1.5.7). Furthermore, in x-space we have

$$A^\mu_{\text{phys}} = A^\mu + \frac{1}{m^2} \partial^\mu \partial_\nu A^\nu. \tag{1.5.48}$$

1.6 Fermionic vector (ghost) fields

These fields occur as ghost fields in spin-2 gauge theories, for example in quantum gravity (see next section). Here we only need massless fields which satisfy the wave equation

$$\Box u^\mu(x) = 0 = \Box \tilde{u}^\mu(x). \tag{1.6.1}$$

As in Sect.1.2 in the case of fermionic scalar fields, we have to consider two vector fields which describe two different kinds of ghost particles

$$u^\mu(x) = (2\pi)^{-3/2} \int \frac{d^3k}{\sqrt{2\omega}} \left(c_2^\mu(\mathbf{k}) e^{-ikx} - g^{\mu\mu} c_1^\mu(\mathbf{k})^+ e^{ikx} \right) \tag{1.6.2}$$

$$\tilde{u}^\mu(x) = (2\pi)^{-3/2} \int \frac{d^3k}{\sqrt{2\omega}} \left(-c_1^\mu(\mathbf{k}) e^{-ikx} - g^{\mu\mu} c_2^\mu(\mathbf{k})^+ e^{ikx} \right) \tag{1.6.3}$$

where the $g^{\mu\mu}$ is necessary to get different signs for $\mu = 0$ and $\mu = 1, 2, 3$ (no sum over μ). If we assume normal anticommutation relations

$$\{c_j^\mu(\mathbf{k}), c_l^\nu(\mathbf{k}')^+\} = \delta_{jl} \delta^{\mu\nu} \delta^3(\mathbf{k} - \mathbf{k}'), \tag{1.6.4}$$

then, first these operators are annihilation and creation operators in a Fock-Hilbert space with positive definite scalar product and, second, the vector fields obey the covariant anticommutation relation

$$\{u^\mu(x), \tilde{u}^\nu(y)\} = ig^{\mu\nu} D_0(x - y) \tag{1.6.5}$$

and the anticommutators between two u's or two $\tilde{u}$'s vanish.

As in the previous sections we define a conjugation K in such a way that $u^\mu(x)$ becomes self-conjugate

$$u^\mu(x)^K = u^\mu(x). \tag{1.6.6}$$

This can be achieved by the definitions

$$c_j^n(\mathbf{k})^K = c_{j'}^n(\mathbf{k})^+, \quad n = 1, 2, 3 \tag{1.6.7}$$

with

$$j' = j \pm 1 \quad \text{for} \quad j = \begin{cases} 1 \\ 2 \end{cases} \tag{1.6.8}$$

and

$$c_j^0(\mathbf{k})^K = -c_{j'}^0(\mathbf{k})^+. \tag{1.6.9}$$

This gives

$$c_j^\mu(\mathbf{k})^K = -g^{\mu\mu} c_{j'}^\mu(\mathbf{k})^+, \tag{1.6.10}$$

so that (1.6.2) can be written in manifestly self-conjugate form

$$u^\mu(x) = (2\pi)^{-3/2} \int \frac{d^3k}{\sqrt{2\omega}} \left(c_2^\mu(\mathbf{k}) e^{-ikx} + c_2^\mu(\mathbf{k})^K e^{ikx} \right). \tag{1.6.11}$$

Then $\tilde{u}^\mu(x)$ becomes skew-conjugate

$$\tilde{u}^\mu(x) = (2\pi)^{-3/2} \int \frac{d^3k}{\sqrt{2\omega}} \left(-c_1^\mu(\mathbf{k}) e^{-ikx} + c_1^\mu(\mathbf{k})^K e^{ikx} \right) = -\tilde{u}^\mu(x)^K. \tag{1.6.12}$$

1.7 Tensor fields

A vector field transforms under the proper Lorentz group $\mathcal{L}_+^\uparrow$ according to the spinor representation $D^{(1/2,1/2)}$ (see FQED, Sect.1.2). Then a tensor field transforms with the tensor product of two such representations which can be decomposed into irreducible representations as follows

$$D^{(1/2,1/2)} \otimes D^{(1/2,1/2)} = D^{(1,1)} \oplus D^{(1,0)} \oplus D^{(0,1)} \oplus D^{(0,0)}. \qquad (1.7.1)$$

The first representation $D^{(1,1)}$ is nine-dimensional and given by symmetric tensors with trace 0. The second one $D^{(1,0)}$ is three-dimensional and given by anti-selfdual tensors, while the third one $D^{(0,1)}$ is also three-dimensional given by antisymmetric selfdual tensors. The last one is the trivial one-dimensional representation represented by a scalar field. The dimensions add up to $4 \times 4 = 16$.

In the following we will consider a symmetric tensor field $h^{\alpha\beta}(x)$ with arbitrary trace. It has 9+1=10 independent components and transforms with the reducible representation $D^{(1,1)} \oplus D^{(0,0)}$. This field is assumed to satisfy the wave equation

$$\Box h^{\alpha\beta}(x) = 0 \qquad (1.7.2)$$

because we are only interested in the massless case in view of gravity. On the other hand, a pure spin-2 field has $2 \times 2 + 1 = 5$ independent components, so that in this case 5 subsidiary conditions must be imposed. Those can be chosen in the following simple Lorentz-covariant way

$$h^{\alpha\beta}{}_{,\beta} = 0 \qquad (1.7.3)$$

$$h^\alpha{}_\alpha = 0. \qquad (1.7.4)$$

The first condition (1.7.3), where the comma notation for derivative is used, is the so-called Hilbert-gauge condition and is completely analogous to the Lorentz condition (1.3.2). It has four components and the trace condition (1.7.4) gives the fifth condition.

We want to consider the symmetric tensor field as a gauge field. For this reason we look for gauge transformations which leave the Hilbert condition (1.7.3) unchanged. It is easy to see that such transformations are of the following form

$$\tilde{h}^{\alpha\beta} = h^{\alpha\beta} + \lambda(u^{\alpha,\beta} + u^{\beta,\alpha} - g^{\alpha\beta}u^\mu{}_{,\mu}). \qquad (1.7.5)$$

Indeed, taking the derivative ∂_β we get

$$\tilde{h}^{\alpha\beta}{}_{,\beta} = h^{\alpha\beta}{}_{,\beta} + \lambda(\Box u^\alpha + u^{\beta,\alpha}{}_{,\beta} - u^{\mu,\alpha}{}_{,\mu}) = h^{\alpha\beta}{}_{,\beta}, \qquad (1.7.6)$$

if

$$\Box u^\alpha = 0 \qquad (1.7.7)$$

fulfills the wave equation. This property also guarantees that the transformed field (1.7.5) satisfies the wave equation, too.

We immediately consider the gauge transformation (1.7.5) as an operator gauge transformation as in Sect.1.4 in case of the vector field. Therefore, we write down the infinitesimal generator Q in complete analogy to (1.4.6)

$$Q = \int d^3x\, h^{\alpha\beta}{}_{,\beta}\, \overleftrightarrow{\partial}_0 u_\alpha, \tag{1.7.8}$$

where we have taken the Hilbert gauge condition (1.7.3) under the integral instead of the Lorentz condition. The vector field u_α must be quantized with anticommutators, in order to get Q nilpotent: $Q^2 = 0$. This follows in the same way as in (1.4.11). The operator Q given by (1.7.8) is the right infinitesimal generator for (1.7.5) if it has the following commutator

$$[Q, h^{\alpha\beta}(x)] = -\frac{i}{2}\left(u^{\alpha,\beta} + u^{\beta,\alpha} - g^{\alpha\beta}u^\mu{}_{,\mu}\right)(x) =$$

$$\stackrel{\text{def}}{=} -ib^{\alpha\beta\mu\nu}u_{\mu,\nu}. \tag{1.7.9}$$

The factor $-i/2$ is convention and the b-tensor

$$b^{\alpha\beta\mu\nu} = \tfrac{1}{2}(g^{\alpha\mu}g^{\beta\nu} + g^{\alpha\nu}g^{\beta\mu} - g^{\alpha\beta}g^{\mu\nu}) \tag{1.7.10}$$

will often appear in connection with tensor fields. As in (1.4.7) the commutator (1.7.9) implies the following commutation relation for the tensor field

$$[h^{\alpha\beta}(x), h^{\mu\nu}(y)] = -ib^{\alpha\beta\mu\nu}D_0(x-y)$$

$$= -\frac{i}{2}(g^{\alpha\mu}g^{\beta\nu} + g^{\alpha\nu}g^{\beta\mu} - g^{\alpha\beta}g^{\mu\nu})D_0(x-y). \tag{1.7.11}$$

Then Q generates the following infinitesimal gauge transformations

$$d_Q h^{\alpha\beta} = [Q, h^{\alpha\beta}] = -\frac{i}{2}(u^\alpha{}_{,\beta} + u^\beta{}_{,\alpha} - g^{\alpha\beta}u^\mu{}_{,\mu}) \tag{1.7.12}$$

$$d_Q u^\alpha = \{Q, u^\alpha\} = 0 \tag{1.7.13}$$

$$d_Q \tilde{u}^\alpha = \{Q, \tilde{u}^\alpha\} = ih^{\alpha\mu}{}_{,\mu}. \tag{1.7.14}$$

The ghost vector fields are quantized as discussed in the last section.

We have still to verify the consistency of the above abstract quantization of the tensor field by constructing an explicit representation in Fock space. For this purpose we return to the irreducible representations of the Lorentz group (1.7.1) and decompose $h^{\alpha\beta}$ according to $D^{(1,1)} \oplus D^{(0,0)}$

$$h^{\alpha\beta}(x) = H^{\alpha\beta}(x) + \frac{1}{4}g^{\alpha\beta}\Phi(x), \tag{1.7.15}$$

where

$$H^{\alpha\beta}(x) = h^{\alpha\beta}(x) - \frac{1}{4}g^{\alpha\beta}h(x), \quad h = h^\alpha{}_\alpha. \tag{1.7.16}$$

Then $H^\alpha{}_\alpha = 0$ is traceless and
$$h^\alpha{}_\alpha = h = \Phi. \tag{1.7.17}$$

¿From (1.7.11) we obtain the following commutation relations
$$[\Phi(x), \Phi(y)] = 4i D_0(x-y) \tag{1.7.18}$$
$$[H^{\alpha\beta}(x), H^{\mu\nu}(y)] = -it^{\alpha\beta\mu\nu} D_0(x-y), \tag{1.7.19}$$
with
$$t^{\alpha\beta\mu\nu} = \tfrac{1}{2}(g^{\alpha\mu}g^{\beta\nu} + g^{\alpha\nu}g^{\beta\mu} - \tfrac{1}{2}g^{\alpha\beta}g^{\mu\nu}) = t^{\mu\nu\alpha\beta} \tag{1.7.20}$$
and
$$[H^{\alpha\beta}(x), \Phi(y)] = 0. \tag{1.7.21}$$

We claim that (1.7.18-19) can be represented as follows
$$H^{\alpha\beta}(x) = (2\pi)^{-3/2} \int \frac{d^3k}{\sqrt{2\omega}} \left(a_{\alpha\beta}(\mathbf{k}) e^{-ikx} + g^{\alpha\alpha} g^{\beta\beta} a^+_{\alpha\beta}(\mathbf{k}) e^{ikx} \right), \tag{1.7.22}$$

where $a_{\alpha\beta} = a_{\beta\alpha}$ is symmetric and satisfies the commutation relation
$$[a_{\alpha\beta}(\mathbf{k}), a^+_{\mu\nu}(\mathbf{k}')] = g^{\alpha\alpha} g^{\beta\beta} t^{\alpha\beta\mu\nu} \delta(\mathbf{k} - \mathbf{k}') \tag{1.7.23}$$
$$\stackrel{\text{def}}{=} \tilde{t}^{\alpha\beta\mu\nu} \delta(\mathbf{k} - \mathbf{k}').$$

we see from the factors $g^{\alpha\alpha} g^{\beta\beta}$ that (1.7.23) is not Lorentz-covariant. The scalar part is given by
$$\Phi(x) = (2\pi)^{-3/2} \int \frac{d^3k}{\sqrt{2\omega}} \left(a(\mathbf{k}) e^{-ikx} - a^+(\mathbf{k}) e^{ikx} \right) \tag{1.7.24}$$
with
$$[a(\mathbf{k}), a^+(\mathbf{k}')] = 4\delta(\mathbf{k} - \mathbf{k}'). \tag{1.7.25}$$

Since the right-hand side is positive, the Φ-sector can be constructed in the usual way by applying products of a^+'s to the vacuum.

The situation is not so simple in the H-sector because $\tilde{t}$ in (1.7.23) is not a diagonal matrix: let $j \neq l$, then we have

$(\alpha,\beta)(\mu,\nu):$	$(0,0)(0,0)$	$(0,0)(j,j)$	$(0,j)(0,j)$	$(j,j)(j,j)$	$(j,j)(l,l)$	$(j,l)(j,l)$
$\tilde{t}^{\alpha\beta\mu\nu}\quad:$	$3/4$	$1/4$	$1/2$	$3/4$	$-1/4$	$1/2$

$$\tag{1.7.26}$$

We perform a linear transformation of the diagonal operators $a_{\alpha\alpha}$ and $a^+_{\alpha\alpha}$
$$a_{\alpha\alpha} = \sum_\beta M_{\alpha\beta} \tilde{a}_{\beta\beta} \tag{1.7.27}$$

$$a^+_{\alpha\alpha} = \sum_\beta M^*_{\alpha\beta} \tilde{a}^+_{\beta\beta} \tag{1.7.28}$$

in such a way that the new operators are usual annihilation and creation operators

$$[\tilde{a}_{\alpha\alpha}(\boldsymbol{k}), \tilde{a}^+_{\beta\beta}(\boldsymbol{k}')] = \delta_{\alpha\beta}\delta(\boldsymbol{k}-\boldsymbol{k}'). \qquad (1.7.29)$$

Substituting this into (1.7.23) we conclude that the matrix M must satisfy the condition

$$\tilde{t}^{\alpha\alpha\mu\mu} = \begin{pmatrix} \frac{3}{4} & \frac{1}{4} & \frac{1}{4} & \frac{1}{4} \\ \frac{1}{4} & \frac{3}{4} & -\frac{1}{4} & -\frac{1}{4} \\ \frac{1}{4} & -\frac{1}{4} & \frac{3}{4} & -\frac{1}{4} \\ \frac{1}{4} & -\frac{1}{4} & -\frac{1}{4} & \frac{3}{4} \end{pmatrix} = (MM^+)_{\alpha\mu}, \qquad (1.7.30)$$

where the rows and columns correspond to (0,0), (1,1), (2,2), (3,3). The matrix (1.7.30) is symmetric and, therefore, can be transformed to diagonal form D by means of an orthogonal transformation O. The eigenvalues are non-negative $D = \text{diag}(0,1,1,1)$. Then M can be determined from

$$\tilde{t} = ODO^{-1} = OD^{1/2}D^{1/2}O^+ = MM^+. \qquad (1.7.31)$$

In this way we find

$$M = \begin{pmatrix} 0 & \frac{1}{2} & \frac{1}{2} & \frac{1}{2} \\ 0 & -\frac{1}{2} & \frac{1}{2} & \frac{1}{2} \\ 0 & \frac{1}{2} & -\frac{1}{2} & \frac{1}{2} \\ 0 & \frac{1}{2} & \frac{1}{2} & -\frac{1}{2} \end{pmatrix}. \qquad (1.7.32)$$

This gives the following result for the transformation (1.7.27)

$$a_{00} = \tfrac{1}{2}(\tilde{a}_{11} + \tilde{a}_{22} + \tilde{a}_{33})$$

$$a_{11} = \tfrac{1}{2}(-\tilde{a}_{11} + \tilde{a}_{22} + \tilde{a}_{33})$$

$$a_{22} = \tfrac{1}{2}(\tilde{a}_{11} - \tilde{a}_{22} + \tilde{a}_{33})$$

$$a_{33} = \tfrac{1}{2}(\tilde{a}_{11} + \tilde{a}_{22} - \tilde{a}_{33}). \qquad (1.7.33)$$

We note that $\tilde{a}_{00}$ does not appear because one pair of absorption and emission operators is superfluous due to the trace condition $H^\alpha{}_\alpha = 0$. In fact, from (1.7.33) we see

$$\sum_{j=1}^{3} a_{jj} = a_{00}.$$

The Fock representation can now be constructed as usual by means of $\tilde{a}^+_{11}, \tilde{a}^+_{22}, \tilde{a}^+_{33}$ and $a^+_{\alpha\beta}$ with $\alpha \neq \beta$. A simple notation of these operators is obtained by writing $\tfrac{1}{2}a_{00} = b_{00}, \tilde{a}_{jj} = b_{jj}, a_{\alpha\beta} = b_{\alpha\beta}, \alpha \neq \beta$. Then we have the commutation relations in normal form

$$[b_{\alpha\beta}(\boldsymbol{k}), b^+_{\mu\nu}(\boldsymbol{k}')] = \tfrac{1}{2}(\delta_{\alpha\mu}\delta_{\beta\nu} + \delta_{\alpha\nu}\delta_{\beta\mu})\delta(\boldsymbol{k}-\boldsymbol{k}'). \qquad (1.7.34)$$

We now want to specify the physical subspace with help of the gauge charge

34   Free fields

$$Q = \int d^3x \left(\Pi^{\alpha\beta}{}_{,\beta} + \frac{1}{4}\Phi^{,\alpha} \right) \overleftrightarrow{\partial}_0 u_\alpha. \qquad (1.7.35)$$

When we substitute (1.6.2) and (1.7.22-24) herein we must be careful in writing the Lorentz indices of k correctly. These are true covariant or contravariant indices with respect to Minkowski metric $g^{\alpha\beta}$, while the indices of the creation and annihilation operators are not. The latter will be always written upstairs. Q in momentum space is then given by

$$Q = \int d^3k \left(A^\alpha(\boldsymbol{k})^+ c_2^\gamma(\boldsymbol{k}) - B^\alpha(\boldsymbol{k}) c_1^\gamma(\boldsymbol{k})^+ \right) g_{\alpha\gamma}, \qquad (1.7.36)$$

where

$$A^\alpha = g^{\alpha\alpha} g^{\beta\beta} a^{\alpha\beta}(\boldsymbol{k}) k^\beta - \frac{k^\alpha}{4} a(\boldsymbol{k}) \qquad (1.7.37)$$

$$B^\alpha = (a^{\alpha\beta}(\boldsymbol{k}) k_\beta + \frac{k^\alpha}{4} a(\boldsymbol{k})) g^{\alpha\alpha}, \qquad (1.7.38)$$

and it is summed over β. As in (1.4.21) the physical subspace $\mathcal{F}_{\text{phys}}$ is the kernel of the anticommutator

$$\{Q, Q^+\} = \int d^3k \, d^3k' \left(A^\alpha(\boldsymbol{k})^+ A^\beta(\boldsymbol{k}'))\{c_2^\gamma(\boldsymbol{k}), c_2^\delta(\boldsymbol{k}')^+\} \right.$$
$$\left. + B^{\beta+} B^\alpha \{c_1^\delta, c_1^{\gamma+}\} + c_2^{\delta+} c_2^\gamma [A^\beta, A^{\alpha+}] + c_1^{\gamma+} c_1^\delta [B^\alpha, B^{\beta+}] \right) g_{\alpha\gamma} g_{\beta\delta}.$$

$$= \int d^3k \sum_{\alpha=0}^{3} \left[A^{\alpha+} A^\alpha + B^{\alpha+} B^\alpha + \right.$$
$$\left. + \omega^2 (c_1^{\alpha+} c_1^\alpha + c_2^{\alpha+} c_2^\alpha) + \tfrac{1}{2}(k_\alpha c_1^{\alpha+}) k_\alpha c_1^\alpha + \tfrac{1}{2}(k_\alpha c_2^{\alpha+}) k_\alpha c_2^\alpha \right]. \qquad (1.7.39)$$

Substituting back

$$A^0 = \omega(a^{00} - a_\|^0 - \frac{a}{4}),$$

$$A^j = \omega(-a^{0j} + a_\|^j - \frac{k^j}{\omega}\frac{a}{4}),$$

$$B^0 = \omega(a^{00} + a_\|^0 + \frac{a}{4}),$$

$$B^j = \omega(-a^{0j} - a_\|^j - \frac{k^j}{\omega}\frac{a}{4}), \qquad (1.7.40)$$

where

$$a_\|^\mu = \frac{k_j}{\omega} a^{\mu j} \qquad (1.7.41)$$

is the absorption operators for the longitudinal mode, we obtain for the graviton terms

$$\sum_{\alpha=0}^{3} \left(A^{\alpha+} A^\alpha + B^{\alpha+} B^\alpha \right) = 2\omega^2 \left(a^{00+} a^{00} + a_\|^{0+} a_\|^0 + \right.$$

$$+\frac{a^+a}{8} + a^{0j+}a^{0j} + a_{\parallel}^{j+}a_{\parallel}^{j}\Big). \tag{1.7.42}$$

Here we must substitute the diagonal a's by $\tilde{a}$'s (1.7.33) because the latter define the states in the Fock space. Then we get for (1.7.42)

$$= \omega^2(\tilde{a}_{11}^+\tilde{a}_{11} + \tilde{a}_{22}^+\tilde{a}_{22} + \tilde{a}_{33}^+\tilde{a}_{33}) + k_3^2(\tilde{a}_{11}^+\tilde{a}_{22} + \tilde{a}_{22}^+\tilde{a}_{11}) +$$
$$+ k_2^2(\tilde{a}_{11}^+\tilde{a}_{33} + \tilde{a}_{33}^+\tilde{a}_{11}) + k_1^2(\tilde{a}_{22}^+\tilde{a}_{33} + \tilde{a}_{33}^+\tilde{a}_{22}) +$$
$$+2 \sum_{\text{cycl.}1,2,3} k_l k_m (\tilde{a}_{nn}^+ a^{lm} + a^{lm+}\tilde{a}_{nn}) + 2 \sum_{\substack{m\neq j \\ n\neq j}} k_m k_n a^{jm+} a^{jn} +$$
$$2\omega^2(a_{\parallel}^{0+}a_{\parallel}^{0} + a^{0j+}a^{0j} + \frac{a^+a}{8}). \tag{1.7.43}$$

To specify the physical states we notice that all states where (1.7.39) does not vanish are unphysical. That means all ghost states are unphysical. The unphysical "graviton" states obviously depend on k. Let us therefore assume $k^\mu = (\omega, 0, 0, \omega)$ parallel to the 3-axis with $k^\mu k_\mu = 0$. Then the following diagonal contribution in (1.7.43) survives

$$\omega^2(\tilde{a}_{11}^+\tilde{a}_{11} + \tilde{a}_{22}^+\tilde{a}_{22} + \tilde{a}_{33}^+\tilde{a}_{33} + \tilde{a}_{11}^+\tilde{a}_{22} + \tilde{a}_{22}^+\tilde{a}_{11}) = \omega^2[(\tilde{a}_{11} + \tilde{a}_{22})^+(\tilde{a}_{11} + \tilde{a}_{22}) +$$
$$+ \tilde{a}_{33}^+\tilde{a}_{33}]. \tag{1.7.44}$$

That means $\tilde{a}_{33}^+\Omega$ and $(\tilde{a}_{11}+\tilde{a}_{22})^+\Omega$ are unphysical, it remains $(\tilde{a}_{11}-\tilde{a}_{22})^+\Omega$ physical. From the non-diagonal terms in (1.7.43) there remains only $a^{12+}\Omega$ physical, because

$$k_3^2(a^{13+}a^{13} + a^{23+}a^{23})$$

survives in (1.7.43) and is therefore unphysical.

The remaining two physical degrees of freedom for fixed $k = (\omega, 0, 0, \omega)$ can be described by two real polarization tensors

$$\varepsilon_1 = \begin{pmatrix} 0 & 0 & 0 & 0 \\ 0 & 1 & 0 & 0 \\ 0 & 0 & -1 & 0 \\ 0 & 0 & 0 & 0 \end{pmatrix}, \quad \varepsilon_2 = \begin{pmatrix} 0 & 0 & 0 & 0 \\ 0 & 0 & 1 & 0 \\ 0 & 1 & 0 & 0 \\ 0 & 0 & 0 & 0 \end{pmatrix}. \tag{1.7.45}$$

The two complex combinations

$$\varepsilon_\pm = \varepsilon_1 \pm i\varepsilon_2 = \begin{pmatrix} 0 & 0 & 0 & 0 \\ 0 & 1 & \pm i & 0 \\ 0 & \pm i & -1 & 0 \\ 0 & 0 & 0 & 0 \end{pmatrix} \tag{1.7.46}$$

are the so-called helicity states. Under rotation

$$R(\varphi) = \begin{pmatrix} 1 & 0 & 0 & 0 \\ 0 & \cos\varphi & \sin\varphi & 0 \\ 0 & -\sin\varphi & \cos\varphi & 0 \\ 0 & 0 & 0 & 1 \end{pmatrix} \tag{1.7.47}$$

around the 3-axis they are transformed as follows

$$\varepsilon'^{\mu\nu}_{\pm} = R^{\mu}{}_{\alpha}R^{\nu}{}_{\beta}\varepsilon^{\alpha\beta}_{\pm} = (R\varepsilon R^T)^{\mu\nu}. \tag{1.7.48}$$

After multiplying out the matrices we find

$$\varepsilon'_{\pm} = e^{\pm 2i\varphi}\varepsilon_{\pm}. \tag{1.7.49}$$

That means we have states with helicities ± 2. They represent free gravitons.

The quantization of the tensor field is not yet finished because we also need the commutation relations for the positive and negative parts separately. They can easily be found from (1.7.22-25):

$$[\Phi^{(-)}(x), \Phi^{(+)}(y)] = 4iD_0^{(+)}(x-y) \tag{1.7.50}$$

$$[H^{\alpha\beta(-)}(x), H^{\mu\nu(+)}(y)] = -it^{\alpha\beta\mu\nu}D_0^{(+)}(x-y), \tag{1.7.51}$$

with $t^{\alpha\beta\mu\nu}$ given by (1.7.20). By (1.7.15) this implies

$$[h^{\alpha\beta(-)}(x), h^{\mu\nu(+)}(y)] = -ib^{\alpha\beta\mu\nu}D_0^{(+)}(x-y). \tag{1.7.52}$$

1.8 Spinor fields

The most important fields in nature beside the gauge fields are spinor fields describing particles with spin 1/2: the electron, muon, tau lepton and the corresponding neutrini, as well as the quarks. The electron and the up- and down-quarks are the main constituents of ordinary matter, where the gauge fields (photon and gluons) supply the glue. Although the quarks only occur as bound states in real physical particles, their coupling to the gluons is one of the fundamental interactions in QCD.

We will mainly consider spinor fields satisfying the Dirac equation

$$i\gamma^{\mu}\partial_{\mu}\psi(x) = m\psi(x). \tag{1.8.1}$$

The γ-matrices obey the anticommutation relation

$$\gamma^{\mu}\gamma^{\nu} + \gamma^{\nu}\gamma^{\mu} = 2g^{\mu\nu}, \tag{1.8.2}$$

and the Dirac spinor $\psi(x)$ transforms under proper Lorentz transformations according to the four-dimensional representation $D^{(1/2,0)} \oplus D^{(0,1/2)}$.

To define the quantized Dirac field we consider a solution of (1.8.1) of the following form

$$\psi(x) = (2\pi)^{-3/2}\int d^3p \sum_{s=\pm 1}[u_s(\boldsymbol{p})e^{-ipx}b_s(\boldsymbol{p}) + v_s(\boldsymbol{p})e^{ipx}d_s^+(\boldsymbol{p})], \tag{1.8.3}$$

$$\stackrel{\text{def}}{=} \psi^{(-)} + \psi^{(+)}.$$

Spinor fields 37

The u- and v-spinors herein are obtained from the Fourier transformed equations

$$(p_\mu\gamma^\mu - m)u_s(\boldsymbol{p}) = 0$$
$$(p_\mu\gamma^\mu + m)v_s(-\boldsymbol{p}) = 0, \tag{1.8.4}$$

with the normalization

$$u_s^+(\boldsymbol{p})u_{s'}(\boldsymbol{p}) = \delta_{ss'} = v_s^+(\boldsymbol{p})v_{s'}(\boldsymbol{p}) \tag{1.8.5}$$

$$u_s^+(\boldsymbol{p})v_{s'}(-\boldsymbol{p}) = 0 = v_s^+(-\boldsymbol{p})u_{s'}(\boldsymbol{p})$$

$$u_s^+(\boldsymbol{p})\gamma^0 u_{s'}(\boldsymbol{p}) = \frac{m}{E}\delta_{ss'} = -v_s(\boldsymbol{p})^+\gamma^0 v_{s'}(\boldsymbol{p}). \tag{1.8.6}$$

The u- and v-spinors span the positive and negative spectral subspaces of the Dirac operator, respectively, which are defined by the projection operators

$$P_+(\boldsymbol{p}) = \sum_s u_s(\boldsymbol{p})u_s^+(\boldsymbol{p}) = \left(\frac{\not{p}+m}{2E}\right)\gamma^0 \tag{1.8.7}$$

$$P_-(\boldsymbol{p}) = \sum_s v_s(-\boldsymbol{p})v_s^+(-\boldsymbol{p}) = \left(\frac{\not{p}-m}{2E}\right)\gamma^0, \tag{1.8.8}$$

where

$$\not{p} = p_\mu\gamma^\mu, \quad p_0 = E = \sqrt{\boldsymbol{p}^2 + m^2}. \tag{1.8.9}$$

The projections are orthogonal

$$P_\pm(\boldsymbol{p})^2 = P_\pm(\boldsymbol{p}), \quad P_+(\boldsymbol{p}) + P_-(\boldsymbol{p}) = \mathbb{1}. \tag{1.8.10}$$

The quantization of the Dirac field is most easily achieved by considering the b's and d's as operator-valued distributions satisfying the anticommutation relations

$$\{b_s(\boldsymbol{p}), b_{s'}^+(\boldsymbol{q})\} = \delta_{ss'}\delta^3(\boldsymbol{p}-\boldsymbol{q}) = \{d_s(\boldsymbol{p}), d_{s'}^+(\boldsymbol{q})\}, \tag{1.8.11}$$

and all other anticommutators vanish. Then, the operators b and d can be interpreted as annihilation and their adjoints b^+, d^+ as creation operators and the Fock space can be constructed from a unique vacuum in the usual way (see FQED, sect.2.2). To get the anticommutation relations for the whole Dirac field we need the adjoint Dirac field

$$\psi^+(x) = (2\pi)^{-3/2}\int d^3p\,[b_s^+(\boldsymbol{p})u_s(\boldsymbol{p})^+ e^{ipx} + d_s(\boldsymbol{p})v_s(\boldsymbol{p})^+ e^{-ipx}].$$

Multiplying by γ^0, we get the so-called Dirac adjoint

$$\overline{\psi}(x) = \psi^+(x)\gamma^0 = \overline{\psi}^{(+)} + \overline{\psi}^{(-)}$$

$$\overline{\psi}^{(+)} = (2\pi)^{-3/2}\int d^3p\,b_s^+(\boldsymbol{p})\overline{u}_s(\boldsymbol{p})e^{ipx}$$

$$\overline{\psi}^{(-)}(x) = (2\pi)^{-3/2} \int d^3p\, d_s(\boldsymbol{p})\overline{v}_s(\boldsymbol{p})e^{-ipx}. \tag{1.8.12}$$

With the aid of (1.8.11) we find

$$\{\psi_a^{(-)}(x), \overline{\psi}_b^{(+)}(y)\} = (2\pi)^{-3}\int d^3p\, u_{sa}(\boldsymbol{p})\overline{u}_{sb}(\boldsymbol{p})e^{-ip(x-y)}. \tag{1.8.13}$$

In the result (1.8.13), the covariant positive spectral projection operator (1.8.7) appears

$$\{\psi^{(-)}(x), \overline{\psi}^{(+)}(y)\} = (2\pi)^{-3}\int \frac{d^3p}{2E}(\not{p}+m)e^{-ip(x-y)} \stackrel{\text{def}}{=} \frac{1}{i}S^{(+)}(x-y)$$
$$= -i(i\not{\partial}+m)D_m^{(+)}(x-y). \tag{1.8.14}$$

In the same way, one obtains the other non-vanishing anticommutator

$$\{\psi^{(+)}(x), \overline{\psi}^{(-)}(y)\} \stackrel{\text{def}}{=} \frac{1}{i}S^{(-)}(x-y) = (2\pi)^{-3}\int \frac{d^3p}{2E}(\not{p}-m)e^{ip(x-y)} \tag{1.8.15}$$
$$= -i(i\not{\partial}+m)D_m^{(-)}(x-y).$$

This gives the anticommutation relation for the total Dirac field

$$\{\psi(x), \overline{\psi}(y)\} = \frac{1}{i}S(x-y), \tag{1.8.16}$$

with

$$S(x) = S^{(-)}(x) + S^{(+)}(x) = (i\not{\partial}+m)D_m(x-y). \tag{1.8.17}$$

The anticommutators between two ψ's and two $\overline{\psi}$'s vanish.

The Dirac field just defined describes charged particles. The charge conjugate field is given by

$$\psi_C(x) = C\overline{\psi}(x)^T, \tag{1.8.18}$$

where T denotes the transposed spinor. The matrix C has the property that

$$\gamma^{\mu T} = -C^{-1}\gamma^\mu C, \quad C^T = -C \tag{1.8.19}$$

for all $\mu = 0, 1, 2, 3$. If the Dirac equation describes neutral particles which are identical with their antiparticles, then we must require $\psi_C = \psi$. This leads to the so-called Majorana field

$$\psi_M(x) = (2\pi)^{-3/2}\int d^3p \sum_{s=\pm 1}[u_s(\boldsymbol{p})e^{-ipx}b_s(\boldsymbol{p}) + sv_s(\boldsymbol{p})e^{ipx}b^+_{-s}(\boldsymbol{p})]. \tag{1.8.20}$$

It satisfies the following anticommutation relation

$$\{\psi_M(x), \psi_M(y)^T\} = -iS(x-y)C. \tag{1.8.21}$$

As we noted above this anticommutator vanishes for the Dirac field.

1.9 Normally ordered products in free fields

Products of numerical distributions are not always defined, the same is true for operator valued distributions. For example, let $\varphi(x)$ be the quantized scalar field of Sect.1.1, then the square $\varphi^2(x)$ is not defined. To see this we start from the well-defined direct product

$$\varphi(x)\varphi(y) = (\varphi^{(-)}(x) + \varphi^{(+)}(x))(\varphi^{(-)}(y) + \varphi^{(+)}(y))$$
$$= \varphi^{(-)}(x)\varphi^{(-)}(y) + \varphi^{(-)}(x)\varphi^{(+)}(y) + \varphi^{(+)}(x)\varphi^{(-)}(y) + \varphi^{(+)}(x)\varphi^{(+)}(y)$$
$$=: \varphi(x)\varphi(y) : + [\varphi^{(-)}(x), \varphi^{(+)}(y)]. \quad (1.9.1)$$

In the last line we have commuted the emission part $\varphi^{(+)}(y)$ to the left in the second term. Then the resulting four terms with all emission operators to the left with respect to the absorption operators are called the normally ordered product or Wick product, denoted by double dots.

The direct product (1.9.1) gives an operator in Fock space after smearing with a test function $f(x,y) \in \mathcal{S}(\mathbb{R}^8)$. However, the limit $y \to x$ does not make sense because the commutator in (1.9.1) is equal to $-iD^{(+)}(x-y)$ and $D^{(+)}(0) \sim \delta(0)$ is not defined. This is the only defect, the limit exists in the normally ordered terms, for example

$$\int d^4x\, f(x) : \varphi^2 : (x) \Omega = \int d^4x\, f(x) : \varphi^{(+)2} : (x) \Omega \quad (1.9.2)$$

is a Fock state containing a pair of particles. Consequently, the normal product can be defined by

$$: \varphi^2 : (x) \stackrel{\text{def}}{=} \lim_{y \to x} \Big(\varphi(x)\varphi(y) + iD^{(+)}(x-y) \Big). \quad (1.9.3)$$

Normal products with more factors are defined recursively: Let $: \varphi(x) := \varphi(x)$, then

$$: \varphi(x_1) \cdot \ldots \cdot \varphi(x_j) := : \varphi(x_1) \cdot \ldots \cdot \varphi(x_{j-1}) : \varphi(x_j)$$
$$+ i \sum_{k=1}^{j-1} D^{(+)}(x_k - x_j) : \varphi(x_1) \cdot \ldots \cdot \overline{\varphi(x_k)} \ldots \varphi(x_{j-1}) :, \quad (1.9.4)$$

where the factor with overline is lacking. In the last term there are no fields with argument x_j and x_k, instead we have the contraction symbol $iD^{(+)}(x_k - x_j)$, one says the two fields are contracted. In the first term on the right side we can again substitute the definition (1.9.4)

$$=: \varphi(x_1) \cdot \ldots \cdot \varphi(x_{j-2}) : \varphi(x_{j-1})\varphi(x_j) +$$
$$+ i \sum_{k=1}^{j-2} D^{(+)}(x_k - x_{j-1}) : \varphi(x_1) \cdot \ldots \cdot \overline{\varphi(x_k)} \ldots \varphi(x_{j-2}) : \varphi(x_j) +$$

$$+i\sum_{k=1}^{j-1} D^{(+)}(x_k - x_j) : \varphi(x_1)\cdot\ldots\overline{\varphi(x_k)}\ldots\varphi(x_{j-1}) : . \qquad (1.9.5)$$

Again by (1.9.4) we write the second term as follows

$$: \varphi(1)\ldots\overline{\varphi(k)}\ldots\varphi(j-2)\overline{\varphi(j-1)} : \varphi(j) =: \varphi(1)\ldots\overline{\varphi(k)}\ldots\overline{\varphi(j-1)}\varphi(j) :$$

$$-i\sum_{k\neq k_1=1}^{j-2} D^{(+)}(k_1 - j) : \varphi(1)\ldots\overline{\varphi(k)}\ldots\overline{\varphi(k_1)}\ldots\overline{\varphi(j-1)}\varphi(j) :, \qquad (1.9.6)$$

to simplify the notation we have always dropped the x in the arguments, i.e. x_j is written as j. Inserting this into (1.9.4) we obtain

$$: \varphi(1)\ldots\varphi(j) := : \varphi(1)\ldots\varphi(j-2) : \varphi(j-1)\varphi(j)+$$

$$+i\sum_{k=1}^{j-2} D^{(+)}(k-(j-1)) : \varphi(1)\ldots\overline{\varphi(k)}\ldots\overline{\varphi(j-1)}\varphi(j) :$$

$$+i\sum_{k=1}^{j-1} D^{(+)}(k-j) : \varphi(1)\ldots\overline{\varphi(k)}\ldots\varphi(j-1)\overline{\varphi(j)} :$$

$$+\sum_{k_1=1}^{j-2}\sum_{k_1\neq k_2=1}^{j-2} D^{(+)}(k_1 - (j-1))D^{(+)}(k_2 - j) : \varphi(1)\ldots\overline{\varphi(k_1)}\ldots\overline{\varphi(k_2)}\ldots$$

$$\overline{\varphi(j-1)\varphi(j)} : . \qquad (1.9.7)$$

On the right side there appear terms with one and two contractions. Continuing this process we can express an ordinary product by normal products and contractions. This is the content of the

Theorem 1.9.1. A product of n scalar field operators is normally ordered as follows:

$$\varphi(1)\ldots\varphi(n) = : \varphi(1)\ldots\varphi(n) : + : \overline{\varphi(1)\varphi(2)}\ldots\varphi(n) : + \text{permutations}$$

$$+ : \overline{\varphi(1)\varphi(2)}\ldots\overline{\varphi(j)}\ldots\varphi(n) : + \ldots + \overline{\varphi(1)\varphi(2)}\,\overline{\varphi(3)\varphi(4)}\ldots$$

$$+ \text{permutations}, \qquad (1.9.8)$$

where the sum contains all normal products with all possible contractions (pairings). The contractions are given by

$$\overline{\varphi(j)\varphi(k)} = -iD^{(+)}(x_j - x_k). \qquad (1.9.9)$$

The theorem of Wick can also be established with Fermi fields. The only difference is that every term with contraction gets a sign $(-1)^P$ where P is the signature of the permutation necessary to bring the two contracted fields together.

Now we want to study normal products in the concrete Fock representation in p-space.

Lemma 1.9.2. Let $f \in \mathcal{S}(\mathbb{R}^{4l})$ and $\Phi = (\Phi_n) \in \mathcal{F}$ then

$$\left(\int dx_1 \ldots dx_l\, f(x_1, \ldots x_l) : D^{\alpha_1}\varphi(x_1) \ldots D^{\alpha_l}\varphi(x_l) : \Phi\right)_n (p_1, \ldots p_n) =$$

$$= (2\pi)^{l/2} \sum_{j=0}^{l} \left[\frac{(n-l+2j)!}{n!}\right]^{1/2} \int \prod_{k=1}^{j} d\mu_m(q_k) \sum_{k_1 < \ldots < k_{l-j}=1}^{n} \frac{1}{j!}$$

$$\sum_P P\Big[(-iq_1)^{\alpha_1} \ldots (-iq_j)^{\alpha_j} (ip_{k_1})^{\alpha_{j+1}} \ldots (ip_{k_{l-j}})^{\alpha_l} \times$$

$$\hat{f}(-q_1, \ldots, -q_j, p_{k_1}, \ldots p_{k_{l-j}}) \Phi_{n-l+2j}(q_1, \ldots q_j, p_1, \ldots \overline{p_{k_1}}, \ldots \overline{p_{k_{l-j}}} \ldots p_n)\Big], \tag{1.9.10}$$

where $\alpha_j = (\alpha_j^0, \alpha_j^1, \alpha_j^2, \alpha_j^3)$ are multi-indices and the sum over permutations P denotes the symmetrization operation.

Proof. The general term in (1.9.10) has j absorption operators with coordinates $q_1, \ldots q_j$ and $l-j$ emission operators with coordinates $p_{k_1}, \ldots p_{k_{l-j}}$. First we consider the simple case where f is a product

$$f(x_1, \ldots x_l) = g_1(x_1) \cdot \ldots \cdot g_l(x_l) \tag{1.9.11}$$

and no derivatives D^{α_l}. We start from the recursive definition (1.9.4)

$$\Big(:\varphi(g_1) \cdot \ldots \cdot \varphi(g_l) : \Phi\Big)_n = \Big(:\varphi(g_1) \cdot \ldots \cdot \varphi(g_{l-1}) : \varphi(g_l)\Phi\Big)_n$$

$$+ i \sum_{s=1}^{l-1} \int dx_s dx_l\, g_s(x_s) D^{(+)}(x_s - x_l) g_l(x_l) \Big(:\varphi(g_1) \cdot \ldots \overline{\varphi(g_s)} \ldots \varphi(g_{l-1}) : \Phi\Big)_n. \tag{1.9.12}$$

Let us abbreviate the two terms on the right side by $A + B$. The proof of (1.9.10) is by induction, that means we can use it for less than l factors. Instead of summing over $k_1 < k_2 < \ldots < k_{l-j-1}$ we sum over $k_1 \neq \ldots \neq k_{l-j-1}$ and divide by $(l-1-j)!$, this gives

$$A = (2\pi)^{(l-1)/2} \sum_{j=0}^{l-1} \left[\frac{(n-l+1+2j)!}{n!}\right]^{1/2} \int \prod_{k=1}^{j} d\mu_m(q_k) \sum_{k_1 \neq \ldots \neq k_{l-1-j}}$$

$$\frac{1}{j!(l-1-j)!} \sum_P P\Big[\prod_{k=1}^{j} \hat{g}_k(-q_k) \prod_{r=1}^{l-1-j} \hat{g}_{j+r}(p_{k_r})\Big]$$

42   Free fields

$$(\varphi(g_l)\Phi)_{n-l+1+2j}(q_1,\ldots q_j, p_1\ldots \overline{p_{k_1}}\ldots \overline{p_{k_{l-1-j}}}\ldots p_n). \quad (1.9.13)$$

¿From (1.1.45) we have

$$(\varphi(g_l)\Phi)_{n-l+1+2j} = \sqrt{2\pi}\Big\{\sqrt{n-l+2+2j}\int d\mu_m(q)\,\hat{g}_l(-q)$$

$$\cdot \Phi_{n-l+2+2j}(q, q_q\ldots q_j, p_1\ldots \overline{p_{k_1}}\ldots \overline{p_{k_{l-1-j}}}\ldots p_n) + \frac{1}{\sqrt{n-l+1+2j}}$$

$$\times\Big[\sum_{r=1}^{j}\hat{g}_l(q_r)\Phi_{n-l+2j}(q_1\ldots\overline{q_r}\ldots q_j, p_1\ldots\overline{p_{k_1}}\ldots\overline{p_{k_{l-1-j}}}\ldots p_n)$$

$$+\sum_{\substack{k_{l-j}=1\\k_{l-j}\neq k_1,\ldots}}^{n}\hat{g}_l(p_{k_{l-j}})\Phi_{n-l+2j}(q_1,\ldots q_j, p_1\ldots\overline{p_{k_1}}\ldots\overline{p_{k_{l-1-j}}}\ldots\overline{p_{k_{l-j}}}\ldots p_n)\Big]\Big\}$$

$$(1.9.14)$$

This we use in (1.9.13)

$$A = (2\pi)^{l/2}\sum_{j=0}^{l-1}\Big\{\Big[\frac{(n-l+2+2j)!}{n!}\Big]^{1/2}\int\prod_{k=1}^{j}d\mu_m(q_k)d\mu_m(q)\sum_{k_1\neq\ldots\neq k_{l-1-j}}$$

$$\frac{1}{j!(l-1-j)!}\sum_{P}P\Big[\prod_{k=1}^{j}\hat{g}_k(-q_k)\prod_{r=1}^{l-1-j}\hat{g}_{j+r}(p_{k_r})\Big]\hat{g}_l(-q)$$

$$\Phi_{n-l+2+2j}(q, q_1,\ldots q_j, p_1\ldots\overline{p_{k_1}}\ldots\overline{p_{k_{l-1-j}}}\ldots p_n)$$

$$+\Big[\frac{(n-l+2j)!}{n!}\Big]^{1/2}\int\prod_{k=1}^{j}d\mu_m(q_k)\sum_{k_1\neq\ldots\neq k_{l-1-j}}$$

$$\frac{1}{j!(l-1-j)!}\sum_{P}P\Big[\prod_{k=1}^{j}\hat{g}_k(-q_k)\prod_{r=1}^{l-1-j}\hat{g}_{j+r}(p_{k_r})\Big]\sum_{s=1}^{j}\hat{g}_l(q_s)$$

$$\Phi_{n-l+2j}(q_1,\ldots\overline{q_s}\ldots q_j, p_1\ldots\overline{p_{k_1}}\ldots\overline{p_{k_{l-1-j}}}\ldots p_n)$$

$$+\Big[\frac{(n-l+2j)!}{n!}\Big]^{1/2}\int\prod_{k=1}^{j}d\mu_m(q_k)\sum_{k_1\neq\ldots\neq k_{l-1-j}}\frac{1}{j!(l-1-j)!}$$

$$\sum_{P}P\Big[\prod_{k=1}^{j}\hat{g}_k(-q_k)\prod_{r=1}^{l-1-j}\hat{g}_{j+r}(p_{k_r})\Big]\sum_{\substack{k_{l-j}=1\\k_{l-j}\neq k_1\ldots k_{l-1-j}}}^{n}\hat{g}_l(p_{k_{l-j}})$$

$$\Phi_{n-l+2j}(q_1,\ldots q_j, p_1\ldots\overline{p_{k_1}}\ldots\overline{p_{k_{l-j}}}\ldots p_n)\Big\}. \quad (1.9.15)$$

Now we compute the second term B in (1.9.12). Since

$$\int dx_s dx_l \, g_s(x_s) i D^{(+)}(x_s - x_l) g_l(x_l) =$$

$$= i(2\pi)^2 \int dk \, \hat{g}_s(-k) \hat{D}^{(+)}(k) \hat{g}_l(k) =$$

$$= -2\pi \int d\mu_m(k) \, \hat{g}_s(-k) \hat{g}_l(k), \tag{1.9.16}$$

where we have used (1.1.12) for the Fourier transform of $D^{(+)}$, we obtain

$$B = -2\pi \sum_{s=1}^{l-1} \int d\mu_m(q) \, \hat{g}_s(-q) \hat{g}_l(q) (2\pi)^{(l-2)/2} \sum_{j=0}^{l-1} \left[\frac{(n-l+2j)!}{n!} \right]^{1/2} \times$$

$$\times \int \prod_{\substack{k=1 \\ k \neq s}} d\mu_m(q_k) \sum_{\substack{k_1 \neq \ldots \neq k_{l-1-j} \\ k_r \neq s}} \frac{1}{j!(l-1-j)!}$$

$$\times \sum_P P \left[\prod_{\substack{k=1 \\ k \neq s}} \hat{g}_k(-q_k) \prod_{r=1}^{l-1-j} \hat{g}_{j+r}(p_{k_r}) \right]$$

$$\Phi_{n-l+2j}(q_1, \ldots \overline{q_s} \ldots q_j, p_1, \ldots \overline{k_{k_1}} \ldots \overline{p_{k_{l-1-j}}} \ldots p_n). \tag{1.9.17}$$

The last term comes from the annihilation operator, therefore we have $s \leq j$ so that the sum over s terminates at $s = j$. Then B exactly compensates the second term in (1.9.15). The first and third term together yield

$$(2\pi)^{l/2} \sum_{j=0}^{l} \left[\frac{(n-l+2j)!}{n!} \right]^{1/2} \int \prod_{k=1}^{j} d\mu_m(q_k) \sum_{k_1 \neq \ldots \neq k_{l-j}} \frac{1}{j!(l-j)!} \sum_P P$$

$$\left[\prod_{k=1}^{j} \hat{g}_k(-q_k) \prod_{r=1}^{l-j} \hat{g}_{j+r}(p_{k_r}) \right] \Phi_{n-l+2j}(q_1, \ldots q_j, p_1, \ldots \overline{p_{k_1}}, \ldots \overline{p_{k_{l-j}}} \ldots p_n) \right]. \tag{1.9.18}$$

To see this one substitutes in the first term $q \to q_{j+1}$ and $j+1 \to j$ which now run from 1 to l, in addition one takes $\hat{g}_l(-q_{j+1})$ under the sum over permutations, giving rise to the factor $1/j$. In the third term we take $\hat{g}_l(p_{k_{l-j}})$ under the sum over permutations and compensate by the factor $1/(l-j)$. Both terms represent the two possibilities that $\hat{g}_l$ has a $-q$ or a p as an argument. With (1.9.18) the induction is proved. Since the products (1.9.11) are dense in Schwartz space, the lemma extends to the whole $\mathcal{S}(\mathbb{R}^{4l})$. The powers on the right side of (1.9.10) come from the derivatives in x-space after Fourier transformation. This completes the proof of the lemma. $\square$

¿From (1.9.10) a Wick monomial is obtained by the formal substitution

$$f(x_1,\ldots x_l) = \prod_{j=2}^{l} \delta(x_j - x_1) g(x_1). \tag{1.9.19}$$

The result

$$\left(:D^{\alpha_1}\varphi\ldots D^{\alpha_l}\varphi:(g)\Phi\right)_n(p_1,\ldots p_n) = (2\pi)^{2-3l/2}$$

$$\times \sum_{j=0}^{l} \left[\frac{(n-l+2j)!}{n!}\right]^{1/2} \int \prod_{k=1}^{j} d\mu_m(q_k) \sum_{k_1<\ldots<k_{l-j}=1}^{n} \frac{1}{j!}$$

$$\sum_{P} P\Big[(-iq_1)^{\alpha_1}\ldots(-iq_j)^{\alpha_j}(ip_{k_1})^{\alpha_{j+1}}\ldots(ip_{k_{l-j}})^{\alpha_l} \times$$

$$\times \hat{g}\Big(\sum_{r=1}^{j} q_r - \sum_{r=1}^{l-j} p_{k_r}\Big)\Big] \Phi_{n-l+2j}(q_1,\ldots q_j, p_1,\ldots \overline{p_{k_1}},\ldots \overline{p_{k_{l-j}}}\ldots p_n) \tag{1.9.20}$$

is indeed well defined. This follows from the next

Lemma 1.9.3. Let $F \in \mathcal{S}(\mathbb{R}^4)$ and $\Phi \in \mathcal{S}(\mathbb{R}^{4(j+s-1)})$, then the function

$$G = \int \prod_{k=1}^{j} d\mu_m(q_k) F\Big(\sum_{k=1}^{j} q_k - \sum_{r=1}^{n-s} p_r\Big) \Phi(q_1,\ldots q_j, p_{n-s+1}\ldots p_n) \tag{1.9.21}$$

is in Schwartz space in the variables $p_1,\ldots,p_n$.

Proof. We first note that $|F|$ is bounded and

$$|\Phi(q_1,\ldots q_j, p_{n-s+1}\ldots p_n)| \le \frac{C_N}{\left(\sum_{k=1}^{j}\sum_{\mu=0}^{3}(q_j^\mu)^2\right)^N + 1} \tag{1.9.22}$$

is polynomially bounded because $\Phi \in \mathcal{S}$. Then the integral (1.9.21) converges together with all derivatives with respect to p_r.

It remains to study the behavior for large $|p_r| \to \infty$. For every integer M there exists a constant D_M such that

$$\Big|F\Big(\sum_{k=1}^{j} q_k - \sum_{r=1}^{n-s} p_r\Big)\Big| < \frac{D_M}{\Big[\big|\sum_k q_k^0 - \sum_r p_r^0\big|+1\Big]^M} \tag{1.9.23}$$

and similarly

$$|\Phi| < \frac{E_{LQ}}{\left(\prod_{k=1}^{j} q_k^0\right)\left((\sum_{k=1}^{j} p_k^0)+1\right)^L \left((\sum_{l=n-s+1}^{n} p_l^0)^Q+1\right)}. \tag{1.9.24}$$

This gives the bound

$$|G| < D_M E_{LQ} \int \prod_{k=1}^{j} \frac{d\mu(q_k)}{q_k^0} \, [|t-u|+1]^{-M}[t+1]^{-L}$$
$$\times [v^Q+1]^{-1}, \qquad (1.9.25)$$

with
$$u = \sum_r p_r^0, \quad v = \sum_{l=n-s+1}^{n} p_l^0, \quad t = \sum_k q_k^0. \qquad (1.9.26)$$

We now use
$$d\mu_m(q) = \frac{4\pi \boldsymbol{q}^2 \, d|\boldsymbol{q}|}{2q_0}, \quad q_0 dq_0 = |\boldsymbol{q}| d|\boldsymbol{q}|$$

and $\|\boldsymbol{q}\| < q_0$, then

$$|G| < D_M E_{LQ} (2\pi)^j \int_m^\infty \prod_{k=1}^{j} dq_k^0 \, [|t-u|+1]^{-M}[t+1]^{-L}$$
$$\times [v^Q+1]^{-1}. \qquad (1.9.27)$$

Here we introduce new integration variables
$$t_1 = q_1^0, \quad t_2 = q_1^0 + q_2^0, \ldots t_{j-1} = \sum_{k=1}^{j-1} q_k^0 \qquad (1.9.28)$$

together with t. Then we get an iterated integral

$$|G| < D_M E_{LQ} (2\pi)^j \int_0^\infty dt \int_0^t dt_{j-1} \ldots \int_0^{t_2} dt_1 \ldots$$
$$\leq \frac{D_M E_{LQ}}{v^Q+1} \frac{(2\pi)^j}{(j-1)!} \int_0^\infty dt \frac{t^{j-1}}{(|t-u|+1)^M (t+1)^L}. \qquad (1.9.29)$$

For simplicity we choose $L = M + j - 1$ and split the integral

$$|G| < \frac{D_M E_{LQ}}{v^Q+1} \frac{(2\pi)^j}{(j-1)!} \left[\int_0^u \frac{dt}{[(u-t+1)(t+1)]^M} + \int_u^\infty \frac{dt}{[(t-u+1)(t+1)]^M} \right]. \qquad (1.9.30)$$

Let us abbreviate the two terms by $G_1 + G_2$. The second term G_2 is easily estimated

$$G_2 \leq \int_u^\infty \frac{dt}{(t+1)^M} = \frac{1}{(M-1)(u+1)^{M-1}}. \qquad (1.9.31)$$

To estimate G_1 we determine the maximum of the integrand which occurs at $t = u/2$, it leads to the bound

$$G_1 < \frac{u}{(1+u/2)^{2M}}. \tag{1.9.31}$$

The total G is then bounded by

$$|G| < \frac{D_M E_{LQ}}{v^Q + 1} \frac{(2\pi)^j}{(j-1)!} \left[\frac{u}{(1+u/2)^{2M}} + \frac{1}{(M-1)(u+1)^{M-1}} \right]. \tag{1.9.32}$$

Consequently, for $|u| \to \infty$ G goes to zero more rapidly than any power of u. Finally we note that all p_r are on the mass shell $(p_r^0)^2 - \boldsymbol{p}_r^2 = m^2$, that means for $|\boldsymbol{p}_r| \to \infty$ it follows $|p_r^0| \to \infty$ and $u \to \infty$ in virtue of (1.9.26). This completes the proof of lemma 2. □

Lemma 1.9.3 shows that Wick monomials (1.9.20) are well defined operators on states with finitely many particles and wave functions in $\mathcal{S}$. This is certainly a dense set in Fock space.

A normally ordered product has the important property that its vacuum expectation value vanishes. This is due to the fact that the absorption operators stand at the right end where they give zero when applied to the vacuum. Similarly, the emission operators at the left end $(\Omega, \varphi^{(+)} \ldots) = (\varphi^{(+)+}\Omega, \ldots)$ give zero when applied to the left vacuum, because they become absorption operators after taking the adjoint. Taking the vacuum expectation value of Wick's theorem (1.9.8), all normal products give no contribution and only the fully contracted terms survive

$$(\Omega, \varphi(1) \ldots \varphi(n)\Omega) = \text{sum of all pairings of } \varphi(1) \ldots \varphi(n). \tag{1.9.33}$$

This allows us to write Wick's theorem (1.9.8) in the following compact form

$$\varphi(1) \ldots \varphi(n) = \sum_{\substack{s_1, \ldots, s_n \\ s_j = 0, 1}} (\Omega, \varphi^{1-s_1}(1) \ldots \varphi^{1-s_n}(n)\Omega) : \varphi^{s_1}(1) \ldots \varphi^{s_n}(n) : . \tag{1.9.34}$$

Note that all terms with an odd number of $s_j = 0$ are zero because the vacuum expectation value vanishes.

Finally we want to generalize this to products of arbitrary Wick monomials. Let us define the so-called Wick sub-monomials

$$F^0(x) =: \varphi^m : (x)$$

$$F^r(x) = \frac{\partial^r}{\partial \varphi^r} : \varphi^m : (x) = \frac{m!}{(m-r)!} : \varphi^{m-r} : (x) \tag{1.9.35}$$

for $0 \le r \le m$. We first consider the normal ordering of two factors

$$F^{r_1}(x_1) F^{r_2}(x_2) = \frac{(m!)^2}{(m-r_1)!(m-r_2)!} : \varphi^{m-r_1}(x_1) :: \varphi^{m-r_2}(x_2) : . \tag{1.9.36}$$

Here we must perform all possible pairings between the many $\varphi(x_1)$ and $\varphi(x_2)$ and the unpaired φ's appear in a normal product. Considering a general term

$\sim$: $\varphi^{s_1}(x_1)\varphi^{s_2}(x_2)$: we observe that there are $\binom{m-r_1}{s_1}$ ways to select the s_1 fields $\varphi(x_1)$ out of the $m - r_1$ φ's present, because the order of the φ's does not matter, and similarly with $\varphi(x_2)$. Then we get

$$= \frac{(m!)^2}{(m-r_1)!(m-r_2)!} \sum_{\substack{s_1,s_2 \\ 0 \le s_j \le m-r_j}} \binom{m-r_1}{s_1}\binom{m-r_2}{s_2}$$

$$\left[\text{all possible pairings}\right] : \varphi^{s_1}(x_1)\varphi^{s_2}(x_2) : . \tag{1.9.37}$$

The possible pairings are between $\varphi^{m-r_1-s_1}(x_1)$ and $\varphi^{m-r_2-s_2}(x_2)$ with $r_1 + s_1 = r_2 + s_2$. As above, the sum of these pairings can be written as a vacuum expectation value and the restriction on s_1, s_2 can be dropped. The combinatorial factors

$$\frac{m!}{(m-r_1)!}\binom{m-r_1}{s_1} = \frac{m!}{(m-r_1)!}\frac{(m-r_1)!}{s_1!(m-r_1-s_1)!} = \frac{m!}{(m-r_1-s_1)!s_1!}$$

lead to the sub-monomial $F^{r_1+s_1}$ so that we arrive at the following result

$$F^{r_1}(x_1)F^{r_2}(x_2) = \sum_{\substack{s_1,s_2 \\ 0 \le s_j \le m-r_j}} (\Omega, F^{r_1+s_1}(x_1)F^{r_2+s_2}(x_2)\Omega)$$

$$\times \frac{: \varphi^{s_1}(x_1)\varphi^{s_2}(x_2) :}{s_1!s_2!}. \tag{1.9.38}$$

This immediately generalizes to n factors

$$F^{r_1}(1)\ldots F^{r_n}(n) = \sum_{\substack{s_1\ldots s_n \\ 0 \le s_j \le m-r_j}} (\Omega, F^{r_1+s_1}(1)\ldots F^{r_n+s_n}(n)\Omega)$$

$$\times \frac{: \varphi^{s_1}(1)\ldots \varphi^{s_n}(n) :}{s_1!\ldots s_n!}. \tag{1.9.39}$$

1.10 Problems

1.1 Write the Jordan-Pauli distribution D_m (1.1.13) as a 3-dimensional Fourier integral and show that

$$D_m(x) = -\frac{1}{8\pi^2 r}\frac{d}{dr}\int_{-\infty}^{\infty}\frac{dp}{E}\left[\sin(Ex_0+pr) + \sin(Ex_0-pr)\right], \tag{1.10.1}$$

where $r = |\boldsymbol{x}|$, $E = +\sqrt{\boldsymbol{p}^2 + m^2}$.

1.2 Show that the integral in (1.10.1) vanish for $-r < x_0 < r$.

Hint: Use the substitutions
$$x_0 = b\sinh\tau \ , \quad r = b\cosh\tau \ , \quad b = \sqrt{r^2 - x_0^2}.$$

1.3 For $x_0 > r$ use the substitutions
$$x_0 = a\cosh\tau \ , \quad r = a\sinh\tau \ , \quad a = \sqrt{x_0^2 - r^2}, \qquad (1.10.2)$$
and similarly for $x_0 < -r$. Show that (1.10.1) is equal to
$$D_m(x) = -\frac{1}{4\pi r}\frac{d}{dr}\mathrm{sgn}\,x_0 \Theta(x^2) J_0\left(m\sqrt{x_0^2 - r^2}\right)$$
$$= \frac{1}{2\pi}\mathrm{sgn}\,x_0 \left[\delta(x^2) - \Theta(x^2)\frac{m}{2\sqrt{x^2}}J_1\left(m\sqrt{x^2}\right)\right], \qquad (1.10.3)$$
where J_0, J_1 are Bessel functions.

1.4 Write $D_m^{(+)}$ (1.1.12) as a 3-dimensional Fourier integral and compute
$$D_1(x) = \frac{i}{(2\pi)^3}\int\frac{d^3p}{2E}\,e^{i\mathbf{p}\cdot\mathbf{x}}\cos Ex_0. \qquad (1.10.4)$$

Result:

$$D_1(x) = \frac{i}{4\pi}\frac{d}{d(x^2)}\left[-\Theta(x^2)N_0\left(m\sqrt{x^2}\right) + \Theta(-x^2)\frac{2}{\pi}K_0\left(m\sqrt{-x^2}\right)\right], \quad (1.10.5)$$

where K_0 and N_0 are modified Bessel functions.

1.5 Prove that the square bracket in (1.10.5) is continuous on the light cone $x^2 = 0$.

1.6 With help of problem 1.4-5 show that
$$D^{(+)}(x) = \frac{1}{4\pi}\left\{\mathrm{sgn}\,x^0\left[\delta(x^2) - \Theta(x^2)\frac{m}{2\sqrt{x^2}}J_1\left(m\sqrt{x^2}\right)\right] + \right.$$
$$\left. + i\,\mathrm{P}\left[\Theta(x^2)\frac{m}{2\sqrt{x^2}}N_1\left(m\sqrt{x^2}\right) + \frac{m}{\pi\sqrt{-x^2}}\Theta(-x^2)K_1\left(m\sqrt{-x^2}\right)\right]\right\}, \quad (1.10.6)$$
where P means the principal value integral.

1.7 Verify:
$$D^{(-)}(x) = -D^{(+)}(-x) = D^{(+)}(x)^*. \qquad (1.10.7)$$

1.8 Show that the unsmeared absorption operator
$$(\tilde{a}(\mathbf{p})\Phi)_n = \sqrt{n+1}\,\varphi_{n+1}(\mathbf{p}_1,\ldots\mathbf{p}_n,\mathbf{p}) \qquad (1.10.8)$$
is a well defined operator, but its formal adjoint $\tilde{a}^+(\mathbf{p})$ is not.

1.9 Verify that
$$(\Psi, \tilde{a}^+(\mathbf{p})\Phi) = (\tilde{a}(\mathbf{p})\Psi, \Phi) \qquad (1.10.9)$$
is a densely defined bilinear form. Show that $(\Psi, \tilde{a}^+(\mathbf{p})a(\mathbf{q})\Phi)$ is also a bilinear form, but $(\Psi, \tilde{a}(\mathbf{q})a^+(\mathbf{p})\Phi)$ is not.

2. Causal perturbation theory

After the discussion of free fields it is high time to come to interactions. We will introduce interactions in the framework of scattering theory. Here the basic object is the scattering matrix (S-matrix). It maps the asymptotically incoming free fields on the outgoing ones; the interaction is switched off at large times and large distances. Then the S-matrix can be expressed by free fields as a formal power series.

In this chapter we introduce the method to construct this series. As the title indicates, causality will be the essential property in the construction. Following the method of Epstein and Glaser, the perturbation series is obtained inductively, order by order, by means of causality and translation invariance, unitarity is not used. The first order, i.e. the coupling, must be given. All steps are under control, consequently, no ultraviolet divergences appear. The naive Feynman rules only hold for tree graphs, not for closed loops in general.

The most delicate step in the construction is the decomposition of distributions with causal support into retarded and advanced parts. If this distribution splitting is carried out without care by multiplication with step functions, then the usual ultraviolet divergences appear. But if it is carefully done by first multiplying with a C^∞ function and then performing the appropriate limit to the step function, everything is finite and well-defined.

As an introduction to the method we first consider scattering in ordinary quantum mechanics.

2.1 The S-matrix in quantum mechanics

We consider a quantum mechanical system described by a (time independent) Hamiltonian H which is a selfadjoint operator on a Hilbert space $\mathcal{H}$. The time evolution of the system is then given by the unitary transformation

$$\psi(t) = e^{-iHt}\psi_0. \qquad (2.1.1)$$

We assume H to be of the following form

$$H = H_0 + V, \qquad (2.1.2)$$

where H_0 is the free Hamiltonian and the interaction V has short range. The latter means that the so-called wave operators

$$W_{\text{out}}^{\text{in}} = s - \lim_{t \to \mp\infty} e^{iHt} e^{-iH_0 t} \qquad (2.1.3)$$

exist as strong limits on $\mathcal{H}$.

In the case of a time-dependent interaction $V = V(t)$, the time evolution is given by a unitary propagator

$$\psi(t) = U(t,s)\psi(s) \quad , \quad U(t,s)^+ = U(s,t) \qquad (2.1.4)$$

$$U(t,s)U(s,r) = U(t,r) \qquad (2.1.5)$$

instead of (2.1.1). The wave operators then are defined as follows

$$\begin{aligned} W_{\text{out}}^{\text{in}} &= s - \lim_{t \to \mp\infty} U(t,0)^+ e^{-iH_0 t} \\ &= s - \lim_{t \to \mp\infty} U(0,t) e^{-iH_0 t}. \end{aligned} \qquad (2.1.6)$$

The plus in the exponent always means the adjoint in Hilbert space, the asterisk $*$ is reserved for complex conjugation, while the bar is used for the Dirac adjoint (see (1.8.12)).

The central object of scattering theory is the scattering matrix (S-matrix)

$$S = W_{\text{out}}^+ W_{\text{in}} = \lim_{\substack{s \to -\infty \\ t \to +\infty}} e^{iH_0 t} U(t,s) e^{-iH_0 s}. \qquad (2.1.7)$$

¿From this definition the physical meaning of the S-matrix can be read of : A normalized initial asymptotic state φ considered at time $t = 0$, say, is first transformed to $s = -\infty$ by the free dynamics, then it is evolved from $-\infty$ to $t = +\infty$ by the full interacting dynamics by means of $U(t,s)$ and finally it is transformed back from $+\infty$ to $t = 0$ by the free dynamics. The resulting state $S\varphi$ is therefore the outgoing scattering state, transformed to $t = 0$ by the free time evolution. It can then be compared with an arbitrary normalized outgoing asymptotic state ψ by calculating the scalar product $(\psi, S\varphi)$. The absolute square of this is the probability for a transition from φ to ψ in the scattering process:

$$P(\varphi \to \psi) = |(\psi, S\varphi)|^2. \qquad (2.1.8)$$

The state $\psi(t)$ (2.1.4) is the solution of the Schrödinger equation

$$i\frac{d}{dt}\psi(t) = (H_0 + V(t))\psi(t) \quad , \quad \hbar = 1. \qquad (2.1.9)$$

We go over to the so-called interaction picture by the substitution

$$\psi(t) = e^{-iH_0 t} \varphi(t). \qquad (2.1.10)$$

$\varphi(t)$ then satisfies the simple equation

$$i\frac{d}{dt}\varphi(t) = \tilde{V}(t)\varphi(t), \qquad (2.1.11)$$

where
$$\tilde{V}(t) = e^{iH_0 t} V(t) e^{-iH_0 t} \tag{2.1.12}$$
is the operator $V(t)$ in the interaction picture. Note that the S-matrix (2.1.7) is just the limit of the time evolution in the interaction picture:
$$S = \lim_{\substack{s \to -\infty \\ t \to +\infty}} \tilde{U}(t, s). \tag{2.1.13}$$

Equation (2.1.11) can be written as an integral equation
$$\varphi(t) = \varphi(s) - i \int_s^t dt_1 \tilde{V}(t_1) \varphi(t_1). \tag{2.1.14}$$

If the interaction $V(t)$ is a bounded operator, equation (2.1.14) can be iterated, leading to the so-called Dyson series
$$\varphi(t) = \left[1 + \sum_{n=1}^{\infty} (-i)^n \int_s^t dt_1 \int_s^{t_1} \cdots \int_s^{t_{n-1}} dt_n \tilde{V}(t_1) \ldots \tilde{V}(t_n) \right] \varphi(s). \tag{2.1.15}$$

This series converges in operator norm, because the n-th order term U_n can be estimated as follows:
$$\|U_n\| \leq \int_s^t dt_1 \cdots \int_s^{t_{n-1}} dt_n \|\tilde{V}(t_1)\| \ldots \|\tilde{V}(t_n)\|$$
$$= \frac{1}{n!} \left[\int_s^t d\tau \|\tilde{V}(\tau)\| \right]^n. \tag{2.1.16}$$

The Dyson series (2.1.15) can be taken as definition of the unitary propagator (2.1.4).

If the time dependent interaction $V(t)$ decreases for large times such that
$$\int_{-\infty}^{+\infty} ds \|V(s)\| < \infty, \tag{2.1.17}$$
then the Dyson series for the S-matrix
$$S = \sum_{n=0}^{\infty} (-i)^n \int_{-\infty}^{+\infty} dt_1 \int_{-\infty}^{t_1} dt_2 \cdots \int_{-\infty}^{t_{n-1}} dt_n \tilde{V}(t_1) \ldots \tilde{V}(t_n) \tag{2.1.18}$$
is also norm-convergent. It then defines a unitary operator in $\mathcal{H}$. Consequently, asymptotic completeness which expresses unitarity of the S-matrix is no problem here. This is in sharp contrast to scattering theory for static

potentials, where unitarity is difficult to prove (see M.Reed, B.Simon, *Methods of modern mathematical physics, vol.III, Academic Press 1979*). At first sight, it seems to be very restrictive to suppose the interaction $V(t)$ to be bounded in norm for almost all t (cf. (2.1.17)). However, external fields with unbounded norm are unphysical in general, for example in quantum electrodynamics (see FQED, chap.2).

The domain of integration in the iterated integral (2.1.18) is a simplex in $\mathbb{R}^n$. Such an integral can be extended to an integral over a cube. However, since the factors $\tilde{V}(t_j)$ do not commute, it is necessary to maintain the time ordering:

$$S = \sum_{n=0}^{\infty} \frac{(-i)^n}{n!} \int_{-\infty}^{+\infty} dt_1 \int_{-\infty}^{+\infty} dt_2 \ldots \int_{-\infty}^{+\infty} dt_n \, T[\tilde{V}(t_1) \ldots \tilde{V}(t_n)] \qquad (2.1.19)$$

$$\stackrel{\text{def}}{=} T \exp -i \int_{-\infty}^{+\infty} dt \, \tilde{V}(t). \qquad (2.1.20)$$

The symbol T before a product means that the factors are arranged with decreasing time coordinate. The order of the factors in such a time-ordered product is immaterial, because the time-ordering T ensures the correct order. Hence, the integrand in (2.1.19) is symmetric in $t_1, \ldots, t_n$; every simplex gives the same contribution. The time-ordered exponential (2.1.20) is a compact notation for the series (2.1.18, 19).

We now give a second derivation of the perturbation series (2.1.19) *without using the Schrödinger equation*. Since in quantum field theory one does not have well-defined dynamical evolution equations, in general, it is this derivation which can be generalized to quantum field theory. In view of this application, let us multiply the interaction by a switching function $g(t)$, which vanishes rapidly for $t \to \pm \infty$. We want to construct the S-matrix as a power series in g of the form

$$S(g) = 1 + \sum_{n=1}^{\infty} \frac{1}{n!} \int dt_1 \ldots dt_n \, T_n(t_1, \ldots, t_n) g(t_1) \ldots g(t_n) \qquad (2.1.21)$$

$$\stackrel{\text{def}}{=} 1 + T.$$

By definition, $T_n(t_1, \ldots, t_n)$ must be symmetric with respect to permutations of the t_j, otherwise the contribution to (2.1.21) would be zero. Then it is appropriate to consider the disordered set of time points $X = \{t_1, \ldots, t_n\}$ as argument of T_n.

Simultaneously with $S(g)$, the inverse $S(g)^{-1}$ can be expanded as follows

$$S(g)^{-1} = 1 + \sum_{n=1}^{\infty} \frac{1}{n!} \int dt_1 \ldots dt_n \, \tilde{T}_n(t_1, \ldots, t_n) g(t_1) \ldots g(t_n) \qquad (2.1.22)$$

$$= (1+T)^{-1} = 1 + \sum_{r=1}^{\infty}(-T)^r. \qquad (2.1.23)$$

The $\tilde{T}_n$ follow from (2.1.23) by expanding the r-th power of $(-T)$

$$\tilde{T}_n(X) = \sum_{r=1}^{n}(-)^r \sum_{P_r} T_{n_1}(X_1)\ldots T_{n_r}(X_r), \qquad (2.1.24)$$

where the second sum runs over all partitions P_r of X into r disjoint subsets

$$X = X_1 \cup \ldots \cup X_r, \, X_r \neq \emptyset, \, |X_j| = n_j.$$

In particular we have

$$\tilde{T}_1(t) = -T_1(t). \qquad (2.1.25)$$

Instead of the Schrödinger equation, we will use a causality property of the S-matrix which follows from (2.1.13). Let us consider two switching functions g_1, g_2 with disjoint support in time

$$\operatorname{supp} g_1 \subset (-\infty, s), \, \operatorname{supp} g_2 \subset (s, +\infty). \qquad (2.1.26)$$

Then, in virtue of (2.1.13) we have

$$S(g_1 + g_2) = U_0(0, \infty)U(+\infty, -\infty)U_0(-\infty, 0)$$
$$= U_0(0, \infty)U(+\infty, s)U_0(s, 0)U_0(0, s)U(s, -\infty)U_0(-\infty, 0) = S(g_2)S(g_1), \qquad (2.1.27)$$

where the arguments $\pm\infty$ stand for the corresponding strong limits in (2.1.7) and U_0 for the free time evolution. Equation (2.1.27) expresses causality in the sense that, what happens for $t < s$ (described by $S(g_1)$) is not influenced by what happens for $t > s$ (described by $S(g_2)$).

We now want to find a perturbative formulation of the causality condition (2.1.27). Substituting (2.1.21) into the left hand side of (2.1.27), we get

$$S(g_1 + g_2) = \sum_n \frac{1}{n!} \int dt_1 \ldots dt_n \, T_n(t_1, \ldots, t_n)$$
$$\times (g_1(t_1) + g_2(t_1)) \ldots (g_1(t_n) + g_2(t_n)). \qquad (2.1.28)$$

By permutation of the integration variables t_j in the 2^n terms, we may arrange the switching functions in the form $g_2(t_1)\ldots g_2(t_m)g_1(t_{m+1})\ldots g_1(t_n)$. Since there are

$$\frac{n!}{m!(n-m)!} \quad \text{permutations,} \quad \sum_{m=0}^{n} \frac{n!}{m!(n-m)!} = 2^n,$$

we arrive at

$$S(g_1 + g_2) = \sum_{n=0}^{\infty} \sum_{m=0}^{n} \frac{1}{m!(n-m)!} \int dt_1 \ldots dt_n \, T_n(t_1, \ldots, t_n)$$

$$\times g_2(t_1)\ldots g_2(t_m)g_1(t_{m+1})\ldots g_1(t_n)$$

$$= S(g_1)S(g_2) = \sum_{n=0}^{\infty}\sum_{m=0}^{n}\frac{1}{m!(n-m)!}\int dt_1\ldots dt_n$$

$$T_m(t_1,\ldots,t_m)T_{n-m}(t_{m+1},\ldots,t_n)g_2(t_1)\ldots g_2(t_m)g_1(t_{m+1})\ldots g_1(t_n). \tag{2.1.29}$$

This leads to the desired perturbative causality condition (problem 2.1)

$$T_n(t_1,\ldots,t_n) = T_m(t_1,\ldots,t_m)T_{n-m}(t_{m+1},\ldots,t_n), \tag{2.1.30}$$

if all $\{t_1,\ldots,t_m\}$ are greater than all $\{t_{m+1},\ldots,t_n\}$.

We now claim that all orders T_n in (2.1.21) can be inductively determined by means of (2.1.30), provided the first order $T_1(t)$ is given. In second order we proceed as follows: We form the two functions

$$A'_2(t_1,t_2) = \tilde{T}_1(t_1)T_1(t_2) = -T_1(t_1)T_1(t_2) \tag{2.1.31}$$

$$R'_2(t_1,t_2) = T_1(t_2)\tilde{T}_1(t_1) = -T_1(t_2)T_1(t_1), \tag{2.1.32}$$

and consider the so-called advanced and retarded functions

$$A_2(t_1,t_2) = A'_2(t_1,t_2) + T_2(t_1,t_2)$$

$$R_2(t_1,t_2) = R'_2(t_1,t_2) + T_2(t_2,t_1). \tag{2.1.33}$$

It follows from (2.1.30) that A_2 vanishes for $t_1 > t_2$ and R_2 vanishes for $t_1 < t_2$. The combination

$$D_2 = R_2 - A_2 = R'_2 - A'_2 \tag{2.1.34}$$

is known to us because T_2 drops out due to the symmetry $T_2(t_1,t_2) = T_2(t_2,t_1)$. Then, R_2 and A_2 can be identified in (2.1.34) by their support properties, for example

$$R_2(t_1,t_2) = \Theta(t_1 - t_2)D_2(t_1,t_2)$$

$$= \Theta(t_1 - t_2)(T_1(t_1)T_1(t_2) - T_1(t_2)T_1(t_1)). \tag{2.1.35}$$

R_2 is uniquely determined up to its value at $t_1 = t_2$. Finally, $T_2(t_1,t_2)$ can be determined from (2.1.33)

$$T_2(t_1,t_2) = R_2(t_1,t_2) - R'_2(t_1,t_2)$$

$$= \Theta(t_1 - t_2)T_1(t_1)T_1(t_2) + \Theta(t_2 - t_1)T_1(t_2)T_1(t_1)$$

$$= T\{T_1(t_1)T_1(t_2)\}. \tag{2.1.36}$$

This is exactly the second order time-ordered product in agreement with (2.1.19). For the induction step from $n-1$ to n we refer to Sect.3.1 in Chap. 3. Although the induction is carried out there for the case of field theory, it is an easy exercise to adapt it to quantum mechanics.

The two constructions of the time ordered exponential (2.1.20), discussed here, are analogous to the different definitions of the exponential function in elementary calculus: either by a differential equation (2.1.9) or by a functional equation (2.1.27). In field theory the second construction is superior.

For later use we give explicit results for the first order S-matrix. We consider the scattering of a nonrelativistic particle with mass m by a potential $V(t, \boldsymbol{x})$. From (2.1.18) we obtain

$$S_1 = -i \int_{-\infty}^{+\infty} dt\, e^{iH_0 t} V(t) e^{-iH_0 t}. \tag{2.1.37}$$

It is even possible to apply this formula to static (time-independent) potentials if one calculates with distributions. Going over to momentum space, the integral kernel of S_1 is equal to

$$S_1(\boldsymbol{p}, \boldsymbol{q}) = -i(2\pi)^{-3/2} \int_{-\infty}^{+\infty} dt \left(e^{iH_0 t}\right)(\boldsymbol{p}) \hat{V}(\boldsymbol{p}-\boldsymbol{q}) \left(e^{iH_0 t}\right)(\boldsymbol{q})$$

$$= -i(2\pi)^{-3/2} \int_{-\infty}^{+\infty} dt\, e^{i(\boldsymbol{p}^2 - \boldsymbol{q}^2)t/2m} \hat{V}(\boldsymbol{p}-\boldsymbol{q})$$

$$= -i(2\pi)^{-1/2} 2m \delta(\boldsymbol{p}^2 - \boldsymbol{q}^2) \hat{V}(\boldsymbol{p}-\boldsymbol{q}). \tag{2.1.38}$$

The hat here means the 3-dimensional Fourier transform. This integral kernel gives the so-called T-matrix

$$\stackrel{\text{def}}{=} -2\pi i \delta(\boldsymbol{p}^2 - \boldsymbol{q}^2) T(\boldsymbol{p}, \boldsymbol{q}),$$

so that

$$T(\boldsymbol{p}, \boldsymbol{q}) = (2\pi)^{-3/2} 2m \hat{V}(\boldsymbol{p}-\boldsymbol{q}). \tag{2.1.39}$$

It is directly related to the scattering amplitude

$$f(\boldsymbol{p}, \boldsymbol{q}) \stackrel{\text{def}}{=} -2\pi^2 T(\boldsymbol{p}, \boldsymbol{q}) = -(2\pi)^{1/2} m \hat{V}(\boldsymbol{p}-\boldsymbol{q}) \tag{2.1.40}$$

and to the differential cross section

$$\frac{d\sigma}{d\Omega} = |f|^2 \tag{2.1.41}$$

(see M.Reed, B.Simon, *loc. cit.*).

2.2 The Method of Epstein and Glaser

The essential difference between quantum mechanics and quantum field theory is the following: the basic objects in quantum mechanics are operators in Hilbert space, whereas quantum field theory deals with operator-valued distributions. As a consequence, the time-ordered products (2.1.18-19) are well defined in case of quantum mechanics where the $\tilde{V}(t)$ are operators. In field theory we would like to define the similar four-dimensional T-product, $x_j = (x_j^0, x_j^1, x_j^2, x_j^3) \in \mathbb{R}^4$,

$$T_n(x_1, \ldots, x_n) = T\{T_1(x_1) \cdot \ldots \cdot T_1(x_n)\}$$

$$\stackrel{\text{def}}{=} \sum_\Pi \Theta(x_{\Pi 1}^0 - x_{\Pi 2}^0) \cdot \ldots \cdot \Theta(x_{\Pi(n-1)}^0 - x_{\Pi n}^0) T_1(x_{\Pi 1}) \cdot \ldots \cdot T_1(x_{\Pi n}),$$

where the sum runs over all $n!$ permutations. But now $T_1(x)$ is an operator-valued distribution and the product with the discontinuous step functions is ill-defined. There are two ways out of this problem. Most people introduce ultraviolet cut-offs in $T_1(x)$, so that the multiplication by Θ-functions is possible. One then has to work hard to remove the cut-offs. The second way is the method of Epstein and Glaser where no ultraviolet cut-off is necessary. Instead, one works more carefully with the Θ-functions: we will first substitute them by C^∞-functions and then carefully discuss the limit.

It is our aim to construct the S-matrix of a quantum field theory by means of perturbation theory. That is to say, we want to express S as a power series in some coupling constant g. We cannot make any statement about the (probably asymptotic) convergence of this series. So the S-matrix is defined as a formal power series. Such series are well-defined mathematical objects if every finite order is well-defined. We can investigate the properties of the S-matrix by working term by term with the series. The individual terms are written down by utilizing the free fields introduced in chap.1 and normally ordered products of them. Since the terms in the perturbation series are distributions, it is necessary to test the S-matrix with a C-number test function $g(x) \in \mathcal{S}(\mathbb{R}^4)$, assumed to be in Schwartz space. The coupling constant g will often be included in the test function. We therefore start from the expression

$$S(g) = 1 + \sum_{n=1}^{\infty} \frac{1}{n!} \int d^4x_1 \ldots d^4x_n \, T_n(x_1, \ldots x_n) g(x_1) \ldots g(x_n)$$

$$\stackrel{\text{def}}{=} 1 + T. \qquad (2.2.1)$$

Since the test function $g(x)$ vanishes at infinity, it switches the interaction on and off in time. This process is completely unphysical. When we calculate physical cross sections, we have to perform the so-called adiabatic limit $g(x) \to 1$ at the end. This procedure has a great advantage. For $g \in \mathcal{S}(\mathbb{R}^4)$, the long range part of the interaction is cut off. Accordingly, $S(g)$ (2.2.1) is free of infrared divergences. The latter only appear in the adiabatic limit $g \to 1$. The operator valued distributions $T_n(x_1, \ldots, x_n)$ in (2.2.1) do not "know" what the test function g is, hence, they are free from infrared divergences. These so-called n-point distributions $T_n(x_1, \ldots, x_n)$ are the basic objects to be constructed, and this construction, therefore, is not plagued by infrared problems. This does not at all mean that such problems do not exist. They appear, however, at a later stage, namely when one calculates observable quantities from the T_n for $g(x)$ going to 1 (or to a coupling constant g).

The method of Epstein and Glaser is a generalization of the simple inductive construction described at the end of the last section. According to the definition (2.2.1), $T_n(x_1, \ldots, x_n)$ is symmetric in $x_1, \ldots, x_n$. It is therefore often convenient to consider the disordered set of n points in Minkowski space $\mathbb{M}$

$$X = \{x_j \in \mathbb{M} \,|\, j = 1, \ldots n\} \quad (2.2.2)$$

as argument of T_n. Later on, we will expand $T_n(X)$ in terms of free field operators in normally ordered form, for example,

$$T_n(X) = \sum_k : \prod_j \overline{\psi}(x_j)\, t_n^k(x_1, \ldots, x_n) \prod_l \psi(x_l) :: \prod_m A(x_m) :, \quad (2.2.3)$$

where t_n^k is a numerical distribution, the arguments of the field operators depend on the term k considered.

Like $S(g)$ (2.2.1), the inverse $S(g)^{-1}$ can be expressed by a perturbation series

$$S(g)^{-1} = 1 + \sum_{n=1}^{\infty} \frac{1}{n!} \int d^4x_1 \ldots d^4x_n\, \tilde{T}_n(x_1, \ldots, x_n) g(x_1) \ldots g(x_n) \quad (2.2.4)$$

$$= (1 + T)^{-1} = 1 + \sum_{r=1}^{\infty} (-T)^r. \quad (2.2.5)$$

The corresponding n-point distributions $\tilde{T}_n$ follow from (2.2.5) as in the formal inversion of a power series

$$\tilde{T}_n(X) = \sum_{r=1}^{n} (-)^r \sum_{P_r} T_{n_1}(X_1) \ldots T_{n_r}(X_r), \quad (2.2.6)$$

where the second sum runs over all partitions P_r of X (2.2.2) into r disjoint subsets

$$X = X_1 \cup \ldots \cup X_r, \quad X_j \neq \emptyset, \quad |X_j| = n_j.$$

All products of distributions in (2.2.6) are well-defined, because the arguments are disjoint sets of points such that the products are direct products of distributions.

Besides (2.2.6), there are two further relations between the T's which are important in the following. We write the product

$$1 = S(g)S(g)^{-1} = 1 + \sum_{n=1}^{\infty} \sum_{n_1+n_2=n} \frac{1}{n_1!n_2!} \int T_{n_1}(x_1,\ldots x_{n_1})$$

$$\times \tilde{T}_{n_2}(y_1,\ldots y_{n_2})g(x_1)\ldots g(x_{n_1})g(y_1)\ldots g(y_{n_2})d^4x_1\ldots d^4y_{n_2} \quad (2.2.7)$$

in symmetrical form by carrying out the $n!/n_1!n_2!$ permutations between the x and y variables

$$= 1 + \sum_{n=1}^{\infty} \frac{1}{n!} \sum_{P_2^0} \int d^4x_1 \ldots d^4x_n \, T_{n_1}(X)\tilde{T}_{n_2}(Y)g(x_1)\ldots g(x_n), \quad (2.2.8)$$

where P_2^0 are all partitions into two subsets

$$P_2^0 : \{x_1,\ldots,x_n\} = X \cup Y, \quad |X| = n_1 \quad (2.2.9)$$

with empty sets $X, Y = \emptyset$ allowed. The trivial expressions

$$T_0(\emptyset) = 1 = \tilde{T}_0(\emptyset) \quad (2.2.10)$$

have to be used. From (2.2.7, 8) we conclude that

$$\sum_{P_2^0} T_{n_1}(X)\tilde{T}_{n-n_1}(Z \setminus X) = 0, \quad (2.2.11)$$

for all fixed Z with $|Z| = n \geq 1$, $|X| = n_1$ and, in different notation

$$\sum_{P_2^0} T_{n-n_2}(Z \setminus Y)\tilde{T}_{n_2}(Y) = 0. \quad (2.2.12)$$

If we exchange S and S^{-1}, we obtain relations with T and $\tilde{T}$ interchanged, for example

$$\sum_{P_2^0} \tilde{T}_{n-n_1}(X)T_{n_1}(Z \setminus X) = 0. \quad (2.2.13)$$

We now proceed to discuss general properties which the S-matrix (2.2.1) should have. The first one is unitarity

$$S(g)^{-1} = S(g)^+. \quad (2.2.14)$$

Using (2.2.4), it can be expressed by means of the n-point distributions in the form

$$\tilde{T}_n(x_1,\ldots,x_n) = T_n(x_1,\ldots,x_n)^+.$$

We shall realize much later in Sect. 4.7 that unitarity is not as simple as this in gauge theories. The second property is translation invariance: Let $U(a, \mathbf{1})$ be the unitary translation operator in the total Fock space $\mathcal{F}$

$$(U(a,\mathbf{1})\varPhi)_j(x) = \varPhi_j(x_1+a,\ldots,x_j+a), \qquad (2.2.15)$$

this transformation law is understood to hold in all sectors j of $\mathcal{F}$. Then one requires

$$U(a,\mathbf{1})S(g)U(a,\mathbf{1})^{-1} = S(g_a) \qquad (2.2.16)$$

where

$$g_a(x) = g(x-a). \qquad (2.2.17)$$

This implies a similar law for the T's

$$U(a,\mathbf{1})T_n(x_1,\ldots,x_n)U(a,\mathbf{1})^{-1} = T_n(x_1+a,\ldots,x_n+a), \qquad (2.2.18)$$

and the same must hold for $\tilde{T}_n$. The next property is Lorentz covariance. If $U(0,\Lambda)$ is the (pseudo-unitary) representation of the proper Lorentz group $\mathcal{L}_+^\uparrow$, defined by the free fields (cf. (1.1.36) and (1.3.37)), then we should have

$$U(0,\Lambda)S(g)U(0,\Lambda)^{-1} = S(g_\Lambda), \qquad (2.2.19)$$

with

$$g_\Lambda(x) = g(\Lambda^{-1}x). \qquad (2.2.20)$$

For the n-point distributions, this means

$$U(0,\Lambda)T_n(x_1,\ldots,x_n)U(0,\Lambda)^{-1} = T_n(\Lambda x_1,\ldots,\Lambda x_n). \qquad (2.2.21)$$

We now come to the most important property which is causality. Let us suppose that there exists a frame of reference in which the test functions g_1, g_2 have disjoint supports in time

$$\operatorname{supp} g_1 \subset \{x \in \mathbb{M} \mid x^0 \in (-\infty, r)\} \quad,\quad \operatorname{supp} g_2 \subset \{x \in \mathbb{M} \mid x^0 \in (r,+\infty)\}. \qquad (2.2.22)$$

Then we require

$$S(g_1+g_2) = S(g_2)S(g_1). \qquad (2.2.23)$$

This causality condition does not have a direct physical foundation because, as discussed above, the switching on and off the interaction is unphysical. But the condition is physical in the external field problem and it works beautifully there (see FQED, Chap.2). In any case, the correctness of (2.2.23) can only be shown by working out its consequences. In an arbitrary Lorentz frame there exists a space-like plane which separates $\operatorname{supp} g_1$ and $\operatorname{supp} g_2$, such that $\operatorname{supp} g_2$ is later than $\operatorname{supp} g_1$. In this situation we shall write

$$\operatorname{supp} g_1 < \operatorname{supp} g_2. \qquad (2.2.24)$$

For arbitrary sets X, Y of points in Minkowski space, the relation $X < Y$ means that all points $x \in X$ are earlier than all $y \in Y$ in some Lorentz frame.

We now proceed to investigate the consequences of (2.2.23) for the T's. The expression

$$S(g_1 + g_2) = \sum_n \frac{1}{n!} \int d^4x_1 \ldots d^4x_n\, T_n(x_1, \ldots, x_n)$$

$$\times (g_1(x_1) + g_2(x_1)) \ldots (g_1(x_n) + g_2(x_n)) \qquad (2.2.25)$$

can be reordered by permutations of the integration variables x_j in the 2^n terms, to have them in the following form

$$g_2(x_1) \ldots g_2(x_m) g_1(x_{m+1}) \ldots g_1(x_n).$$

Since there are

$$\frac{n!}{m!(n-m)!} \quad \text{permutations,} \quad \sum_{m=0}^{n} \frac{n!}{m!(n-m)!} = 2^n, \qquad (2.2.26)$$

we arrive at

$$S(g_1 + g_2) = \sum_{n=0}^{\infty} \sum_{m=0}^{n} \frac{1}{m!(n-m)!} \int d^4x_1 \ldots d^4x_n$$

$$\times T_n(x_1, \ldots, x_n) g_2(x_1) \ldots g_2(x_m) g_1(x_{m+1}) \ldots g_1(x_n)$$

$$= S(g_2) S(g_1) = \sum_{n=0}^{\infty} \sum_{m=0}^{n} \frac{1}{m!(n-m)!} \int d^4x_1 \ldots d^4x_n$$

$$\times T_m(x_1, \ldots, x_m) T_{n-m}(x_{m+1}, \ldots, x_n) g_2(x_1) \ldots g_2(x_m) g_1(x_{m+1}) \ldots g_1(x_n).$$
$$(2.2.27)$$

This leads to the condition

$$T_n(x_1, \ldots, x_n) = T_m(x_1, \ldots, x_m) T_{n-m}(x_{m+1}, \ldots, x_n) \qquad (2.2.28)$$

if $\{x_1, \ldots, x_m\} > \{x_{m+1}, \ldots, x_n\}$. Similarly, the causality condition for $S^{-1}(g)$

$$S(g_1 + g_2)^{-1} = S(g_1)^{-1} S(g_2)^{-1}$$

with g_1, g_2 satisfying (2.2.24), implies

$$\tilde{T}_n(x_1, \ldots, x_n) = \tilde{T}_m(x_1, \ldots, x_m) \tilde{T}_{n-m}(x_{m+1}, \ldots, x_n), \qquad (2.2.29)$$

if $\{x_1, \ldots, x_m\} < \{x_{m+1}, \ldots, x_n\}$.

The basic causality condition (2.2.28) shows that the T_n are time-ordered products. If all x_j have different temporal components x_j^0, the arguments of T_n can be permuted such that $x_1^0 > x_2^0 > \ldots > x_n^0$, say, are ordered in time. Repeated application of (2.2.28) then gives

$$T_n(x_1, \ldots, x_n) = T_1(x_1) T_1(x_2) \ldots T_1(x_n). \qquad (2.2.30)$$

The letter T has been chosen to denote the time-ordered products. They are the basic objects of the theory. For shortness the notion "n-point function" is also used for T_n.

Now we are ready to turn to the inductive construction of $T_n(x_1, \ldots, x_n)$. Suppose all $T_m(x_1, \ldots, x_m)$ for $1 \leq m \leq n-1$ are known and have the above properties. Then, according to (2.2.6), the $\tilde{T}_m(X)$ can be calculated for all $1 \leq m = |X| \leq n-1$. From this it is possible to form the following distributions

$$A'_n(x_1, \ldots, x_n) = \sum_{P_2} \tilde{T}_{n_1}(X) T_{n-n_1}(Y, x_n) \tag{2.2.31}$$

$$R'_n(x_1, \ldots, x_n) = \sum_{P_2} T_{n-n_1}(Y, x_n) \tilde{T}_{n_1}(X), \tag{2.2.32}$$

where the sums run over all partitions

$$P_2 : \{x_1, \ldots, x_{n-1}\} = X \cup Y, \quad X \neq \emptyset \tag{2.2.33}$$

into disjoint subsets with $|X| = n_1 \geq 1$, $|Y| \leq n-2$. We also introduce

$$D_n(x_1, \ldots, x_n) = R'_n - A'_n. \tag{2.2.34}$$

If the sums are extended over all partitions P_2^0, including the empty set $X = \emptyset$, then we get the distributions

$$A_n(x_1, \ldots, x_n) = \sum_{P_2^0} \tilde{T}_{n_1}(X) T_{n-n_1}(Y, x_n)$$

$$= A'_n + T_n(x_1, \ldots, x_n), \tag{2.2.35}$$

$$R_n(x_1, \ldots, x_n) = \sum_{P_2^0} T_{n-n_1}(Y, x_n) \tilde{T}_{n_1}(X)$$

$$= R'_n + T_n(x_1, \ldots, x_n). \tag{2.2.36}$$

These two distribution are not known by the induction assumption because they contain the unknown $T_n(x_1, \ldots, x_n)$. Only the difference

$$D_n = R'_n - A'_n = R_n - A_n \tag{2.2.37}$$

is known (2.2.34). What remains to be done is to determine R_n (or A_n) in (2.2.37) separately. This is achieved by investigating the support properties of the various distributions.

Theorem 2.2.1. Let $Y = P \cup Q$, $P \neq \emptyset$, $P \cap Q = \emptyset$, $|Y| = n_1 \leq n-1$ and $x \notin Y$. If $\{Q, x\} > P$, $|Q| = n_2$, then we have

$$R'_{n_1+1}(Y, x) = -T_{n_2+1}(Q, x) T_{n_1-n_2}(P). \tag{2.2.38}$$

If $\{Q, x\} < P$, then we have

$$A'_{n_1+1}(Y, x) = -T_{n_1-n_2}(P) T_{n_2+1}(Q, x). \tag{2.2.39}$$

Proof. We start from (2.2.32)

$$R'_{n_1+1}(Y, x) = \sum_{P_2} T_{n_1+1-n_3}(Y', x)\tilde{T}_{n_3}(X), \qquad (2.2.40)$$

where P_2 are the partitions of Y

$$P_2: Y = X \cup Y', \quad |X| = n_3 \neq 0. \qquad (2.2.41)$$

Let
$$Y' = Y_1 \cup Y_2, \quad Y_1 = Y' \cap P, \quad Y_2 = Y' \cap Q$$
$$X = X_1 \cup X_2, \quad X_1 = X \cap P, \quad X_2 = X \cap Q, \qquad (2.2.42)$$

then we obviously have

$$Y_1 < Y_2, \quad X_1 < X_2, \quad \{Y_2, x\} > Y_1. \qquad (2.2.43)$$

Hence, causality implies

$$R'_{n_1+1}(Y, x) = \sum_{P_4^0} T(Y_2, x)T(Y_1)\tilde{T}(X_1)\tilde{T}(X_2), \qquad (2.2.44)$$

where the subscripts of the T's have been dropped for simplicity, the latter are always equal to the number of points in the argument. P_4^0 are all partitions of the form

$$P_4^0: P = X_1 \cup Y_1, \quad Q = X_2 \cup Y_2, \quad X_1 \cup X_2 \neq \emptyset. \qquad (2.2.45)$$

However, for $X_2 \neq \emptyset$, we can have $X_1 = \emptyset$. Then it follows from (2.2.12) that

$$\sum_{P_2^0} T(Y_1)\tilde{T}(X_1) = 0. \qquad (2.2.46)$$

Consequently, in (2.2.44) only the terms with $X_2 = \emptyset$, $Y_2 = Q$ remain

$$R'_{n_1+1}(Y, x) = T(Q, x)\sum_{P_2} T(Y_1)\tilde{T}(X_1). \qquad (2.2.47)$$

The partitions P_2 are

$$P_2: P = X_1 \cup Y_1, \quad X_1 \neq \emptyset,$$

with the empty set excluded. If one includes the empty set, one gets 0 according to (2.2.46). One may, therefore, rewrite (2.2.47) as

$$R'_{n_1+1}(Y, x) = -T(Q, x)T(P),$$

which proves (2.2.38). The proof of (2.2.39) is the same. □

By
$$\overline{V^+}(x) = \{y \mid (y-x)^2 \geq 0, \, y^0 \geq x^0\} \qquad (2.2.48)$$

we denote the closed forward cone of x, and by

$$\overline{V^-}(x) = \{y \mid (y-x)^2 \geq 0, \, y^0 \leq x^0\} \tag{2.2.49}$$

the closed backward cone. The n-dimensional generalizations are

$$\Gamma_n^\pm(x) = \{(x_1,\ldots,x_n) \mid x_j \in \overline{V^\pm}(x), \, \forall j = 1,\ldots n\}. \tag{2.2.50}$$

For $|Y| = n_1 \leq n-2$, it follows from (2.2.36) that

$$R_{n_1+1}(Y, x) = R'_{n_1+1}(Y, x) + T_{n_1+1}(P \cup Q, x)$$
$$= -T_{n_2+1}(Q, x)T_{n_1-n_2}(P) + T_{n_2+1}(Q, x)T_{n_1-n_2}(P) = 0, \tag{2.2.51}$$

where (2.2.38) and causality (2.2.28) was used. Since this holds for arbitrary Y as in Theorem 1.1, it follows that R (2.2.51) vanishes, if there exists a point earlier than x in some Lorentz frame. Consequently, R has a retarded support:

Corollary 2.2.2.
$$\operatorname{supp} R_{n_1+1}(Y, x) \subseteq \Gamma_{n_1+1}^+(x) \tag{2.2.52}$$

and similarly

$$\operatorname{supp} A_{n_1+1}(Y, x) \subseteq \Gamma_{n_1+1}^-(x). \tag{2.2.53}$$

Because of these support properties, R and A are called retarded and advanced distributions, respectively. The distribution D (2.2.37) then has a causal support:

Corollary 2.2.3. If $|Y| = n_1 \leq n-2$, then

$$\operatorname{supp} D_{n_1+1}(Y, x) \subseteq \Gamma_{n_1+1}^+(x) \cup \Gamma_{n_1+1}^-(x). \tag{2.2.54}$$

These support properties must be preserved in the step from $n-1$ to n. In particular, corollary 2.2.3 must hold for $n_1 = n-1$, too. But $D_n(Y, x)$ is known according to (2.2.37) and the induction assumption. Hence, the support property (2.2.54) must follow without using causality of T_n.

Theorem 2.2.4. If $n \geq 3$, then

$$\operatorname{supp} D_n(x_1,\ldots x_{n-1}, x_n) \subseteq \Gamma_n^+(x_n) \cup \Gamma_n^-(x_n). \tag{2.2.55}$$

Proof. We divide the proof into two parts.
1) According to theorem 2.2.1, we have

$$R'_n(x_1,\ldots, x_{n-1}, x_n) = -T_{n_2+1}(Q, x_n)T_{n-n_2-1}(P), \tag{2.2.56}$$

if $\{Q, x_n\} > P$, $|Q| = n_2$, and

$$A'_n(x_1,\ldots,x_{n-1},x_n) = -T_{n-n_2-1}(P)T_{n_2+1}(Q,x_n), \qquad (2.2.57)$$

if $\{Q,x_n\} < P$. Let Ω be the set of all points $x = (x_1,\ldots,x_n) \in \mathbb{M}^n$, such that in some Lorentz frame (which may depend on x) the n points of x can be decomposed as follows

$$\{x_1,\ldots,x_n\} = P' \cup Q' \cup S\,, \ P', Q' \neq \emptyset \qquad (2.2.58)$$

with
$$x_j^0 > x_n^0 \quad \text{for} \quad \forall x_j \in P'$$
$$x_j^0 < x_n^0 \quad \text{for} \quad \forall x_j \in Q'$$
$$x_j^0 = x_n^0 \quad \text{for} \quad \forall x_j \in S. \qquad (2.2.59)$$

We obviously have $P' > Q'$. Then, by means of causality, we get from (2.2.56)

$$R'_n(x_1,\ldots,x_{n-1},x_n) = -T(P' \cup S)T(Q')$$
$$= -T(P')T(S)T(Q'), \qquad (2.2.60)$$

and similarly for (2.2.57)

$$A'_n(x_1,\ldots,x_{n-1},x_n) = -T(P')T(Q' \cup S)$$
$$= -T(P')T(S)T(Q'). \qquad (2.2.61)$$

Consequently,
$$D_n = R'_n - A'_n = 0 \qquad (2.2.62)$$

vanishes in the open set Ω.

2) Suppose now
$$x = \{x_1,\ldots,x_{n-1},x_n\} \notin \Gamma^+_{n-1}(x_n) \cup \Gamma^-_{n-1}(x_n) \qquad (2.2.63)$$

is not in the support (2.2.55). This is possible in the following ways:

a) One point x_1 is in $\overline{V^+}(x_n)$ and another one, say x_2, is in $\overline{V^-}(x_n)$. Then $x \in \Omega$ and $D_n = 0$ according to (2.2.62).

b) One point, say x_1, is space-like with respect to x_n, $(x_1 - x_n)^2 < 0$. Then we choose a frame of reference such that $x_1^0 = x_n^0$. If there are two points x_j, x_k with $x_j^0 > x_n^0$ and $x_k^0 < x_n^0$, then x (2.2.63) is in Ω, hence, $D_n = 0$. We therefore assume

$$x_j^0 \geq x_n^0 \quad , \quad \forall j = 2,\ldots,n-1.$$

The case $x_j^0 \leq x_n^0$ is similar. If there is a point, say x_2, with $x_2^0 > x_n^0$, then it is possible by a small Lorentz transformation to arrive at a situation with $x_1^0 < x_n^0$, but still $x_2^0 > x_n^0$. Once more, we find $x \in \Omega$, hence, $D_n = 0$.

c) There remains the case where all x_j are simultaneous

$$x_j^0 = x_n^0 \quad , \quad \forall j = 1,\ldots,n-1. \qquad (2.2.64)$$

Then we select a point, say x_1, with maximal spatial distance $|x_1-x_n|$. Let P be the rest $\{x_2,\ldots,x_n\}$. It is now possible by small Lorentz transformations to get $x_1 > P$ or $x_1 < P$. In the first case, one finds according to (2.2.56)

$$R'_n(x_1,\ldots,x_n) = -T_1(x_1)T_{n-1}(P),$$

and in the second case from (2.2.57)

$$A'_n(x_1,\ldots,x_n) = -T_{n-1}(P)T_1(x_1).$$

¿From P we select another point, say x_2, which is space-like with respect to the rest P'. By suitable choice of Lorentz frames we can achieve that $x_2 > P'$ or $x_2 < P'$. Then causality of T_{n-1} implies

$$T_{n-1}(P) = T_1(x_2)T_{n-2}(P') = T_{n-2}(P')T_1(x_2).$$

For the same reason all three factors in

$$R'_n = -T_1(x_1)T_1(x_2)T_{n-2}(P')$$

commute which leads to $R'_n = A'_n$. Hence, $D_n = R'_n - A'_n = 0$, which completes the proof of theorem 2.2.4. □

For $n \leq 2$, the support property (2.2.55) of D_n must be verified explicitly. If theorem 2.2.4 would not be true, then the inductive construction of the T_n by means of causality would be impossible. Now we see this construction clearly before us: From the known $T_m(x_1,\ldots,x_m)$, $m \leq n-1$ one computes $A'_n(x_1,\ldots,x_n)$ (2.2.31) and $R'_n(x_1,\ldots,x_n)$ (2.2.32), and then $D_n = R'_n - A'_n$ (2.2.37). One decomposes D_n with respect to the supports (2.2.55)

$$D_n(x_1,\ldots,x_n) = R_n(x_1,\ldots,x_n) - A_n(x_1,\ldots,x_n), \quad (2.2.65)$$

$$\text{supp } R_n \subseteq \Gamma^+_{n-1}(x_n) \quad , \quad \text{supp } A_n \subseteq \Gamma^-_{n-1}(x_n).$$

Finally, T_n is found from (2.2.35) or (2.2.36)

$$T_n(x_1,\ldots,x_n) = R_n(x_1,\ldots,x_n) - R'_n(x_1,\ldots,x_n) \quad (2.2.66a)$$

$$= A_n(x_1,\ldots,x_n) - A'_n(x_1,\ldots,x_n). \quad (2.2.66b)$$

The only non-trivial step in this construction is the distribution splitting (2.2.65). This will be investigated in the following section.

To complete the inductive step we have to verify that T_n (2.2.66) satisfies all properties used in the inductive construction, in particular the causality condition (2.2.28). Assuming $\{x_1,\ldots,x_n\} = P \cup Q$ with $P < Q$, there are two cases to be examined: If $x_n \in P$, A_n vanishes and (2.2.66b) gives the causality condition

$$T_n = -A'_n = T(Q)T(P),$$

using (2.2.57). On the other hand, if $x_n \in Q$, R_n is zero and from (2.2.66a) and (2.2.56) we get the same result. Furthermore, T_n obeys translation invariance (2.2.18), because it cannot be destroyed in the distribution splitting

(2.2.65). Finally, if the distribution splitting is unique, $T_n(x_1,\ldots,x_n)$ must also be totally symmetric in virtue of (2.2.36), although R'_n and R_n separately are not symmetrical with respect to x_n. We will learn in the next section that R_n is unique up to local terms. But the latter can be symmetrized, so that from an appropriate splitting solution R_n, after symmetrization, we get a T_n with all desired properties.

The n-point distribution T_n so constructed can be regarded as a well-defined time ordered product of n factors T_1:

$$T_n(x_1,\ldots,x_n) = T\{T_1(x_1)\cdot\ldots\cdot T_1(x_n)\}. \qquad (2.2.67)$$

Then the S-matrix assumes the usual form

$$S(g) = \sum_{n=0}^{\infty} \frac{1}{n!} \int d^4x_1\ldots d^4x_n\, T\{T_1(x_1)\cdot\ldots\cdot T_1(x_n)\} g(x_1)\cdot\ldots\cdot g(x_n). \qquad (2.2.68)$$

Because of the similarity with the ordinary exponential series the following short notation is widely used

$$S(g) = T\exp\int d^4x\, T_1(x) g(x). \qquad (2.2.69)$$

The whole inductive construction can obviously be generalized to the situation with more than one first order coupling $T_1^0(x),\ldots T_1^m(x)$. Then, for a clear book-keeping it is convenient to use different test functions $g_0(x),\ldots g_m(x)$ for the $m+1$ vertices. The switched S-matrix is now of the following form

$$S(g_0,\ldots g_m) = \sum_{n=0}^{\infty} \frac{1}{n!} \int d^4x_1\ldots d^4x_n \sum_{j_1\ldots j_n} T_n^{j_1\ldots j_n}(x_1,\ldots,x_n)$$

$$\times g_{j_1}(x_1)\ldots g_{j_n}(x_n), \qquad (2.2.70)$$

where the T_n's are the well-defined time-ordered products

$$T_n^{j_1\ldots j_n}(x_1,\ldots,x_n) = T\{T_1^{j_1}(x_1)\ldots T_1^{j_n}(x_n)\}. \qquad (2.2.71)$$

They are obtained by using in step l of the inductive construction the coupling $T_1^{j_l}(x_l)$. Since after each step the result is symmetrized we still have the following permutation symmetry

$$T_n^{j_1\ldots j_n}(x_1,\ldots,x_n) = T_n^{Pj_1\ldots Pj_n}(x_{P1},\ldots x_{Pn}) \qquad (2.2.72)$$

for arbitrary permutations $P\in\mathcal{S}_n$.

2.3 Splitting of causal distributions in x-space

The operator-valued causal distributions which we want to split are always expanded in terms of normal products in free fields, as in (2.2.3) for example,

$$D_n(x_1,\ldots,x_n) = \sum_k : \prod_j \overline{\psi}(x_j) d_n^k(x_1,\ldots,x_n) \prod_l \psi(x_l) :: \prod_m A(x_m) : . \qquad (2.3.1)$$

We claim that the numerical distributions d_n^k have causal support. This is due to the fact that the numerical distributions are vacuum expectation values of operators $D_n^{j_1,\ldots,j_n}$ with causal support, constructed with Wick submonomials as above (2.2.71). This fact is a consequence of the generalized Wick's theorem (1.9.39), indeed, it is easily seen that A'_n, R'_n and D_n are of the form (1.9.39). It is our aim to preserve this form in the retarded distribution. Then we must only split the numerical distributions as follows

$$d_n^k(x) = r_n(x) - a_n(x)$$

$$\operatorname{supp} r_n \subseteq \Gamma_{n-1}^+(x_n) \quad , \quad \operatorname{supp} a_n \subseteq \Gamma_{n-1}^-(x_n), \qquad (2.3.2)$$

where $x = (x_1,\ldots,x_n)$. All distributions are assumed to be tempered $\in \mathcal{S}'(\mathbb{R}^{4n})$, so that we can use distributional Fourier transformation. The simplest way of splitting would be

$$r_n(x) = \chi_n(x) d_n^k(x), \quad \text{with} \qquad (2.3.3)$$

$$\chi_n(x) = \prod_{j=1}^{n-1} \Theta(x_j^0 - x_n^0). \qquad (2.3.4)$$

The difficulty is that (2.3.4) is discontinuous, and then, if d_n^k is singular at $x = 0$, r_n (2.3.3) is generally not a tempered distribution. Because of translation invariance, it is sufficient to put $x_n = 0$ and to consider

$$d(x) \stackrel{\text{def}}{=} d_n^k(x_1,\ldots,x_{n-1},0) \in \mathcal{S}'(\mathbb{R}^m), \quad m = 4n-4. \qquad (2.3.5)$$

The intersection of the discontinuity surface of (2.3.4) and supp d is the origin $x = 0$. Therefore, the behavior of $d(x)$ in the neighborhood of $x = 0$ is essential for the splitting procedure. For this reason we introduce the following definition:

Definition 3.1. *The distribution $d(x) \in \mathcal{S}'(\mathbb{R}^m)$ has a quasi-asymptotics $d_0(x)$ at $x = 0$ with respect to a positive continuous function $\rho(\delta)$, $\delta > 0$, if the limit*

$$\lim_{\delta \to 0} \rho(\delta) \delta^m d(\delta x) = d_0(x) \neq 0 \qquad (2.3.6)$$

exists in $\mathcal{S}'(\mathbb{R}^m)$.

The quasi-asymptotics probes the vicinity of $x = 0$, only: If

$$d(x) = d_1(x) + d_2(x) \tag{2.3.7}$$

where d_1 has a compact support K_0 containing $x = 0$ and supp d_2 is bounded away from 0, it follows that

$$\lim_{\delta \to 0} \rho(\delta) \delta^m \langle d_2(\delta x), \varphi_0 \rangle = \lim_{\delta \to 0} \rho(\delta) \langle d_2(x), \varphi_0\left(\frac{x}{\delta}\right) \rangle = 0 \tag{2.3.8}$$

for every $\varphi_0 \in C_0^\infty(\mathbb{R}^m)$. Since C_0^∞ is dense in $\mathcal{S}$, the distribution (2.3.8) vanishes on $\mathcal{S}$ also, hence

$$\lim_{\delta \to 0} \rho(\delta) \delta^m \langle d_1(\delta x), \varphi(x) \rangle = \langle d_0, \varphi \rangle \tag{2.3.9}$$

for all $\varphi \in \mathcal{S}$. In

$$\lim_{\delta \to 0} \rho(\delta) \langle d(x), \varphi\left(\frac{x}{\delta}\right) \rangle = \langle d_0, \varphi \rangle. \tag{2.3.10}$$

we go over to momentum space to find an equivalent condition for the Fourier transform $\hat{d}(p)$. Since

$$\langle d(x), \varphi\left(\frac{x}{\delta}\right) \rangle = \langle \hat{d}(p), \left(\varphi\left(\frac{x}{\delta}\right)\right)\check{\ }(p) \rangle = \delta^m \langle \hat{d}(p), \check{\varphi}(\delta p) \rangle$$

$$= \langle \hat{d}\left(\frac{p}{\delta}\right), \check{\varphi}(p) \rangle, \tag{2.3.11}$$

where $\check{\varphi}$ denotes the inverse Fourier transform, we get the following equivalent definition:

Definition 2.3.2. *The distribution $\hat{d}(p) \in \mathcal{S}'(\mathbb{R}^m)$ has quasi-asymptotics $\hat{d}_0(p)$ at $p = \infty$ if*

$$\lim_{\delta \to 0} \rho(\delta) \langle \hat{d}\left(\frac{p}{\delta}\right), \check{\varphi}(p) \rangle = \langle \hat{d}_0, \check{\varphi} \rangle \tag{2.3.12}$$

exists for all $\check{\varphi} \in \mathcal{S}(\mathbb{R}^m)$.

In momentum space the quasi-asymptotics controls the ultraviolet behavior of the distribution. Let us consider a scaling transformation

$$\lim_{\delta \to 0} \rho(\delta) \langle \hat{d}\left(\frac{p}{\delta}\right), \check{\varphi}(ap) \rangle = \langle \hat{d}_0(p), \check{\varphi}(ap) \rangle$$

$$= a^{-m} \lim_{\delta \to 0} \rho(\delta) \langle \hat{d}\left(\frac{p}{a\delta}\right), \check{\varphi}(p) \rangle = a^{-m} \lim_{\delta \to 0} \frac{\rho(\delta)}{\rho(a\delta)} \rho(a\delta) \langle \hat{d}\left(\frac{p}{a\delta}\right), \check{\varphi}(p) \rangle. \tag{2.3.13}$$

Since

$$\lim_{\delta \to 0} \rho(a\delta) \langle \hat{d}\left(\frac{p}{a\delta}\right), \check{\varphi}(p) \rangle = \langle \hat{d}_0(p), \check{\varphi}(p) \rangle$$

exists, we may conclude that the limit

$$\lim_{\delta \to 0} \frac{\rho(a\delta)}{\rho(\delta)} = a^{-m} \frac{\langle \hat{d}_0(p), \check{\varphi}(p) \rangle}{\langle \hat{d}_0(p), \check{\varphi}(ap) \rangle} \stackrel{\text{def}}{=} \rho_0(a) \tag{2.3.14}$$

exists, too, assuming that the denominator is different from 0. By another scaling transformation it follows

$$\rho_0(ab) = \rho_0(a)\rho_0(b), \tag{2.3.15}$$

which implies $\rho_0(a) = a^\omega$ with some real ω. We therefore call $\rho(\delta)$ the power-counting function.

Definition 2.3.3. *The distribution $d \in \mathcal{S}'(\mathbb{R}^m)$ is called singular of order ω, if it has a quasi-asymptotics $d_0(x)$ at $x = 0$, or its Fourier transform has quasi-asymptotics $\hat{d}_0(p)$ at $p = \infty$, respectively, with power-counting function $\rho(\delta)$ satisfying*

$$\lim_{\delta \to 0} \frac{\rho(a\delta)}{\rho(\delta)} = a^\omega, \tag{2.3.16}$$

for each $a > 0$.

Equation (2.3.14) implies

$$a^m \langle \hat{d}_0(p), \check{\varphi}(ap) \rangle = \langle \hat{d}_0(\frac{p}{a}), \check{\varphi}(p) \rangle = a^{-\omega} \langle \hat{d}_0(p), \check{\varphi}(p) \rangle$$

$$= \langle d_0(x), \varphi(\frac{x}{a}) \rangle = a^m \langle d_0(ax), \varphi(x) \rangle = a^{-\omega} \langle d_0(x), \varphi(x) \rangle, \tag{2.3.17}$$

i.e. $\hat{d}_0$ is homogeneous of degree ω:

$$\hat{d}_0(\frac{p}{a}) = a^{-\omega} \hat{d}_0(p) \tag{2.3.18}$$

$$d_0(ax) = a^{-(m+\omega)} d_0(x). \tag{2.3.19}$$

This implies that d_0 has quasi-asymptotics $\rho(\delta) = \delta^\omega$ and the singular order ω, too. A positive measurable function $\rho(\delta)$, satisfying (2.3.16), is called regularly varying at zero by mathematicians (E. Senata, *Regularly Varying Functions, Lecture Notes in Mathematics 508, Springer-Verlag 1976*). The most important properties of those functions are collected in Appendix B of FQED. In particular, we have the following estimates: If $\varepsilon > 0$ is an arbitrarily small number, then there exist constants C, C' and δ_0, such that

$$C\delta^{\omega+\varepsilon} \geq \rho(\delta) \geq C'\delta^{\omega-\varepsilon}, \tag{2.3.20}$$

for $\delta < \delta_0$.

We want to apply the definitions to the following examples:
1) $d = 1$: From (2.3.6) we get $\rho(\delta) = \delta^{-m}$ and $\omega = -m$.
2) $d(x) = D^a \delta(x)$ where

$$D^a \stackrel{\text{def}}{=} \frac{\partial^{a_1+\ldots+a_m}}{\partial x_1^{a_1}\ldots\partial x_m^{a_m}}, \quad |a|=a_1+\ldots+a_m.$$

Since
$$\hat{d}(p) = (2\pi)^{-m/2}(ip)^a,$$

we obtain $\rho(\delta) = \delta^{|a|}$ from (2.3.12) and $\omega = |a|$.

3) Let us consider the Jordan-Pauli distribution (1.10.3)

$$D(x) = \frac{\operatorname{sgn} x^0}{2\pi}\Big[\delta(x^2) - \frac{m}{2}\frac{\Theta(x^2)}{\sqrt{x^2}}J_1(m\sqrt{x^2})\Big]. \tag{2.3.21}$$

The one-dimensional δ-distribution satisfies

$$\delta(\delta^2 x^2) = \frac{\delta(x^2)}{\delta^2},$$

whereas the term with the Bessel function stays bounded for $\delta\sqrt{x^2} \to 0$. Hence

$$\lim_{\delta\to 0}\delta^2 D(\delta x) = \frac{\operatorname{sgn} x^0}{2\pi}\delta(x^2) = D_0(x) \tag{2.3.22}$$

which is just the mass zero Jordan-Pauli distribution. This illustrates the general fact that the quasi-asymptotics d_0 is given by the corresponding mass zero distribution. Since the Jordan-Pauli distribution is considered in $\mathbb{R}^4$ ($m=4$), we find $\rho(\delta) = \delta^{-2}$ and $\omega(D) = -2$.

4) The positive frequency part (1.1.12)

$$\hat{D}^{(+)}(p) = \frac{i}{2\pi}\Theta(p^0)\delta(p^2-m^2) \tag{2.3.23}$$

is best considered in momentum space. Since

$$\int \Theta\Big(\frac{p_0}{\delta}\Big)\delta\Big(\frac{p^2}{\delta^2}-m^2\Big)\varphi(p)\,d^4p = \delta^2\int\Theta(p_0)\delta(p^2-\delta^2 m^2)\varphi(p)\,d^4p$$

$$= \delta^2\int\frac{d^3p}{2\sqrt{\mathbf{p}^2+\delta^2 m^2}}\varphi(\sqrt{\mathbf{p}^2+\delta^2 m^2},\mathbf{p}), \tag{2.3.24}$$

we find

$$\lim_{\delta\to 0}\delta^{-2}\hat{D}^{(+)}\Big(\frac{p}{\delta}\Big) = \hat{D}_0^{(+)}(p) \tag{2.3.25}$$

which implies $\omega(D^{(+)}) = -2$, in agreement with the foregoing example. We obviously have $\omega(D^{(-)}) = -2$, too.

We now turn to the splitting problem, where we have to distinguish two cases:

a) $\omega < 0$: In this case, the power-counting function goes to infinity

$$\rho(\delta) \to \infty \quad \text{for} \quad \delta \to 0. \tag{2.3.26}$$

This implies

Splitting of causal distributions in x-space

$$\left\langle d(x), \varphi\left(\frac{x}{\delta}\right)\right\rangle \to \frac{\langle d_0, \varphi\rangle}{\rho(\delta)} \to 0. \tag{2.3.27}$$

We choose a monotonous C^∞-function χ_0 over $\mathbb{R}^1$ with

$$\chi_0(t) = \begin{cases} 0 & \text{for } t \leq 0 \\ < 1 & \text{for } 0 < t < 1 \\ 1 & \text{for } t \geq 1. \end{cases} \tag{2.3.28}$$

In addition we choose a vector $v = (v_1, \ldots v_{n-1}) \in \Gamma^+$, which means that all four-vectors v_j are inside the forward cone V^+. Then

$$v \cdot x = \sum_{j=1}^{n-1} v_j \cdot x_j = 0 \tag{2.3.29}$$

is a space-like hyperplane that separates the causal support: All products $v_j \cdot x_j$ are either ≥ 0 for $x \in \Gamma^+$ or ≤ 0 for $x \in \Gamma^-$. It is our aim to prove that the limit

$$\lim_{\delta \to 0} \chi_0\left(\frac{v \cdot x}{\delta}\right) d(x) \stackrel{\text{def}}{=} \Theta(v \cdot x) d(x) = r(x) \tag{2.3.30}$$

exists.

To prove this we have to show that the difference

$$\left[\chi_0\left(a\frac{v \cdot x}{\delta}\right) - \chi_0\left(\frac{v \cdot x}{\delta}\right)\right] d(x) \stackrel{\text{def}}{=} \psi_0\left(\frac{x}{\delta}\right) d(x) \tag{2.3.31}$$

tends to 0 for $\delta \to 0$ uniformly in $a \geq a_1 > 1$. Here ψ_0 makes the causal support of d shrinking in the time directions when δ is going to 0. By the cone - property of $\Gamma^+ \cup \Gamma^-$ the support also shrinks in the spatial directions. Therefore we can choose an auxiliary real function $\psi_1(x) \in C^\infty(\mathbb{R}^m)$ that is $= 1$ in $\Gamma^+ \cup \Gamma^-$ and vanishes outside a certain neighborhood of $\Gamma^+ \cup \Gamma^-$ (which depends on ψ_0). Then we have

$$\left\langle \psi_0\left(\frac{x}{\delta}\right) d(x), \varphi(x) \right\rangle = \left\langle \psi_0\left(\frac{x}{\delta}\right) d(x), \psi_1\left(\frac{x}{\delta}\right) \varphi(x) \right\rangle$$
$$= \left\langle \varphi(x) d(x), \psi_0\left(\frac{x}{\delta}\right) \psi_1\left(\frac{x}{\delta}\right) \right\rangle. \tag{2.3.32}$$

This vanishes for $\delta \to 0$ due to (2.3.27) for all $\varphi \in S(\mathbb{R}^m)$ because (i) $\psi_0(x)\psi_1(x)$ is a test function in $C_0^\infty(\mathbb{R}^m)$, (ii) if $d(x)$ has singular order < 0 then also $\varphi(x)d(x)$, which is a simple consequence of def.2.3.1. The convergence is obviously uniform in $a_1 \geq a \geq 1$ for some finite fixed $a_1 > 1$. To show the uniformity in $a \geq a_1$, we consider instead of ψ_0 (2.3.31) the function

$$\chi_0\left(a^n \frac{v \cdot x}{\delta}\right) - \chi_0\left(\frac{v \cdot x}{\delta}\right) = \sum_{j=0}^{n-1} \psi_0\left(a^j \frac{x}{\delta}\right).$$

Since $d_0(x)$ in (2.3.27) is homogeneous of order ω, it follows

$$\left\langle d_0, \sum_{j=0}^{n-1} \psi_0\left(a^j \frac{x}{\delta}\right)\psi_1\left(a^j \frac{x}{\delta}\right)\right\rangle =$$

$$= \left(\sum_{j=0}^{n-1} a^{j\omega}\right)\left\langle d_0, \psi_0\left(\frac{x}{\delta}\right)\psi_1\left(\frac{x}{\delta}\right)\right\rangle.$$

The sum is bounded for $n \to \infty$ which proves the uniformity in $a \geq a_1$.

The same construction can be carried through with $1 - \chi_0(t)$ which leads to the advanced distribution

$$\lim_{\delta \to 0}\left(1 - \chi_0\left(\frac{v \cdot x}{\delta}\right)\right) d(x) \stackrel{\text{def}}{=} -a(x). \qquad (2.3.33)$$

Since

$$d = \tilde{r} - a = r - a, \qquad (2.3.34)$$

we have a solution of the splitting problem. By construction we can apply r and a on discontinuous test functions

$$\langle r, \varphi \rangle = \langle r(x), \Theta(v \cdot x)\varphi(x)\rangle, \quad \langle r, (1 - \Theta)\varphi\rangle = 0$$

$$\langle a, \varphi \rangle = \langle a, (1 - \Theta)\varphi\rangle, \quad \langle a, \Theta\varphi\rangle = 0. \qquad (2.3.35)$$

This allows us to extend d (2.3.34) to such test functions also:

$$\langle d, \Theta\varphi\rangle = \langle r, \Theta\varphi\rangle - \langle a, \Theta\varphi\rangle = \langle r, \varphi\rangle$$

$$\langle d, (1 - \Theta)\varphi\rangle = -\langle a, \varphi\rangle. \qquad (2.3.36)$$

The same is true for the quasi-asymptotics (2.3.6)

$$\langle d_0, \varphi\rangle = \lim_{\delta \to 0}\rho(\delta)\delta^m\langle d(\delta x), \varphi\rangle.$$

Choosing here $\Theta\varphi$ instead of φ it follows that

$$\langle d_0, \Theta\varphi\rangle = \lim_{\delta \to 0}\rho(\delta)\delta^m\langle r(\delta x), \varphi\rangle. \qquad (2.3.37)$$

This shows that r and similarly a have the same singular order ω as d. If $\tilde{r}$ is another retarded solution of the splitting problem, then $\tilde{r} - r$ must be a tempered distribution with point support $x = 0$:

$$\tilde{r} - r = \sum_a C_a D^a \delta(x). \qquad (2.3.38)$$

It follows from example 2) above that $|a| \leq \omega$. For $\omega < 0$, which is the case here, all C_a must vanish, hence, $\tilde{r} = r$. Thus, in case a) the solution of the splitting problem is unique, in particular, independent of the time-like vector v in (2.3.30) and independent of the choice of χ_0. We see in (2.3.37) that the case $\omega(d) < 0$ is the case of trivial splitting by multiplication with Θ-function.

The non-trivial case is
b) $\omega \geq 0$: Now the power-counting function satisfies

$$\frac{\rho(\delta)}{\delta^{\omega+1}} \to \infty \quad \text{for} \quad \delta \to 0. \tag{2.3.39}$$

To get a vanishing scaling limit as in (2.3.27) we choose a multi-index b with $|b| = \omega + 1$ and consider

$$\langle d(x)x^b, \psi(\frac{x}{\delta})\rangle = \langle d(\delta y)y^b, \psi(y)\rangle \delta^{m+\omega+1}$$

$$\to \langle d_0(y), y^b\psi\rangle \frac{\delta^{\omega+1}}{\rho(\delta)} \to 0. \tag{2.3.40}$$

If ω is not integer (which seems not to occur in quantum field theory), we use the largest integer $[\omega] < \omega$ instead of ω. It follows that the splitting as in case a) is possible if the test function φ satisfies

$$(D^a\varphi)(0) = 0 \quad \text{for} \quad |a| \leq \omega. \tag{2.3.41}$$

To achieve that, we introduce an auxiliary function $w(x) \in \mathcal{S}(\mathbb{R}^m)$ with

$$w(0) = 1, \; (D^a w)(0) = 0 \quad \text{for} \quad 1 \leq |a| \leq \omega, \tag{2.3.42}$$

and define

$$(W\varphi)(x) \stackrel{\text{def}}{=} \varphi(x) - w(x) \sum_{|a|=0}^{\omega} \frac{x^a}{a!}(D^a\varphi)(0) \tag{2.3.43}$$

$$= \sum_{|b|=\omega+1} x^b \psi_b(x).$$

The function $w(x)$ serves for the purpose of getting rapid decrease for $|x| \to \infty$. Now the decomposition according to a) (2.3.37) is possible

$$\langle r(x), \varphi\rangle \stackrel{\text{def}}{=} \langle d, \Theta(v \cdot x)W\varphi\rangle, \tag{2.3.44}$$

$$a(x) = r - d.$$

After construction $r(x)$ defines a tempered distribution with $\operatorname{supp} r \subseteq \Gamma^+(0)$. It agrees with $d(x)$ on $\Gamma^+(0) \setminus \{0\}$ in the sense of distributions, because a test function $\varphi \in \mathcal{S}$ with $\operatorname{supp}\varphi \subset \Gamma^+(0) \setminus \{0\}$ vanishes at $x = 0$, together with all its derivatives, so that the additional subtracted terms in (2.3.43) are 0. But without these terms, there is no splitting of $d(x)$ which makes sense for arbitrary $\varphi \in \mathcal{S}$, because the limit (2.3.30) exists on subtracted test functions only. *If one does the splitting incorrectly by simple multiplication with $\Theta(v \cdot x)$ as in a), one is punished by the well-known ultraviolet divergences in field theory.* These divergences appear in loop graphs which have $\omega \geq 0$. For those graphs the naive splitting with $\Theta(v \cdot x)$ is impossible and, as a consequence, the Feynman rules do not hold.

Again we have
$$\omega(r) = \omega(d) = \omega(a), \qquad (2.3.45)$$
This is a direct consequence of the definitions (2.3.43-44), because the limit
$$\lim_{\delta \to 0} \rho(\delta)\langle r(x), \varphi\left(\frac{x}{\delta}\right)\rangle = \lim_{\delta \to 0} \rho(\delta)\langle d(x), \Theta W\left(\varphi\left(\frac{x}{\delta}\right)\right)\rangle$$
$$= \lim_{\delta \to 0} \rho(\delta)\langle d(x), (\Theta W \varphi)(\frac{x}{\delta})\rangle = \langle d_0(x), (\Theta W \varphi)(x)\rangle$$
exists with the same power counting function as $d(x)$. But in sharp contrast to case a), the splitting b) is not unique. If $\tilde{r}(x)$ is the retarded part of another decomposition, then the difference
$$\tilde{r} - r = \sum_{|a|=0}^{\omega} \tilde{C}_a D^a \delta(x) \qquad (2.3.46)$$
is again a distribution with point support. Since $\omega > 0$, this time the splitting is only determined up to a finite sum of local terms (2.3.46). These undetermined local terms are not fixed by causality, additional physical normalization conditions are necessary to fix them.

2.4 Splitting in momentum space

All explicit calculations in quantum field theory are usually done in momentum space. For this reason, we must investigate the splitting procedure in p-space. We need the distributional Fourier transforms
$$F^{-1}[\Theta(v \cdot x)] \stackrel{\text{def}}{=} \check{\chi}(k) \qquad (2.4.1)$$
$$F^{-1}[x^a w](p) = (iD_p)^a \check{w}(p). \qquad (2.4.2)$$
Since
$$(D^a \varphi)(0) = \langle (-)^a D^a \delta, \varphi \rangle = (-)^a \langle \widehat{D^a \delta}, \check{\varphi} \rangle$$
$$= (-)^a (2\pi)^{-m/2} \langle (-ip)^a, \check{\varphi} \rangle = (2\pi)^{-m/2} \langle (ip)^a, \check{\varphi} \rangle, \qquad (2.4.3)$$
we conclude from (2.3.44) that
$$\langle \hat{r}, \check{\varphi} \rangle = \langle \hat{d}, (\Theta W \varphi)\check{\ } \rangle = (2\pi)^{-m/2} \langle \hat{d}, \check{\chi} * \left[\check{\varphi} - \right.$$
$$\left. - \sum_{|a|=0}^{\omega} \frac{1}{a!}(iD_p)^a \check{w}(p)(2\pi)^{-m/2} \langle (ip')^a, \check{\varphi} \rangle \right] \rangle_p \qquad (2.4.4)$$
$$= (2\pi)^{-m/2} \langle \check{\chi} * \hat{d}, \check{\varphi} - \sum_{|a|=0}^{\omega} \ldots \rangle, \qquad (2.4.5)$$

where the asterisk means convolution. We stress the fact that the convolution $\check{\chi} * \hat{d}$ is only defined on subtracted test functions, not on $\check{\varphi}$ alone. Interchanging p' and p in the subtraction terms, we may write

$$\langle \hat{r}, \check{\varphi} \rangle = (2\pi)^{-m/2} \int dk\, \hat{\chi}(k) \langle \hat{d}(p-k)$$

$$-(2\pi)^{-m/2} \sum_a \frac{(-)^a}{a!} p^a \int dp'\, \hat{d}(p'-k) D^a_{p'} \check{w}(p'),\, \check{\varphi} \rangle_p. \quad (2.4.6)$$

After partial integration in the p'-integral this is equivalent to the following result for the retarded distribution

$$\hat{r}(p) = (2\pi)^{-m/2} \int dk\, \hat{\chi}(k) \Big[\hat{d}(p-k)$$

$$-(2\pi)^{-m/2} \sum_{|a|=0}^{\omega} \frac{p^a}{a!} \int dp'\, \big(D^a_{p'} \hat{d}(p'-k)\big) \check{w}(p') \Big]. \quad (2.4.7)$$

Here the k-integral is understood in the sense of distributions as in (2.4.6).

By Fourier transformation of (2.3.46) we see that $\hat{r}(p)$ is only determined up to a polynomial in p of degree ω. Consequently the general result for the retarded distribution reads

$$\hat{\tilde{r}}(p) = \hat{r}(p) + \sum_{|a|=0}^{\omega} C_a p^a \quad (2.4.8)$$

with $\hat{r}(p)$ given by (2.4.7). We now assume that there exists a point $q \in \mathbb{R}^m$ where the derivatives $D^b \hat{r}(q)$ exist in the usual sense of functions for all $|b| \le \omega$. Let us define

$$\hat{r}_q(p) = \hat{r}(p) - \sum_{|b|=0}^{\omega} \frac{(p-q)^b}{b!} D^b \hat{r}(q). \quad (2.4.9)$$

This is another retarded distribution because we have only added a polynomial in p of degree ω. Furthermore, this solution of the splitting problem is *uniquely* specified by the normalization condition

$$D^b \hat{r}_q(q) = 0, \quad |b| \le \omega. \quad (2.4.10)$$

We compute

$$D^b \hat{r}(q) = (2\pi)^{-m/2} \int dk\, \hat{\chi}(k) \Big[(D^b \hat{d})(q-k)$$

$$-(2\pi)^{-m/2} \sum_{b \le a} \frac{a!\, q^{a-b}}{(a-b)!\, a!} \int dp'\, \check{w}(p') D^a_{p'} \hat{d}(p'-k) \Big] \quad (2.4.11)$$

from (2.4.7) and substitute this into (2.4.9). Since

$$\sum_{b\le a}\frac{(p-q)^b}{b!}\frac{q^{a-b}}{(a-b)!}=\frac{1}{a!}\sum_{b\le a}\binom{a}{b}(p-q)^b q^{a-b}=\frac{p^a}{a!},$$

the subtracted terms in (2.4.7) drop out

$$\hat r_q(p)=(2\pi)^{-m/2}\int dk\,\hat\chi(k)\Big[\hat d(p-k)-\sum_{|b|=0}^{\omega}\frac{(p-q)^b}{b!}(D^b\hat d)(q-k)\Big]. \quad (2.4.12)$$

This is the splitting solution with normalization point q. It is uniquely specified by (2.4.10), that means it does not depend on the time-like vector v in (2.4.1). The subtracted terms are the beginning of the Taylor series at $p=q$. This is an ultraviolet "regularization" in the usual terminology. It should be stressed, however, that here this is a consequence of the causal distribution splitting and not an ad hoc recipe.

It is well-known that causality can be expressed in momentum space by dispersion relations. Therefore we look for a connection of the result (2.4.12) with dispersion relations. We take $q=0$ in (2.4.12), which is possible if all fields are massive, for example, and consider time-like $p\in\varGamma^+$. We choose a special coordinate system such that $p=(p_1^0,\mathbf{0},0,\ldots)$. Note that this coordinate system is not obtained by a Lorentz transformation from the original one, but by an orthogonal transformation in $\mathbb{R}^m$. Furthermore we take v parallel to p, i.e. $v=(1,\mathbf{0},0,\ldots)$. Then v varies with p, but this is admissible because (2.4.12) is actually independent of v. We now have $\Theta(v\cdot x)=\Theta(x_1^0)$ and the Fourier transform (2.4.1) is given by

$$\hat\chi(k)=(2\pi)^{m/2-1}\delta(\mathbf{k}_1,k_2,\ldots k_{n-1})\frac{i}{k_1^0+i0}.$$

We always use the mathematical notation $i0$ for $i\varepsilon$ with the subsequent distributional limit $\varepsilon\to 0$. Using this result in (2.4.12) we shall obtain

$$\hat r_0(p_1^0)=\frac{i}{2\pi}\int_{-\infty}^{+\infty}dk_1^0\,\frac{1}{k_1^0+i0}\Big[\hat d(p_1^0-k_1^0,0,\ldots)$$

$$-\sum_{a=0}^{\omega}\frac{(p_1^0)^a}{a!}(-)^a D_{k_1^0}^a\hat d(q_1^0-k_1^0,0,\ldots)\Big|_{q_1^0=0}\Big]. \quad (2.4.13)$$

The transformation of this result to the usual form of a dispersion integral is contained in the following proposition.

Proposition 2.4.1. The retarded distribution $\hat r_0$ (2.4.13) is equal to the dispersion integral

$$r_0(p_1^0)=\frac{i}{2\pi}(p_1^0)^{\omega+1}\int_{-\infty}^{+\infty}dk_0\,\frac{\hat d(k_0)}{(k_0-i0)^{\omega+1}(p_1^0-k_0+i0)}. \quad (2.4.14)$$

Proof. We simplify the notation and consider the following integral over a finite range

$$F = \int_{-A}^{B} \frac{dk}{k+i0}\left[d(p-k) - \sum_{a=0}^{\omega}(-)^a\frac{p^a}{a!}\partial_k^a d(-k)\right].$$

In the first term we introduce the new integration variable $k' = k - p$ and the second term is integrated by parts

$$F = \int_{-A-p}^{B-p} dk' \frac{d(-k')}{p+k'+i0} - \sum_{a=0}^{\omega}\frac{p^a}{a!}\int_{-A}^{B} d(-k)\partial_k^a\frac{1}{k+i0}dk$$

$$-\sum_{a=1}^{\omega}(-)^a\frac{p^a}{a!}\sum_{b=0}^{a-1}(-)^b\partial_k^b\frac{1}{k+i0}\partial_k^{a-b-1}d(-k)\Big|_{k=-A}^{B}.$$

In the first term we restore the old limits of integration $[-A, B]$

$$F = \int_{-A}^{B} dk'\left[\frac{1}{p+k'+i0} - \sum_{a=0}^{\omega}\frac{p^a}{a!}\partial_k'^a\frac{1}{k'+i0}\right]d(-k')$$

$$+\left(\int_{-A-p}^{-A} - \int_{B-p}^{B}\right)\frac{d(-k')}{p+k'+i0} - \cdots\Big|_{-A}^{B},$$

where the dots stand for the above boundary terms.

We claim that the correction terms in the second line cancel each other in the limits $A, B \to \infty$. To show this we study the expression

$$R(-A) = f(p) + \sum_{a=1}^{\omega}(-)^a\frac{p^a}{a!}\sum_{b=0}^{a-1}(-)^b\partial_k^b\frac{1}{k+i0}\partial_k^{a-b-1}d(-k)\Big|_{k=-A},$$

with

$$f(p) = \int_{-A-p}^{-A} dk\, \frac{d(-k)}{p+k+i0}.$$

Simplifying the boundary terms we get

$$R(-A) = f(p) + \sum_{a=1}^{\omega}\frac{p^a}{a!}\sum_{b=1}^{a}\frac{(b-1)!}{A^b}d^{(a-b)}(A),$$

where $d^{(c)}$ denotes the derivatives of $d(k)$. The crucial fact is that the polynomial in p appearing in the remainder $R(-A)$ is just the Taylor polynomial of $f(p)$ of order ω. To verify this we have to compute the derivatives of $f(p)$.

Since p appears in the lower limit of integration and under the integral we obtain

$$f^{(a)}(p) = -\sum_{b=1}^{a} \frac{(b-1)!}{A^b} d^{(a-b)}(A+p)$$

$$+(-)^a a! \int_{-A-p}^{-A} dk \frac{d(-k)}{(p+k+i0)^{a+1}},$$

which can easily be proved by induction. For $p = 0$ this is indeed equal to the coefficient of $p^a/a!$ in $R(-A)$, up to a minus sign. Consequently, the polynomial in $R(-A)$ drops out, so that we must only estimate the remainder in the Taylor formula.

For this purpose we need an estimate for the derivatives $d^{(a)}(p)$ for large p. This is obtained from (2.3.12) which yields

$$\hat{d}(tq) \longrightarrow \frac{\hat{d}_0(q)}{\rho(1/t)}, \quad |t| \to \infty,$$

and, since the power counting function ρ is bounded by

$$\frac{1}{\rho(1/t)} < \frac{|t|^{\omega+\varepsilon}}{C(\varepsilon)}, \quad \varepsilon < 1,$$

we may conclude that

$$|\hat{d}(p)| < C |p|^{\omega+\varepsilon}, \quad |p| \to \infty. \tag{2.4.15}$$

Similarly, we have

$$|\hat{d}^{(a)}(p)| < C |p|^{\omega+\varepsilon-a}, \quad |p| \to \infty,$$

because differentiation lowers the singular order. The remainder in the Taylor formula goes with $|f^{(\omega+1)}| \sim A^{\varepsilon-1}$ which tends to 0 for $A \to \infty$. Note that the integral in $f^{(\omega+1)}$ also vanishes for $A \to \infty$ by means of (2.4.15).

Returning to our old notation, the retarded distribution can now be written as

$$\hat{r}_0(p_1^0) = \frac{i}{2\pi} \int_{-\infty}^{+\infty} dk_0' \left[\frac{1}{p_1^0 + k_0' + i0} - \sum_{a=0}^{\omega} \frac{(p_1^0)^a}{a!} \frac{\partial^a}{\partial k_0'^a} \frac{1}{k_0' + i0} \right] \hat{d}(-k_0').$$

The square bracket can be easily computed

$$[\ldots] = \left(-\frac{p_1^0}{k_0' + i0} \right)^{\omega+1} \frac{1}{p_1^0 + k_0' + i0}.$$

Changing the variable of integration from k_0' to $-k_0$, we obtain the dispersion integral (2.4.14). □

The expression (2.4.14) is a subtracted dispersion relation like (2.8.52). We shall often apply it to the case of one four-momentum $p \in \mathbb{R}^4$. To write down the result for arbitrary $p \in \Gamma^+$, we use the variable of integration $t = k_0/p_1^0$ and arrive at

$$\hat{r}_0(p) = \frac{i}{2\pi} \int_{-\infty}^{+\infty} dt \, \frac{\hat{d}(tp)}{(t-i0)^{\omega+1}(1-t+i0)}. \qquad (2.4.16)$$

For later reference we call this the central splitting solution, because it is normalized at the origin ($q = 0$ in (2.4.10)). The latter fact has two important consequences. (i) The central splitting solution does not introduce a mass scale into the theory. If $q \neq 0$, then $|q^2| = M^2$ defines such a scale. (ii) Most symmetry properties of $\hat{d}(p)$ are preserved under central splitting, as we will see later, because the origin $q = 0$ is a very symmetrical point.

It is easy to verify by means of (2.4.15) that the dispersion integral (2.4.16) is convergent for $|t| \to \infty$. But it would be ultraviolet divergent, if ω in (2.4.16) is chosen too small. Consequently, *the correct distribution splitting with the right singular order ω is terribly important. Incorrect distribution splitting leads to ultraviolet divergences.* This is the origin of the ultraviolet problem in field theory.

The dispersion relation (2.4.16) is only valid for $p \in \Gamma^+$. For $p \in \Gamma^-$ we get

$$r_0(p) = -\frac{i}{2\pi} \int_{-\infty}^{+\infty} dt \, \frac{\hat{d}(tp)}{(t+i0)^{\omega+1}(1-t-i0)}. \qquad (2.4.17)$$

For arbitrary p we may either return to (2.4.12)

$$\hat{r}_q(p) = \frac{i}{2\pi} \int_{-\infty}^{+\infty} \frac{dt}{t+i0} \left[\hat{d}(p-tv) - \sum_{|a|=0}^{\omega} \frac{p^a}{a!} \left(D_q^a \hat{d} \right)(q-tv) \right], \qquad (2.4.18)$$

where $v \in \Gamma^+$ is fixed, or we may determine $\hat{r}(p)$ by analytic continuation. The latter method is based on the following well-known theorem (see e.g. M. Reed, B. Simon, *Methods of Modern Mathematical Physics, Vol.2 (1978)*, p.23 and Problem 23, p.124).

Theorem 2.4.5. *The retarded distribution $\hat{r}(p)$ is the boundary value of an analytic function, regular in $\mathbb{R}^m + i\Gamma^+$.*

We give the main idea of the proof in the simple case where $r(x)$ is an L^1-function. Then the ordinary Fourier integral

$$\hat{r}(p) = (2\pi)^{-m/2} \int r(x) \, e^{ipx} \, d^m x \qquad (2.4.19)$$

is valid. Here $\operatorname{supp} r(x) \subset \Gamma^+$. Since $p \in \Gamma^+$ implies $p \cdot x \geq 0$ for all $x \in \Gamma^+$, it follows that

$$\operatorname{Re}(ipx) < 0 \quad \text{if} \quad p \in \mathbb{R}^m + i\Gamma^+. \qquad (2.4.20)$$

Consequently, the integral (2.4.19) is exponentially convergent and, hence, analytic in p. Theorem 3.5 allows to continue $\hat{r}(p)$ from time-like to space-like p.

Finally we want to determine the analytic function in the tube $\Gamma^+ + i\Gamma^+$ from the dispersion integral (2.4.16). We take $p = \lambda e$, $e \in \Gamma^+$, $\lambda > 0$

$$\hat{r}_0(\lambda e) = \frac{i}{2\pi} \int_{-\infty}^{+\infty} dt \, \frac{\hat{d}(t\lambda e)}{(t-i0)^{\omega+1}(1-t+i0)},$$

and use $t\lambda = s$ as a new variable of integration

$$\hat{r}_0(\lambda e) = \frac{i}{2\pi} \lambda^{\omega+1} \int_{-\infty}^{+\infty} ds \, \frac{\hat{d}(se)}{(s-i0)^{\omega+1}(\lambda - s + i0)}. \qquad (2.4.21)$$

Here λ can be chosen complex with $\operatorname{Im} \lambda > 0$. In this case the $i0$ of the second factor in the denominator is superfluous.

2.5 Calculation of tree graphs

For application to quantum gravity in Chap.5 we consider a massless tensor field $h^{\mu\nu}(x)$ coupled to an energy-momentum tensor $T_m^{\mu\nu}$, the subscript m stands for matter. The tensor field is quantized according to (1.7.11)

$$\Box h^{\mu\nu}(x) = 0,$$

$$[h^{\alpha\beta}(x), h^{\mu\nu}(y)] = -ib^{\alpha\beta\mu\nu} D_0(x-y)$$

$$= -\frac{i}{2}(g^{\alpha\mu}g^{\beta\nu} + g^{\alpha\nu}g^{\beta\mu} - g^{\alpha\beta}g^{\mu\nu})D_0(x-y), \qquad (2.5.1)$$

$$[h^{\alpha\beta(-)}(x), h^{\mu\nu(+)}(y)] = -ib^{\alpha\beta\mu\nu} D_0^{(+)}(x-y). \qquad (2.5.2)$$

The coupling is also given by the b-tensor (cf. Sect.5.9)

$$T_1(x) = \frac{i}{2}\kappa h^{\alpha\beta} b_{\alpha\beta\mu\nu} T_m^{\mu\nu}. \qquad (2.5.3)$$

The coupling constant κ will be related to Newton's gravitational constant later on.

For simplicity we assume that the energy-momentum tensor $T_{\mu\nu}^m$ comes from a complex free scalar field with classical Lagrangian

$$L = \varphi_{,\mu}^+ \varphi^{,\mu} - m^2 \varphi^+ \varphi. \qquad (2.5.4)$$

This describes the usual situation where the particle is different from its antiparticle, the often used real scalar field is not realistic in this respect. We write "+" for the complex conjugate because we soon go over to quantum fields where + is the adjoint. The classical energy-momentum tensor is then given by

$$T^m_{\mu\nu} = \frac{\partial L}{\partial \varphi^+_{,\nu}} \varphi^+_{,\mu} + \frac{\partial L}{\partial \varphi_{,\mu}} \varphi_{,\nu} - g_{\mu\nu} L$$

$$= \varphi^+_{,\mu} \varphi_{,\nu} + \varphi^+_{,\nu} \varphi_{,\mu} - g_{\mu\nu}(\varphi^+_{,\alpha} \varphi^{,\alpha} - m^2 \varphi^+ \varphi). \qquad (2.5.5)$$

To use this expression in quantum field theory, the products of field operators at the same space-time point x must be considered as normally ordered, otherwise they would not be well defined. Then the coupling (2.5.3) assumes the following form

$$T^m_1(x) = \frac{i}{2}\kappa\Big(2h^{\alpha\beta} : \varphi^+_{,\alpha} \varphi_{,\beta} : -m^2 h : \varphi^+ \varphi : \Big), \qquad (2.5.6)$$

where $h = h^\alpha_\alpha$ is the trace of the tensor field.

In addition to φ we consider a second scalar field $\psi(x)$ with mass M which couples in the same way (2.5.6) with the tensor field. We are interested in the gravitational force between the two particles due to graviton- or h-exchange to second order. According to the causal method we must form the products

$$A'_2(x,y) = -T^m_1(x) T^M_1(y) \qquad (2.5.7)$$

$$R'_2(x,y) = -T^M_1(y) T^m_1(x) \qquad (2.5.8)$$

and the causal distribution

$$D_2(x,y) = R'_2 - A'_2 = [T^m_1(x), T^M_1(y)] \sim -i D_0(x-y). \qquad (2.5.9)$$

Here the commutator is calculated between the two h-fields by means of (2.5.1). In tree graphs the causal splitting is always trivial, then we arrive at the retarded distribution

$$R_2(x,y) \sim -i D_0^{\text{ret}}(x-y). \qquad (2.5.10)$$

Finally, R'_2 (2.5.8) is normally ordered by Wick's theorem performing the contraction between the two h-fields with the help of (2.5.2)

$$R'_2(x,y) \sim i D_0^{(+)}(y-x) = -i D_0^{(-)}(x-y). \qquad (2.5.11)$$

Then we obtain

$$T_2(x,y) = R_2 - R'_2 \sim -i(D_0^{\text{ret}} - D_0^{(-)}) = -i D_0^F(x-y). \qquad (2.5.12)$$

We see that the ordinary Feynman rules hold for tree graphs. In all details the second order T-product is equal to (problem 2.3)

$$T_2(x,y) = \frac{i}{2}\kappa^2 : \Big\{\varphi^+_{,\mu}(x)\psi^{+,\mu}(y)\varphi_{,\nu}(x)\psi^{,\nu}(y) +$$

$$+\varphi_{,\mu}^+(x)\psi^{,\mu}(y)\varphi_{,\nu}(x)\psi^{+,\nu}(y) - \varphi_{,\mu}^+(x)\varphi^{,\mu}(x)\psi_{,\alpha}^+(y)\psi^{,\alpha}+$$
$$+M^2\varphi_{,\mu}^+(x)\varphi^{,\mu}(x)\psi^+(y)\psi(y) + m^2\psi_{,\alpha}^+(y)\psi^{,\alpha}(y)\varphi^+(x)\varphi(y)-$$
$$-2m^2M^2\varphi^+(x)\varphi(x)\psi^+(y)\psi(y)\Big\}: D_0^F(x-y). \quad (2.5.13)$$

The explicit representation of the scalar fields reads

$$\varphi(x) = (2\pi)^{-3/2}\int \frac{d^3p}{\sqrt{2E}}\left(a(\boldsymbol{p})e^{-ipx} + \tilde{a}^+(\boldsymbol{p})e^{ipx}\right) \quad (2.5.14)$$

$$\psi(y) = (2\pi)^{-3/2}\int \frac{d^3p}{\sqrt{2E}}\left(b(\boldsymbol{p})e^{-ipy} + \tilde{b}^+(\boldsymbol{p})e^{ipy}\right). \quad (2.5.15)$$

We want to calculate the S-matrix element $(\Phi_f, S\Phi_i)$ between two in- and outgoing particles

$$\Phi_i = a^+(\boldsymbol{p}_i)b^+(\boldsymbol{q}_i)\Omega \quad (2.5.16)$$

and similarly for Φ_f. Then the first term in (2.5.13) gives the following contribution

$$S_2^{(1)} = \frac{i}{2}\kappa^2\int d^4x\,d^4y\,D_0^F(x-y)$$
$$(\Omega, b(q_f)a(p_f):\varphi_{,\mu}^+(x)\psi^{+,\mu}(y)\varphi_{,\nu}(x)\psi^{,\nu}(y): a^+(p_i)b^+(q_i)\Omega) =$$
$$= \frac{i}{2}\kappa^2(2\pi)^{-6}\int d^4x\,d^4y\,D_0^F(x-y)\frac{1}{4\sqrt{E(p_i)E(q_i)E(p_f)E(q_f)}}$$
$$\times e^{i(p_f-p_i)x+i(q_f-q_i)y}(p_fq_f)(p_iq_i) = \quad (2.5.17)$$
$$\stackrel{\text{def}}{=} \delta^4(p_f+q_f-p_i-q_i)M_1,$$

with

$$M_1 = -\frac{i\kappa^2}{8(2\pi)^2\sqrt{E(p_i)E(q_i)E(p_f)E(q_f)}}\frac{(p_fq_f)(p_iq_i)}{(p_f-p_i)^2+i0}. \quad (2.5.18)$$

In the same way we compute the other five terms in (2.5.13). The total result is given by

$$S_2(\boldsymbol{p}_i,\boldsymbol{q}_i,\boldsymbol{p}_f,\boldsymbol{q}_f) = \delta^4(p_f+q_f-p_i-q_i)M$$
$$M = -\frac{i\kappa^2}{8(2\pi)^2\sqrt{E(p_i)E(q_i)E(p_f)E(q_f)}}\frac{Z}{(p_f-p_i)^2+i0}, \quad (2.5.19)$$

where

$$Z = (p_fq_f)(p_iq_i) + (p_fq_i)(p_iq_f) - (p_fp_i)(q_fq_i) + M^2p_fp_i + m^2q_fq_i - 2m^2M^2. \quad (2.5.20)$$

To relate the S-matrix element between continuum states to a transition probability we must calculate with wave packets

$$\Phi_i = \int d^3p_1 d^3q_1 \Phi(\boldsymbol{p}_1)\Psi(\boldsymbol{q}_1)a^+(\boldsymbol{p}_1)b^+(\boldsymbol{q}_1)\Omega \quad (2.5.21)$$

and similarly for the final state Φ_f. Φ and Ψ are wave functions $\in L^2(\mathbb{R}^3)$ normalized to one. Then

$$S_{fi} = \int d^3p_1 d^3q_1 d^3p_2 d^3q_2 \Phi_f(\boldsymbol{p}_2,\boldsymbol{q}_2)^* S_2(\boldsymbol{p}_1,\boldsymbol{q}_1,\boldsymbol{p}_2,\boldsymbol{q}_2)\Phi(\boldsymbol{p}_1)\Psi(\boldsymbol{q}_1), \quad (2.5.22)$$

and the transition probability is given by

$$p_{fi} \stackrel{\text{def}}{=} |S_{fi}|^2. \quad (2.5.23)$$

We sum over a complete set of two-particle final states by means of the completeness relation

$$\sum_f \Phi_f(\boldsymbol{p}_2,\boldsymbol{q}_2)^* \Phi_f(\boldsymbol{p}_2',\boldsymbol{q}_2') = \delta^3(\boldsymbol{p}_2-\boldsymbol{p}_2')\delta^3(\boldsymbol{q}_2-\boldsymbol{q}_2')$$

and obtain

$$\sum_f p_{fi} = \int d^3p_1 d^3q_1 d^3p_1' d^3q_1' d^3p_2 d^3q_2 M(p_1,q_1,p_2,q_2)\delta^4(p_1+q_1-p_2-q_2)$$

$$\times M(p_1',q_1',p_2,q_2)^* \delta^4(p_1'+q_1'-p_2-q_2)\Phi(\boldsymbol{p}_1)\Psi(\boldsymbol{q}_1)\Phi(\boldsymbol{p}_1')^*\Psi(\boldsymbol{q}_1')^*. \quad (2.5.24)$$

In a scattering experiment the wave functions Φ, Ψ are sharply peaked around the initial momenta $\boldsymbol{p}_i, \boldsymbol{q}_i$, respectively, compared with the distance of variation of M. Then (2.5.24) may be simplified as follows

$$\sum_f p_{fi} = \int d^3p_2 d^3q_2 |M(p_i,q_i,p_2,q_2)|^2 \int d^3p_1 d^3q_1 d^3p_1' d^3q_1'$$

$$\times \delta^4(p_1+q_1-p_2-q_2)\delta^4(p_1'+q_1'-p_2-q_2)\Phi(\boldsymbol{p}_1)\Psi(\boldsymbol{q}_1)\Phi(\boldsymbol{p}_1')^*\Psi(\boldsymbol{q}_1')^*. \quad (2.5.25)$$

The second integral herein only depends on the initial state. We denote it by $F(p), p = p_2+q_2$. Substituting

$$\delta^4(p) = (2\pi)^{-4}\int e^{\pm ipx} d^4x,$$

we can write it in the form

$$F(p) = (2\pi)^{-8}\int d^4x_1 d^4x_2 \int d^3p_1 \ldots d^3q_1' e^{-i(p_1+q_1-p)x_1}$$

$$\times e^{i(p_1'+q_1'-p)x_2}\Phi(\boldsymbol{p}_1)\Psi(\boldsymbol{q}_1)\Phi(\boldsymbol{p}_1')^*\Psi(\boldsymbol{q}_1')^*. \quad (2.5.26)$$

Let

$$\hat{\Phi}(x) = (2\pi)^{-3/2}\int d^3p\, e^{-ipx}\Phi(\boldsymbol{p}) \quad (2.5.27)$$

be the free wave packet in x-space, then (omitting the hat in the following)

$$F(p) = (2\pi)^{-2} \int d^4x_1 d^4x_2 \, \Phi(x_1)\Psi(x_1)\Phi(x_2)^*\Psi(x_2)^* e^{ip(\tau_1-\tau_2)}. \quad (2.5.28)$$

The positive function $F(p)$ is normalized according to

$$\int d^4p \, F(p) = (2\pi)^2 \int d^4x \, |\Phi(x)|^2 |\Psi(x)|^2, \quad (2.5.29)$$

and it is concentrated around $p = p_2 + q_2 = p_1 + q_1 \approx p_i + q_i$. In the limit of infinitely sharp wave packets we, therefore, may represent it by

$$F(p) = \delta(p - p_i - q_i)(2\pi)^2 \int d^4x \, |\Phi(x)|^2 |\Psi(x)|^2. \quad (2.5.30)$$

Then we get the following result for the sum of the transition probabilities

$$\sum_f p_{fi} = \int d^3p_2 d^3q_2 \, |M|^2 \delta(p_2 + q_2 - p_i - q_i) \times$$

$$\times (2\pi)^2 \int d^4x \, |\Phi(x)|^2 |\Psi(x)|^2. \quad (2.5.31)$$

The spreading of the wave packets in the course of time can be neglected, if they are sharply concentrated in momentum space. The free wave packet is then shifted with the velocity $\boldsymbol{v}$ of the particle without change of the shape

$$\Phi(t, \boldsymbol{x}) = \varphi(\boldsymbol{x} + \boldsymbol{x}_1 + \boldsymbol{v}t). \quad (2.5.32)$$

Particle 2 is assumed to be the target which is at rest

$$\Psi(x) = \psi(\boldsymbol{x}). \quad (2.5.33)$$

Scattering is only possible, if we consider a beam of incoming particles. Therefore, the expression (2.5.31) must be averaged over a cylinder of radius R parallel to $\boldsymbol{v}$

$$\sum_f p_{fi}(R) = \frac{1}{\pi R^2} \int_{|\boldsymbol{x}_{1\perp}| \leq R} d^2 x_{1\perp} \int d^4x \, |\varphi(\boldsymbol{x} + \boldsymbol{x}_1 + \boldsymbol{v}t)|^2 |\psi(\boldsymbol{x})|^2 \ldots \quad (2.5.34)$$

The scattering cross section in the laboratory frame is then given by

$$\sigma = \lim_{R \to \infty} \pi R^2 \sum_f p_{fi}(R). \quad (2.5.35)$$

Since

$$\int d^2 x_{1\perp} \int dt \int d^3x \, |\varphi(\boldsymbol{x} + \boldsymbol{x}_1 + \boldsymbol{v}t)|^2 |\psi(\boldsymbol{x})|^2 =$$

$$= \frac{1}{|\boldsymbol{v}|} \int d^3x_1 \int d^3x \, |\varphi(\boldsymbol{x} + \boldsymbol{x}_1)|^2 |\psi(\boldsymbol{x})|^2 = \frac{1}{|\boldsymbol{v}|} = \frac{E_i}{|\boldsymbol{p}_i|},$$

we arrive at

$$\sigma = (2\pi)^2 \frac{F_i}{|\boldsymbol{p}_i|} \int d^3p_2 d^3q_2\, \delta(p_2 + q_2 - p_i - q_i)|M|^2. \tag{2.5.36}$$

This result can be written in Lorentz invariant form by substituting

$$|\boldsymbol{p}_i| = \sqrt{E_i^2 - m^2} = \sqrt{(p_i q_i)^2 \frac{1}{M^2} - m^2}$$

$$= \frac{1}{M}\sqrt{(p_i q_i)^2 - m^2 M^2},$$

and $m = E(q_i)$:

$$\sigma = (2\pi)^2 \frac{E(p_i)E(q_i)}{\sqrt{(p_i q_i)^2 - m^2 M^2}} \int d^3 p_f d^3 q_f$$

$$\times \delta^4(p_f + q_f - p_i - q_i)|M|^2, \tag{2.5.37}$$

where p_f has been written instead of p_2. Indeed, taking the denominator $\sqrt{E(p_1)E(q_i)E(p_f)E(q_f)}$ in (2.5.19) into account, which is the only non-covariant factor, we see that $E(p_i)E(q_i)$ cancels in (2.5.37) and the remaining denominator $E(p_f)E(q_f)$ combines with the three-dimensional measures in (2.5.37) to give the Lorentz invariant measures on the mass shells.

One integration in (2.5.37) can immediately be carried out with help of the δ-distribution and the remaining $\delta^1(E(p_f)+E(q_f)-E(p_i)-E(q_i))$ allows to do the energy integral

$$d^3 p_f = \boldsymbol{p}_f^2 d|p_f| d\Omega = dE(p_f)E(p_f)|\boldsymbol{p}_f|d\Omega.$$

Then only the angular integral is left over, so that we can identify the differential cross section

$$\frac{d\sigma}{d\Omega} = (2\pi)^2 \frac{E(p_i)E(q_i)E(p_f)|\boldsymbol{p}_f|}{\sqrt{(p_i q_i)^2 - m^2 M^2}}|M|^2. \tag{2.5.38}$$

It is our aim now to make contact with non-relativistic potential scattering (2.1.41). By taking the square root of the differential cross section we can identify the scattering amplitude

$$f(\boldsymbol{p}_i, \boldsymbol{p}_f) = \frac{2\pi}{\lambda^{1/4}}\sqrt{E(p_i)E(q_i)E(p_f)|\boldsymbol{p}_f|}|M| \tag{2.5.39}$$

up to the sign, where we have introduced the abbreviation

$$\lambda = (p_i q_i)^2 - m^2 M^2. \tag{2.5.40}$$

Substituting the value of $|M|$ (2.5.19) we have

$$f(\boldsymbol{p}_i, \boldsymbol{p}_f) = \frac{\kappa^2}{16\pi}\sqrt{\frac{|\boldsymbol{p}_f|}{E(q_f)\lambda^{1/2}}} \frac{Z}{(p_f - p_i)^2 + i0}. \tag{2.5.41}$$

In the non-relativistic limit we put

$$(p_f - p_i)^2 = -\boldsymbol{k}^2,$$

and $p_f q_f = mM = p_i q_i$, $p_f q_i = mM = p_i q_f$ and $p_f p_i = m^2$, $q_f q_i = M^2$. Then the numerator Z (2.5.20) becomes $Z = m^2 M^2$. We assume the particle with mass M to be at rest initially, $\boldsymbol{q}_i = 0$, so that

$$\lambda = (E(p_i)E(q_i) - \boldsymbol{p}_i \boldsymbol{q}_i)^2 - m^2 M^2 = E(p_i)^2 M^2 - m^2 M^2 = M^2 \boldsymbol{p}_i^2$$

and

$$f(\boldsymbol{p}_i, \boldsymbol{p}_f) = \frac{\kappa^2}{16\pi} \sqrt{\frac{|\boldsymbol{p}_f|}{ME(q_f)|\boldsymbol{p}_i|}} \frac{m^2 M^2}{k^2}. \qquad (2.5.42)$$

For comparison with potential scattering we can neglect the recoil of the target, i.e. $|\boldsymbol{p}_i| = |\boldsymbol{p}_f| \ll M$, hence

$$f = \frac{\kappa^2}{16\pi} \frac{m^2 M}{k^2} = -\sqrt{2\pi} m \hat{V}(\boldsymbol{k}) \qquad (2.5.43)$$

according to (2.1.40). This allows to identify the static potential between the two particles

$$\hat{V}(\boldsymbol{k}) = -\frac{\kappa^2}{16\sqrt{2}\pi^{3/2}} \frac{mM}{k^2} \stackrel{\text{def}}{=} -G\sqrt{\frac{2}{\pi}} \frac{mM}{k^2}. \qquad (2.5.44)$$

The factor after G is chosen because the three-dimensional Fourier transform of Newton's potential is given by

$$F\left[\frac{1}{|\boldsymbol{x}|}\right] = \sqrt{\frac{2}{\pi}} \frac{1}{k^2}.$$

Thus, the second order tree graph considered leads to the Newtonian potential, if we choose the coupling constant

$$\kappa^2 = 32\pi G, \qquad (2.5.45)$$

where G is Newton's constant.

2.6 Calculation of loop graphs

In this section we consider pure Yang-Mills theory with the following first order coupling

$$T_1 = ig f_{abc} (: A_{\mu a} A_{\nu b} \partial^\nu A_c^\mu : - : A_{\mu a} u_b \partial^\mu \tilde{u}_c :). \tag{2.6.1}$$

The first term is a cubic self-coupling of vector fields. In addition one needs a coupling to ghost fields, in order to get first order gauge invariance. We will derive this coupling in the following chapter (sect.3.2). The vector fields $A_{\mu a}(x), a = 1, \ldots N$ are the (free) gauge potentials, satisfying the commutation relations

$$[A_a^{(-)\mu}(x), A_b^{(+)\nu}(y)] = i\delta_{ab} g^{\mu\nu} D_0^{(+)}(x-y), \tag{2.6.2}$$

where $A^{(\pm)}$ are the emission and absorption parts of A and $D_0^{(\pm)}$ the (mass zero) Jordan-Pauli distributions. $u_a(x)$ and $\tilde{u}_a(x)$ are the free massless fermionic ghost fields fulfilling the anti-commutation relations

$$\{u_a^{(\pm)}(x), \tilde{u}_b^{(\mp)}(y)\} = -i\delta_{ab} D_0^{(\mp)}(x-y). \tag{2.6.3}$$

f_{abc} denote the real antisymmetric structure constants of the gauge group, say $SU(N)$ (see any book on Lie theory, e.g. V.S.Varadarajan, *Lie groups, Lie algebras and their representations*, Prentice-Hall, Inc. 1974). The time dependence of A, u and $\tilde{u}$ is given by the wave equation

$$\Box A_a^\mu(x) = 0, \quad \Box u_a(x) = 0, \quad \Box \tilde{u}_a(x) = 0. \tag{2.6.4}$$

As usual we define

$$F_a^{\mu\nu} \stackrel{\text{def}}{=} \partial^\mu A_a^\nu - \partial^\nu A_a^\mu. \tag{2.6.5}$$

Let us call the index "a" the colour index which is the terminology in quantum chromodynamics (QCD) where the gauge group is $SU(3)$. But the gauge group will play no role in the following, only the structure constants f_{abc} in (2.6.1) are important.

We now want to calculate second order loop graphs. In normally ordering the direct product

$$R_2'(x;y) = -T_1(y)T_1(x), \tag{2.6.6}$$

we select all terms with two contractions:

$$R_2'(x;y) = r_{AA}'^{\nu\mu}(x-y) : A_\nu(x) A_\mu(y) : + r_{AF}'^{\nu\mu\kappa}(x-y) : A_\nu(x) F_{\mu\kappa}(y) : +$$

$$+ r_{FA}'^{\mu\kappa\nu}(x-y) : F_{\mu\kappa}(x) A_\nu(y) : + r_{FF}'^{\mu\kappa\nu\tau}(x-y) : F_{\mu\kappa}(x) F_{\nu\tau}(y) : +$$

$$+ r_{u\tilde{u}}'^\mu(x-y) : u(x) \partial_\mu \tilde{u}(y) : + r_{\tilde{u}u}'^\mu(x-y) : \partial_\mu \tilde{u}(x) u(y) : + \ldots, \tag{2.6.7}$$

where the two free field operators have the same colour index and there is a sum over this index. r_{AA}' has two contributions: $r_{AA}'^1$ containing two times the gluon vertex : $AA\partial A$: of T_1 and $r_{AA}'^2$ containing two times the ghost vertex : $Au\partial\tilde{u}$:. Note that $R'_2(x;y)$ has no 2-legs diagram with one gluon and one ghost vertex. We now give the results for the numerical distributions (problem 2.5):

$$r'^{1\,\nu\mu}_{AA}(x-y) = g^2 N[2(\partial^\nu\partial^\mu D_0^-)(x-y)D_0^-(x-y)- \qquad$$
$$-3(\partial^\nu D_0^-)(x-y)(\partial^\mu D_0^-)(x-y)], \qquad (2.6.8)$$

$$r'^{2\,\nu\mu}_{AA}(x-y) = g^2 N(\partial^\nu D_0^-)(x-y)(\partial^\mu D_0^-)(x-y), \qquad (2.6.9)$$

$$r'^{\nu\mu\kappa}_{AF}(x-y) = -r'^{\mu\kappa\nu}_{FA}(x-y) =$$

$$= \frac{g^2}{2} N[g^{\nu\kappa}(\partial^\mu D_0^-)(x-y)D_0^-(x-y) - g^{\nu\mu}(\partial^\kappa D_0^-)(x-y)D_0^-(x-y)], \qquad (2.6.10)$$

$$r'^{\mu\kappa\nu\tau}_{FF}(x-y) = -\frac{g^2}{2} N g^{\mu\nu} g^{\kappa\tau} D_0^-(x-y)D_0^-(x-y), \qquad (2.6.11)$$

$$r'^\mu_{u\tilde{u}}(x-y) = r'^\mu_{\tilde{u}u}(x-y) = g^2 N \partial^\mu D_0^-(x-y) D_0^-(x-y), \qquad (2.6.12)$$

where N is the order of the group. The corresponding numerical distributions of $A'_2(x;y) = -T_1(x)T_1(y)$ are obtained (in this special case) by replacing D_0^- by D_0^+.

One may wonder why the products $\partial^a D_0^- \partial^b D_0^-$, $|a|,|b|=0,1,2$, appearing in (2.6.8-12) exist. The reason is that in the Fourier transformed expression

$$\frac{1}{(2\pi)^2}\int d^4q\, (-i(p-q))^a \hat{D}_0^-(p-q)(-iq)^b \hat{D}_0^-(q) \qquad (2.6.13)$$

the intersection of the supports of the two $\hat{D}_0^-$ is a compact set. Inserting the explicit expression of $\hat{D}_0^-$ in (2.6.13), we see that we have to deal with the integrals

$$I^\pm(p) \stackrel{\text{def}}{=} \int d^4q\, \delta((p-q)^2)\delta(q^2)\theta(\pm(p_0-q_0))\theta(\pm q_0)\{1,\, q_\nu,\, q_\nu q_\mu\}. \qquad (2.6.14)$$

Starting from (2.6.8-12) and the corresponding a'-distributions one can easily express the $\hat{d}$-distributions in terms of the integrals (2.6.14) (problem 2.6):

$$\hat{d}^{1\,\nu\mu}_{AA}(p) = -\frac{g^2 N}{(2\pi)^4}[2(I^{+\nu\mu}(p) - I^{-\nu\mu}(p)) - 3(p^\nu I^{+\mu}(p) - I^{+\nu\mu}(p)-$$
$$-p^\nu I^{-\nu\mu}(p) + I^{-\nu\mu}(p))], \qquad (2.6.15)$$

$$\hat{d}^{2\,\nu\mu}_{AA}(p) = -\frac{2g^2 N}{(2\pi)^4}[p^\nu I^{+\mu}(p) - I^{+\nu\mu}(p) - p^\nu I^{-\mu}(p) + I^{-\nu\mu}(p)], \qquad (2.6.16)$$

$$\hat{d}^{\nu\mu\kappa}_{AF}(p) = \frac{ig^2 N}{2(2\pi)^4}[g^{\nu\mu}(I^{+\kappa} - I^{-\kappa}) - g^{\nu\kappa}(I^{+\mu} - I^{-\mu})], \qquad (2.6.17)$$

$$\hat{d}^{\mu\kappa\nu\tau}_{FF}(p) = -\frac{g^2 N}{2(2\pi)^4}[I^+ - I^-]g^{\mu\nu}g^{\kappa\tau}, \qquad (2.6.18)$$

$$\hat{d}^{\mu}_{u\bar{u}}(p) = -\frac{ig^2 N}{(2\pi)^4}[I^{+\mu} - I^{-\mu}]. \tag{2.6.19}$$

It remains to compute the integrals (2.6.14). Due to the two $\delta\theta$ factors they vanish, except for $p^2 > 0$ and $p^0 > 0$ for $I^+(p)$ resp. $p^0 < 0$ for $I^-(p)$. Therefore, we may choose a special Lorentz frame with $p = (p^0, \mathbf{0})$. Especially for the scalar integral I^+ we obtain there (assuming $p^0 > 0$)

$$I^+(p) = \int d^4q\, \delta(p_0^2 - 2p_0 q_0)\theta(p_0 - q_0)\frac{1}{2|\mathbf{q}|}\delta(q_0 - |\mathbf{q}|) =$$

$$= \int d\Omega d|\mathbf{q}|\, |\mathbf{q}|^2 \frac{1}{2p_0}\delta(|\mathbf{q}| - \frac{p_0}{2})\frac{1}{2|\mathbf{q}|}\theta(p_0 - |\mathbf{q}|) = \frac{\pi}{2}. \tag{2.6.20}$$

I^- can be calculated analogously and the result is in an arbitrary Lorentz frame

$$I^{\pm}(p) = \frac{\pi}{2}\theta(p^2)\theta(\pm p^0). \tag{2.6.21}$$

Computing the vector integral I^+_ν for $p = (p^0, \mathbf{0})$, $p^0 > 0$, we have a non-vanishing contribution for $\nu = 0$ only. In comparison to (2.6.20) we obtain for I^+_0 an additional factor q_0 in the integrand, which is set equal to $\frac{p_0}{2}$. This leads us to

$$I^{\pm}_\nu(p) = \frac{\pi}{4}p_\nu \theta(p^2)\theta(\pm p^0) \tag{2.6.22}$$

in an arbitrary Lorentz frame.

The covariant decomposition of the tensor integral $I^{\pm}_{\nu\mu}$ is

$$I^{\pm}_{\nu\mu}(p) = A^{\pm}(p^2)p_\nu p_\mu + B^{\pm}(p^2)g_{\nu\mu}. \tag{2.6.23}$$

A glance at (2.6.14) shows $I^{\pm\nu}_\nu = 0$. This implies

$$B^{\pm}(p^2) = -\frac{p^2}{4}A^{\pm}(p^2). \tag{2.6.24}$$

Calculating $I^{\pm}_{\nu\mu}(p)p^\nu p^\mu$ for $p = (p^0, \mathbf{0})$, $p^0 > 0$ an additional factor $(p_0 q_0)^2$ appears in the integrand of (2.6.20). Therefore, we obtain (in an arbitrary Lorentz frame)

$$I^{\pm}_{\nu\mu}(p)p^\nu p^\mu = \frac{\pi}{8}p^4\theta(p^2)\theta(\pm p^0). \tag{2.6.25}$$

Taking (2.6.23-24) into consideration we end up with

$$I^{\pm}_{\nu\mu}(p) = [p_\nu p_\mu - \frac{p^2}{4}g_{\nu\mu}]\frac{\pi}{6}\theta(p^2)\theta(\pm p^0). \tag{2.6.26}$$

We now substitute the values of these integrals into (2.6.15-19) and obtain

$$\hat{d}^{1\,\nu\mu}_{AA}(p) = -\frac{g^2 N\pi}{(2\pi)^4 24}[2p^\nu p^\mu - 5p^2 g^{\nu\mu}]\theta(p^2)\,\text{sgn}\,p^0, \tag{2.6.27}$$

$$\hat{d}_{AA}^{2\,\nu\mu}(p) = -\frac{g^2 N\pi}{(2\pi)^4 24}[2p^\nu p^\mu + p^2 g^{\nu\mu}]\theta(p^2)\,\mathrm{sgn}\,p^0, \qquad (2.6.28)$$

$$\hat{d}_{AF}^{\nu\mu\kappa}(p) = -\hat{d}_{FA}^{\mu\kappa\nu}(p) = -\frac{g^2 N\pi i}{(2\pi)^4 8}[g^{\nu\kappa}p^\mu - g^{\nu\mu}p^\kappa]\theta(p^2)\,\mathrm{sgn}\,p^0, \qquad (2.6.29)$$

$$\hat{d}_{FF}^{\mu\kappa\nu\tau}(p) = -\frac{g^2 N\pi}{(2\pi)^4 4}g^{\mu\nu}g^{\kappa\tau}\theta(p^2)\,\mathrm{sgn}\,p^0, \qquad (2.6.30)$$

$$\hat{d}_{u\tilde{u}}^{\mu}(p) = \hat{d}_{\tilde{u}u}^{\mu}(p) = -\frac{g^2 N\pi i}{(2\pi)^4 4}p^\mu \theta(p^2)\,\mathrm{sgn}\,p^0. \qquad (2.6.31)$$

Note the transversality relations

$$p_\nu(\hat{d}_{AA}^{1\,\nu\mu}(p) + \hat{d}_{AA}^{2\,\nu\mu}(p)) = 0,$$

$$p_\nu \hat{d}_{AF}^{\nu\mu\kappa}(p) = 0. \qquad (2.6.32)$$

The first one would be violated without the ghost term $\hat{d}_{AA}^2$.

The next step in the causal construction is the distribution splitting. It is somewhat tricky here because we are in the massless case. The central solution r^0 (2.4.16), which exists for $m > 0$ only, fulfills

$$(D^a \hat{r}^0)(0) = 0, \quad \forall \mid a \mid \le \omega \qquad (2.6.33)$$

i.e. it has the normalization or subtraction point $p = 0$. For $m = 0$ no splitting solution is ω-times differentiable (in momentum space) at $p = 0$. Therefore, the central solution does not exist. We, therefore, proceed in a different way.

Let $d(x)$ be the causal mass-zero distribution which we want to split into retarded minus advanced part. Then the distribution

$$d_q(x) \stackrel{\mathrm{def}}{=} e^{iqx} d(x) \qquad (2.6.34)$$

has causal support, too. Note the relation

$$\hat{d}_q(p) = \hat{d}(p+q) \qquad (2.6.35)$$

for the Fourier transforms. If r_q is a retarded part of d_q, the reader may easily convince himself (in x-space) that

$$r(x) \stackrel{\mathrm{def}}{=} e^{-iqx} r_q(x) \qquad (2.6.36)$$

is a retarded part of $d(x)$.

Now we choose $q \ne 0$ in such a way that the central solution r_q^0 of d_q exists. This is the case if $q = (q_1, ..., q_l)$ is totally space-like (i.e. $q_I^2 < 0$, where $q_I \stackrel{\mathrm{def}}{=} \sum_{j \in I} q_j$ and I runs through all subsets of $\{1, 2, ..., l\}$). The central solution may be expressed by the dispersion integral (2.4.16)

$$\hat{r}_q^0(p) = \frac{i}{2\pi} \int dt \frac{\hat{d}_q(tp)}{(t-i0)^{\omega+1}(1-t+i0)} = \frac{i}{2\pi} \int dt \frac{\hat{d}(tp+q)}{(t-i0)^{\omega+1}(1-t+i0)} \quad (2.6.37)$$

for $p = (p_1, ..., p_l)$ with $p_j \in V^+ \stackrel{\text{def}}{=} \{k \in \mathbb{R}^4 \mid k^2 > 0, k^0 > 0\}$, $\forall 1 \leq j \leq l$. Let us remark that this integral (2.6.37) even exists if $\hat{d}(p)$ is ω-times continuously differentiable at $p = q$ and the derivatives $(D^a \hat{d})(p)$, $\forall \mid a \mid = \omega$ are continuous at $p = q$.

The retarded part r^q of d belonging to r_q^0 according to (2.6.36), fulfills in momentum space

$$\hat{r}^q(p) = \hat{r}_q^0(p-q), \quad (2.6.38)$$

and for $p_j - q_j \in V^+$, $\forall 1 \leq j \leq l$ it can be computed by

$$\hat{r}^q(p) = \frac{i}{2\pi} \int dt \frac{\hat{d}(tp+(1-t)q)}{(t-i0)^{\omega+1}(1-t+i0)}. \quad (2.6.39)$$

The result for arbitrary p is obtained by analytic continuation over $p_j \in \mathbb{R}^4 + iV^+$, $\forall 1 \leq j \leq l$ according to (2.4.21). But this result has the drawback of not being Lorentz covariant. In fact, the covariance of d: $d(\Lambda x) = D(\Lambda)d(x)$ leads to $d_{\Lambda q}(\Lambda x) = D(\Lambda)d_q(x)$. This relation holds for the retarded part, too, if we choose the central splitting solution r_q^0

$$r_{\Lambda q}^0(\Lambda x) = D(\Lambda)r_q^0(x). \quad (2.6.40)$$

Using (2.6.36) in addition, we obtain

$$r^{\Lambda q}(\Lambda x) = e^{-i\Lambda q \Lambda x} r_{\Lambda q}^0(\Lambda x) = e^{-iqx} D(\Lambda) r_q^0(x) = D(\Lambda) r^q(x). \quad (2.6.41)$$

We see that r^q is in general not covariant because the normalization point q is transformed with Λ.

To obtain a covariant splitting solution we shall take the limit $q \to 0$. However, $\lim_{q \to 0} \hat{r}^q$ and $\lim_{q \to 0} \hat{r}_q^0$ are both diverging. We have to be more careful: For an arbitrary retarded part $r(x)$ of $d(x)$ we define

$$r_q(x) \stackrel{\text{def}}{=} r(x)e^{iqx}. \quad (2.6.42)$$

Obviously r_q is a splitting solution of d_q defined in (2.6.34). We assume q to be totally space-like. Then the central solution r_q^0 (2.6.37) of d_q exists and

$$P_q(p) \stackrel{\text{def}}{=} \hat{r}_q(p) - \hat{r}_q^0(p) \quad (2.6.43)$$

is a q-depending polynomial of degree ω in p. Because of (2.6.42) $\hat{r}(p)$ can be obtained from $\hat{r}_q(p)$ by taking the limit

$$\hat{r}(p) = \lim_{q \to 0} \hat{r}_q(p). \quad (2.6.44)$$

Inserting (2.6.43) we arrive at

$$\hat{r}(p) = \lim_{q \to 0}[\hat{r}_q^0(p) + P_q(p)]. \qquad (2.6.45)$$

By taking the limit $q \to 0$ of $\hat{r}_q^0$, we must add a q-depending polynomial P_q in such a way that the limit exists and obtain with this procedure a splitting solution $\hat{r}$ of d.

Now we apply this splitting method to the causal d-distributions (2.6.27-31). They all have the form

$$\hat{d}(p) = P(p)\hat{d}_1(p) \qquad (2.6.46)$$

with

$$\hat{d}_1(p) \stackrel{\text{def}}{=} \theta(p^2)\operatorname{sgn} p^0 \qquad (2.6.47)$$

and $P(p)$ is a covariant polynomial in p. It suffices to find a Lorentz invariant retarded part $\hat{r}_1(p)$ of $\hat{d}_1(p)$. Then

$$\hat{r}(p) \stackrel{\text{def}}{=} P(p)\hat{r}_1(p) \qquad (2.6.48)$$

is a covariant splitting solution of $\hat{d}(p)$ (2.6.46). Inserting (2.6.37), (2.6.47) in (2.6.45) and taking $\omega(\hat{d}_1) = 0$ into consideration, we obtain for $p \in V^+$, $q \to 0$ in such a way that $p - q \in V^+$, $q^2 < 0$

$$\hat{r}_1(p) = \lim_{q \to 0}\left\{\frac{i}{2\pi}\int\frac{dt}{(t-i0)(1-t+i0)}\theta((tp+q)^2)\operatorname{sgn}(tp^0+q^0) + P_q(p)\right\}, \qquad (2.6.49)$$

where the polynomial $P_q(p)$ is constant (in p). The zeros of $(tp+q)^2$ are

$$t_{1,2} = \frac{1}{p^2}(-pq \pm \sqrt{N}) \qquad (2.6.50)$$

with

$$N \stackrel{\text{def}}{=} (pq)^2 - p^2q^2 > (pq)^2 \qquad (2.6.51)$$

(remember $p^2 > 0$, $q^2 < 0$). Note $t_1 > 0$, $t_2 < 0$. The integral in (2.6.49) may be simplified to

$$\left\{\frac{-i}{2\pi}\int_{-\infty}^{t_2}dt + \frac{i}{2\pi}\int_{t_1}^{\infty}dt\right\}\left\{\frac{1}{t} - P\frac{1}{t-1} + i\pi\delta(t-1)\right\} =$$

$$= \frac{i}{2\pi}(-\log|t_2| + \log|t_2-1| - \log|t_1| + \log|t_1-1| + i\pi). \qquad (2.6.52)$$

Since $t_{1,2} \to 0$ for $q \to 0$ the logarithms $\log|t_{1,2} - 1|$ vanish in this limit and it remains in (2.6.52)

$$-\frac{i}{2\pi}(\log|t_1 t_2| - i\pi) = \frac{i}{2\pi}\log\frac{p^2}{q^2}. \qquad (2.6.53)$$

Adding the polynomial $P_q(p) - \frac{i}{2\pi}\log\frac{q^2}{M^2}$ in (2.6.49), where $M > 0$ is an arbitrary scale parameter which has the dimension of a mass, we obtain an invariant splitting solution of d_1

$$\hat{r}_1(p) = \frac{i}{2\pi}\log\frac{p^2}{M^2}. \qquad (2.6.54)$$

So far this holds for $\lim_{q\to 0}(p-q) = p \in V^+$. By analytic continuation in $\mathbb{R}^4 + iV^+$, we obtain for $p \in \mathbb{R}^4$

$$\hat{r}_1(p) = \frac{i}{2\pi}\log\frac{p^2 + ip^0 0}{M^2}. \qquad (2.6.55)$$

Note that scale invariance of the massless theory is violated or broken after distribution splitting due to the presence of M.

2.7 Normalizability

In causal perturbation theory "normalizability" means that the singular order ω of the numerical distributions in T_n depend only on the corresponding Wick monomials, not on the order n of perturbation theory. Then only a finite number of external field configurations give rise to finitely many free normalization constants. They can be fixed by the symmetries of the theory and by a few physical normalization conditions. In non-normalizable theories ω increases with n. If ω decreases with n, the theory is called super-normalizable.

We again consider pure Yang-Mills theory with the coupling (2.6.1)

$$T_1 = igf_{abc}(:A_{\mu a}A_{\nu b}\partial^\nu A_c^\mu: - :A_{\mu a}u_b\partial^\mu \tilde{u}_c:) \qquad (2.7.1)$$

between "gluon" fields A_a^μ, ghost fields u_a and anti-ghost fields $\tilde{u}_a$. We are going to prove by induction that

$$\omega \leq 4 - b - n_u - n_{\tilde{u}} - d, \qquad (2.7.2)$$

where b is the number of external gluons, n_u (resp. $n_{\tilde{u}}$) the number of external ghosts u (anti-ghosts $\tilde{u}$) and d the number of derivatives on external gluons and anti-ghosts. We have $n_u = n_{\tilde{u}}$ because u and $\tilde{u}$ always appear in pairs. In the inductive construction of T_n from the $T_m, m \leq n-1$, one must first consider tensor products of two distributions

$$T_r^1(x_1,\ldots x_r)T_v^2(y_1,\ldots y_v)$$

with singular orders, say, ω_1 and ω_2 which obey (2.7.2) by induction hypothesis. This product is then normally ordered and we assume that l contractions (bosonic or fermionic) arise in this process. Then, taking translation invariance into account, the numerical part of the contracted expression is of the form

$$t_1(x_1 - x_r, \ldots x_{r-1} - x_r) \prod_{j=1}^{l} \partial^{a_j} D_0^+(x_{r_j} - y_{v_j}) t_2(y_1 - y_v, \ldots y_{v-1} - y_v)$$

$$\stackrel{\text{def}}{=} t(\xi_1, \ldots \xi_{r-1}, \eta_1, \ldots \eta_{v-1}, \eta), \tag{2.7.2}$$

where $a_j \in \mathbb{N}_0^4, |a_j| = 0, 1, 2$. Here, $\{x_{r_j}\}$ is a subset of $\{x_1, \ldots x_r\}$ and $\{y_{v_j}\}$ is a subset of $\{y_1, \ldots y_v\}$, and we have introduced relative coordinates

$$\xi_j = x_j - x_r, \quad \eta_j = y_j - y_v, \quad \eta = x_r - y_v.$$

The contraction function is given by

$$D_0^+(x) = \frac{i}{(2\pi)^3} \int d^4 p \, \delta^{(1)}(p^2) \Theta(p^0) e^{-ipx}, \tag{2.7.3}$$

where $\delta^{(n)}$ denotes the n-dimensional δ-distribution. We compute the Fourier transform (omitting powers of i and 2π)

$$\hat{t}(p_1, \ldots p_{r-1}, q_1, \ldots q_{v-1}, q) = \int t(\xi, \eta) e^{ip\xi + iq\eta} d^{4r-4}\xi \, d^{4v}\eta. \tag{2.7.4}$$

Since products go over into convolutions, we get

$$\hat{t}(\cdots) = \int \prod_{j'=1}^{l} d\kappa_{j'} \, \delta^{(4)}\left(q - \sum_{h=1}^{l} \kappa_h\right)$$

$$\times \hat{t}_1(\ldots p_{r_i} - \kappa_i \ldots) \prod_{j=1}^{l} \kappa_j^{a_j} \hat{D}_0^+(\kappa_j) \, \hat{t}_2(\ldots q_{v_i} + \kappa_i \ldots).$$

Here, κ_i does not appear in the argument of $\hat{t}_1$ (resp. $\hat{t}_2$) if $r_i = r$ (resp. if $v_i = v$). If e.g. $k = r_i = r_s$ we have $\hat{t}_1(\ldots p_k - \kappa_i - \kappa_s \ldots)$. Applying $\hat{t}$ to a test function $\varphi \in \mathcal{S}(\mathbb{R}^{4(r+v-1)})$, we obviously have

$$\langle \hat{t}, \varphi \rangle = \int d^{4r-4} p' \, d^{4v-4} q' \, \hat{t}_1(p') \hat{t}_2(q') \psi(p', q'), \tag{2.7.5}$$

with

$$\psi = \int d^4 q \prod_{j'} d\kappa_{j'} \, \delta^{(4)}\left(q - \sum_h \kappa_h\right) \varphi(\ldots p'_{r_i} + \kappa_i \ldots, \ldots q'_{v_i} - \kappa_i \ldots q)$$

$$\times \prod_{j=1}^{l} \kappa_j^{a_j} \hat{D}_0^+(\kappa_j). \tag{2.7.6}$$

In order to determine the singular order of $\hat{t}$ in p-space, we have to consider the scaled distribution

$$\left\langle \hat{t}\left(\frac{p}{\delta}\right), \varphi \right\rangle = \delta^m \langle \hat{t}(p), \varphi(\delta p) \rangle$$

$$= \delta^m \int d^{4r-4}p'\, d^{4v-4}q'\, \hat{t}_1(p')\hat{t}_2(q')\psi_\delta(p', q'), \qquad (2.7.7)$$

where

$$\psi_\delta(p', q') = \int d^4q \prod_{j'} d\kappa_{j'}\, \delta^{(4)}\left(q - \sum_h \kappa_h\right) \times$$

$$\times \varphi(\ldots \delta(p'_{r_i} + \kappa_i)\ldots, \ldots \delta(q'_{v_i} - \kappa_i) \ldots \delta q) \prod_j \kappa_j^{a_j} \hat{D}_0^+(\kappa_j) \qquad (2.7.8)$$

and $m = 4(r + v - 1)$. We introduce scaled variables $\tilde{\kappa}_j = \delta\kappa_j$, $\tilde{q} = \delta q$ and note that

$$\hat{D}_0^+\left(\frac{\tilde{\kappa}}{\delta}\right) = \delta^{(1)}\left(\frac{\tilde{\kappa}^2}{\delta^2}\right)\Theta\left(\frac{\tilde{\kappa}^0}{\delta}\right) = \delta^2 \hat{D}_0^+(\tilde{\kappa}), \quad \sum_{j=1}^l |a_j| = a. \qquad (2.7.9)$$

This implies

$$\psi_\delta(p', q') = \frac{\delta^{2l-a}}{\delta^{4l}} \int d^4\tilde{q} \prod_{j'} d\tilde{\kappa}_{j'}\, \delta^{(4)}\left(\tilde{q} - \sum_h \tilde{\kappa}_h\right) \times$$

$$\times \varphi(\ldots \delta p'_{r_i} + \tilde{\kappa}_i \ldots, \ldots \delta q'_{v_i} - \tilde{\kappa}_i \ldots, \tilde{q}) \prod_j \tilde{\kappa}_j^{a_j} \hat{D}_0^+(\tilde{\kappa}_j)$$

$$= \frac{1}{\delta^{2l+a}} \psi(\delta p', \delta q').$$

Using again scaled variables $\delta p' = \tilde{p}$, $\delta q' = \tilde{q}$, we find

$$\left\langle \hat{t}\left(\frac{p}{\delta}\right), \varphi \right\rangle = \frac{\delta^4}{\delta^{2l+a}} \int d^{4r-4}\tilde{p}\, d^{4v-4}\tilde{q}\, \hat{t}_1\left(\frac{\tilde{p}}{\delta}\right)\hat{t}_2\left(\frac{\tilde{q}}{\delta}\right)\psi(\tilde{p}, \tilde{q}). \qquad (2.7.10)$$

By the induction hypothesis, $\hat{t}_1$ and $\hat{t}_2$ have singular orders ω_1, ω_2 with power counting functions $\rho_1(\delta), \rho_2(\delta)$, respectively. Then the following limit exists:

$$\lim_{\delta \to 0} \delta^{2l+a-4} \rho_1(\delta) \rho_2(\delta) \left\langle \hat{t}\left(\frac{p}{\delta}\right), \varphi \right\rangle = \langle \hat{t}_0(p), \varphi \rangle.$$

Hence, the singular order of $\hat{t}(p)$ is

$$\omega \leq \omega_1 + \omega_2 + 2l + a - 4. \qquad (2.7.11)$$

It remains to check that this result satisfies (2.7.2). Substituting

$$\omega_j \leq 4 - b_j - n_{uj} - n_{\tilde{u}j} - d_j, \quad j = 1, 2$$

we find

$$\omega \leq 4 - (b_1 + b_2 - 2l_b) - (n_{u1} + n_{u2} - l_g) - (n_{\tilde{u}1} + n_{\tilde{u}2} - l_g) - (d_1 + d_2 - a),$$

where l_b (resp. l_g) is the number of gluon-contractions (ghost- contractions) and the $l = l_b + l_g$ contractions have totally a derivatives. Since the first bracket is just the number of gluon operators after the l_b gluon-contractions and the second bracket is the number of ghost operators u after the l_g ghost-contractions etc., (2.7.2) is proved for the r', a' and d distributions.

The final step of the inductive construction is the splitting of the causal distribution into a retarded and advanced part. In this process the singular order is not changed (Sect.2.3). Hence, (2.7.2) is true in general. This implies that there are only finitely many cases with non-negative ω that require normalization. This is the normalizability of Yang-Mills theories.

The above method of proof can be applied to any other quantum field theory. It is a rigorous version of power counting in momentum space (the usual power counting lacks rigour because it deals with divergent Feynman integrals). Note that normalizability has nothing to do with gauge invariance. In the usual renormalization theory renormalizability and gauge invariance are entangled (see the biographical notes) which complicates the proofs.

Instead of counting powers in momentum space we now want to count powers of Planck's constant $\hbar$ which we no longer set $=1$ for this purpose. There are two places where $\hbar$ appears: The coupling T_1 (2.7.1) gets a factor $\hbar^{-1}$ and all (anti-) commutators have a factor $\hbar$. Let us consider a connected diagram of order n with l contractions, then the total power of $\hbar$ is $\hbar^{l-n}$, apart from the dimension of the external fields which we keep fixed. The number of contractions l is related to the number of loops n_L by

$$n_L = l - n + 1. \tag{2.7.12}$$

Consequently, the power $\hbar^{n_L - 1}$ gives immediately the number of loops. For this reason the expansion in powers of $\hbar$ is called loop-expansion.

2.8 Problems

2.1 Prove the perturbative causality condition (2.1.30) from the global condition (2.1.29) using symmetry under permutations.

2.2 What is the singular order of $\log(p^2/m^2) \in \mathcal{S}'(\mathbb{R}^4)$?

2.3 Prove the following identity for normal products by induction

$$: A_1 \ldots A_n : B =: A_1 \ldots A_{n-1} A_n B :$$
$$+ \sum_{k=1}^n : A_1 \ldots \overline{A_k \ldots A_n B} : . \tag{2.8.1}$$

2.4 Give an inductive proof of Wick's theorem by means of (2.8.1).

2.5 Compute the amplitude T_2 (2.5.13) for graviton exchange in detail.

2.6 Compute the total cross section for electron-electron scattering from the QED coupling

$$T_1(x) = i\,e : \overline{\psi}(x)\gamma^\mu\psi(x) : A_\mu(x) = -\tilde{T}_1(x). \qquad (2.8.2)$$

Hint: Make all steps in complete analogy with Sect.2.5. The electron field $\psi(x)$ is given by (1.8.3) and the photon field $A_\mu(x)$ by (1.3.27) with commutation rules (1.3.17-18).

2.7 Verify the results (2.6.8-12) using the relation

$$f_{abc}f_{abd} = N\delta_{cd}. \qquad (2.8.3)$$

2.8 Verify the results (2.6.15-19).

3. Spin-1 gauge theories: massless gauge fields

It is our aim to construct the S-matrix $S(g)$ as a formal power series (2.2.1)

$$S(g) = 1 + \sum_{n=1}^{\infty} \frac{1}{n!} \int d^4x_1 \ldots d^4x_n \, T_n(x_1,\ldots x_n) g(x_1) \ldots g(x_n). \quad (3.1.0)$$

Using the causal method of the last chapter, it is sufficient to find the first order $T_1(x)$, all higher orders can then be constructed by causality, essentially. In this chapter we discuss how $T_1(x)$ can be found by quantum gauge invariance in the case of massless gauge fields. The concept of formal power series (3.1.0) is useful for quantum gauge theories only if gauge invariance of the S-matrix $S(g)$ can be defined for the power series, that means in terms of the T_n's. Such a perturbative definition of gauge invariance is indeed possible, it is given in the following section. This causal gauge invariance will turn out to be a universal principle of nature, because it determines the structure of the interactions which are mediated by gauge fields. The electroweak, strong and gravitational interactions are of this kind.

3.1 Causal gauge invariance

Since the time-ordered products T_n are expressed by free fields, it is clear that we must use the gauge variation d_Q previously defined for free fields in order to formulate gauge invariance. Let $A_a^\mu, a = 1, \ldots N$ be a collection of vector fields, then the gauge variations are given by (1.4.29)

$$d_Q A_a^\mu = i\partial^\mu u_a, \quad d_Q u_a = 0 \quad (3.1.1)$$

$$d_Q \tilde{u}_a = -i\partial_\mu A_a^\mu. \quad (3.1.2)$$

Here u_a and $\tilde{u}_a$ are the fermionic ghost fields discussed in Sect.1.2. All fields are massless quantum fields satisfying the wave equations

$$\Box A_a^\mu = 0, \quad \Box u_a = 0, \quad \Box \tilde{u}_a = 0 \quad (3.1.3)$$

and the usual (anti-)commutation relations

$$[A_a^\mu(x), A_b^\nu(y)] = i\delta_{ab} g^{\mu\nu} D_0(x-y) \quad (3.1.4)$$

$$\{u(x), \tilde{u}(y)\} = -iD_0(x-y). \quad (3.1.5)$$

All other (anti-)commutators vanish and vector and ghost fields commute. The gauge variation $d_Q F$ is defined as the commutator or anticommutator with the gauge charge

$$Q = \int d^3x (\partial_\nu A_a^\nu \overleftrightarrow{\partial}_0 u_a), \qquad (3.1.6)$$

respectively, depending upon whether F has bosonic or fermionic character.

According to the causal construction the time-ordered products T_n come out in normally ordered form. Therefore, we need a lemma which ensures that d_Q commutes with the normal ordering:

Lemma 3.1.1. Let G be a Wick monomial containing vector fields A and ghost fields $u, \tilde{u}$, then we have

$$d_Q : G := : d_Q G : . \qquad (3.1.7)$$

Proof. The proof is by induction on the number of fields. The start with one field is trivial. Consider now one additional field A

$$: AG := A^{(+)} : G : + : G : A^{(-)}.$$

As always, $A^{(\pm)}$ denotes the emission and absorption parts of A. We now apply d_Q, using the induction assumption that the lemma holds for G:

$$d_Q : AG := d_Q A^{(+)} : G : + A^{(+)} : d_Q G : + : d_Q G : A^{(-)}$$

$$+ (-1)^{n_G} : G : d_Q A^{(-)}.$$

Since d_Q preserves the $\pm$-structure (1.4.29-30), the second and third term combine to the normal product $: A d_Q G :$ and similarly the first and last term

$$=: d_Q G : + : A d_Q G := : d_Q(AG) : . \qquad (3.1.8)$$

The addition of a ghost field requires a little more reasoning about signs:

$$: uG := u^{(+)} : G : + (-1)^{n_G} : G : u^{(-)}. \qquad (3.1.9)$$

Applying d_Q and using the induction hypothesis we get

$$d_Q : uG := d_Q u^{(+)} : G : - u^{(+)} : d_Q G : + (-1)^{n_G} : d_Q G : u^{(-)} + : G : d_Q u^{(-)}$$

$$=: d_Q uG : - : u d_Q G := : d_Q(uG) : . \qquad (3.1.10)$$

The minus sign in (3.1.10) follows from the fact that $d_Q G$ is a Bose operator if n_G is odd and a Fermi operator if n_G is even, respectively. In the same way one proves

$$d_Q : \tilde{u}G := : d_Q \tilde{u}G : . \qquad (3.1.11)$$

□

Now we are ready to define causal gauge invariance. One is tempted to define it simply by $d_Q T_n = 0$, but this is not correct. To find the right definition let us consider QED where we certainly know what gauge invariance means. Ordinary spinor quantum electrodynamics is constructed from

$$T_1(x) = ie : \overline{\psi}(x)\gamma^\mu \psi(x) : A_\mu(x). \tag{3.1.12}$$

The *free* Dirac fields $\psi, \overline{\psi}$ have zero gauge variation $d_Q\psi = 0 = d_Q\overline{\psi}$, but $d_Q A_\mu = i\partial_\mu u$. Then we obtain

$$d_Q T_1 = -e : \overline{\psi}\gamma^\mu \psi : \partial_\mu u = ie\partial_\mu(i : \overline{\psi}\gamma^\mu \psi : u). \tag{3.1.13}$$

Here we have used current conservation

$$\partial_\mu : \overline{\psi}\gamma^\mu \psi := 0 \tag{3.1.14}$$

which follows from the *free* Dirac equations. We see that $d_Q T_1$ is not zero, but a divergence

$$d_Q T_1 = i\partial_\mu T^\mu_{1/1}, \tag{3.1.15}$$

where

$$T^\mu_{1/1} = ie : \overline{\psi}\gamma^\mu \psi : u \tag{3.1.16}$$

is called Q-vertex in the following and (3.1.15) is first order gauge invariance.

It is not hard to generalize this to higher orders. If we freely interchange d_Q and the time ordering we can write

$$d_Q T_n = d_Q T\{T_1(x_1) \cdot \ldots \cdot T_1(x_n)\}$$

$$= \sum_{l=1}^n T\{T_1(x_1) \ldots d_Q T_1(x_l) \cdot \ldots T_1(x_n)\}$$

$$= \sum_{l=1}^n T\{T_1(x_1) \ldots i\partial_\mu T^\mu_{1/1}(x_l) \cdot \ldots T_1(x_n)\}. \tag{3.1.17}$$

The time ordered products herein have to be constructed correctly by the causal method, using the Q-vertex (3.1.16) at x_l instead of the ordinary QED vertex (3.1.12). Again formally taking the derivative out of the T-product we get

$$d_Q T_n = i\sum_{l=1}^n \frac{\partial}{\partial x_l^\mu} T\{T_1(x_1) \ldots T^\mu_{1/1}(x_l) \cdot \ldots T_1(x_n)\}$$

$$\stackrel{\text{def}}{=} i\sum_{l=1}^n \frac{\partial}{\partial x_l^\mu} T^\mu_{n/l}(x_1 \ldots x_n). \tag{3.1.18}$$

This equation certainly holds for $x_j \ne x_k$, for all $j \ne k$, because there we can calculate with the T-product in the same way as with an ordinary product. But the extension to the diagonal $x_1 = \ldots = x_n$ produces local terms in general, both in (3.1.17) and in (3.1.18). If it is possible to absorb such local terms by suitable normalization of the distributions T_n and $T^\mu_{n/l}$, then we call the theory gauge invariant to n-th order. We want to emphasize that causal gauge invariance not only means that $d_Q T_n$ is a divergence, the divergence must also be of the specific form (3.1.18) involving the Q-vertex.

To check the usefulness of the definition (3.1.18) we work out its consequences for QED. Let us consider a fixed vertex x_l and concentrate on all contributions to $T_n(x_1, \ldots, x_n)$ containing the external field operator $A_\mu(x_l)$

$$T_n(x_1, \ldots, x_n) =: T_l^\mu(x_1, \ldots, x_n) A_\mu(x_l) : + \ldots \qquad (3.1.19)$$

The dots represent terms without $A_\mu(x_l)$, it is an important property of QED that only one external leg $A_\mu(x_l)$ can appear. The gauge variation can be calculated in the simple manner

$$d_Q T_n = i : T_l^\mu(x_1, \ldots, x_n) \partial_\mu u(x_l) : + \ldots, \qquad (3.1.20)$$

and this is equal to a divergence

$$= i \partial^{x_l}_\mu \Big(T_l^\mu(x_1, \ldots, x_n) u(x_l) \Big) + \ldots, \qquad (3.1.21)$$

if and only if

$$\frac{\partial}{\partial x_l^\mu} T_l^\mu(x_1, \ldots, x_n) = 0. \qquad (3.1.22)$$

These are the so-called Ward-Takahashi identities of QED (see FQED (4.6.9)). Since they express gauge invariance of QED, our definition (3.1.18) is completely satisfactory in this case.

Now we want to study what causal gauge invariance (3.1.18) means for the total S-matrix (3.1.0). Applying the gauge variation d_Q to the formal power series we obtain

$$d_Q S(g) = \sum_{n=1}^\infty \frac{i}{n!} \int d^4 x_1 \ldots d^4 x_n \sum_{l=1}^n (\partial^{x_l}_\mu T^\mu_{n/l})$$

$$\times g(x_1) \ldots g(x_n). \qquad (3.1.23)$$

Since the test function $g(x)$ is in Schwartz space $\mathcal{S}(\mathbb{R}^4)$, we can integrate by parts

$$= -\sum_{n=1}^\infty \frac{i}{n!} \int d^4 x_1 \ldots d^4 x_n \sum_{l=1}^n T^\mu_{n/l} g(x_1) \ldots (\partial_\mu g)(x_l) \ldots g(x_n). \qquad (3.1.24)$$

If it is possible to perform the adiabatic limit $g \to 1$ here, then the right-hand side goes to zero and we get the naive definition of gauge invariance of the S-matrix

$$\lim_{g \to 1} d_Q S(g) = 0. \qquad (3.1.25)$$

The adiabatic limit exists if all gauge fields are massive. This has been proved by Epstein and Glaser (in *Renormalization Theory, p.193-254, edited* by *G. Velo, A.S. Wightman, D. Reidel Publ. Comp. 1976*). It does not exist for the time-ordered products if some gauge field is massless. In this case (3.1.25) is meaningless and we must use the perturbative definition (3.1.18). The latter is really at the heart of gauge theory because it determines the possible couplings T_1. This we are now going to show for massless spin-1 gauge fields.

3.2 Self-coupling of gauge fields

We consider a collection of vector fields $A_a^\mu(x), a = 1, \ldots N$ and ghost fields $u_a(x)$ with anti-ghost fields $\tilde{u}_a(x)$ quantized in the usual manner (3.1.3-4). It is our goal to find all possible gauge invariant self-couplings $T_1(x)$ of these fields. Since the gauge variation d_Q generates ghost fields from vector fields, it is pretty clear that T_1 must involve ghost and anti-ghost fields as well. But we assume ghost number $=0$, so that u and $\tilde{u}$ must appear in pairs. It is sufficient to consider trilinear couplings proportional to a coupling constant g, quadrilinear ones proportional to g^2 correspond to T_2 and should come out automatically in the causal construction. We therefore start from the following general ansatz

$$T_1(x) = ig\{f_{abc}^1 : A_{\mu a}A_{\nu b}\partial^\nu A_c^\mu : +f_{abc}^2 : A_{\mu a}A_b^\mu \partial^\nu A_{\nu c} :$$
$$+f_{abc}^3 : A_{\mu a}u_b\partial^\mu \tilde{u}_c : +f_{abc}^4 : (\partial^\mu A_{\mu a})u_b\tilde{u}_c : +f_{abc}^5 : A_{\mu a}(\partial^\mu u_b)\tilde{u}_c :\}. \quad (3.2.1)$$

Here we have further assumed that T_1 is a Lorentz scalar. For this reason we need an odd number of derivatives in each term. We only consider one derivative because with three the theory is not normalizable. The f_{abc}^j are arbitrary constants, but pseudo-unitarity requires a skew-conjugate T_1

$$T_1^K(x) = -T_1(x), \quad (3.2.2)$$

so that the f's and g must be real. This was the reason for the imaginary i in (3.2.1). Since the Wick monomial in the second term is symmetric in a and b, we assume

$$f_{abc}^2 = f_{bac}^2 \quad (3.2.3)$$

without loss of generality. The reader easily convinces himself that there is no further possibility to contract the Lorentz indices and place the derivative. All double indices including a, b, c are summed over.

Next we calculate the gauge variation

$$d_Q T_1 = -\{f_{abc}^1(\partial_\mu u_a A_{\nu b}\partial^\nu A_c^\mu + A_{\mu a}\partial_\nu u_b \partial^\nu A_c^\mu + A_{\mu a}A_{\nu b}\partial^\nu \partial^\mu u_c)$$
$$+f_{abc}^2(2\partial_\mu u_a A_b^\mu \partial_\nu A_c^\nu + A_{\mu a}A_b^\mu \partial_\nu \partial^\nu u_c)+$$
$$+f_{abc}^3(\partial_\mu u_a u_b\partial^\mu \tilde{u}_c + A_{\mu a}u_b\partial^\mu \partial_\nu A_c^\nu)+$$

$$+f^4_{abc}(\partial^\mu\partial_\mu u_a u_b \tilde{u}_c + (\partial^\mu A_{\mu a})u_b \partial_\nu A^\nu_c)+$$
$$+f^5_{abc}(\partial_\mu u_a \partial^\mu u_b \tilde{u}_c + A_{\mu a}\partial^\mu u_b \partial_\nu A^\nu_c)\bigg\}. \tag{3.2.4}$$

The last term in the second and the first one in the fourth line vanish due to the wave equation. To simplify the notation we don't write the double dots for normal ordering anymore, *all products of field operators with the same argument are normally ordered if nothing else is said.*

For gauge invariance the expression (3.2.4) must be a divergence

$$= i\partial_\mu T^\mu_{1/1}(x).$$

We therefore write down a general ansatz for $T^\mu_{1/1}$ as well:

$$iT^\mu_{1/1} = g\bigg\{g^1_{abc}\partial^\mu u_a A_{\nu b}A^\nu_c + g^2_{abc}u_a A_{\nu b}\partial^\mu A^\nu_c+$$
$$+g^3_{abc}\partial_\nu u_a A^\nu_b A^\mu_c + g^4_{abc}u_a \partial_\nu A^\nu_b A^\mu_c+$$
$$+g^5_{abc}u_a A^\nu_b \partial_\nu A^\mu_c + g^6_{abc}u_a u_b \partial^\mu \tilde{u}_c + g^7_{abc}\partial^\mu u_a u_b \tilde{u}_c\bigg\}. \tag{3.2.5}$$

The symmetry in the first and antisymmetry in the sixth term give the relations

$$g^1_{abc} = g^1_{acb}, \quad g^6_{abc} = -g^6_{bac}. \tag{3.2.6}$$

This ansatz for $T^\mu_{1/1}$ can be further restricted using the property

$$id_Q \partial_\mu T^\mu_{1/1} = d^2_Q T_1 = 0. \tag{3.2.7}$$

Substituting (3.2.5) and collecting terms with the same field operators we obtain the following homogeneous relations: from

$$\partial^\mu u_a \partial_\mu \partial_\nu u_b A^\nu_c : \quad 2g^1_{abc} + g^2_{acb} + g^3_{abc} - g^3_{bac} - g^3_{bca} + g^5_{acb} = 0 \tag{3.2.8}$$

$$\partial^\mu u_a \partial_\mu A^\nu_b \partial_\nu u_c : \quad 2g^1_{abc} + g^2_{acb} - g^3_{cba} - g^5_{cab} = 0 \tag{3.2.9}$$

$$\partial_\mu u_a \partial^\mu u_b \partial_\nu A^\nu_c : \quad g^3_{abc} - g^3_{bac} + g^4_{acb} - g^4_{bac} + g^7_{abc} - g^7_{bac} = 0 \tag{3.2.10}$$

$$u_a \partial^\mu u_b \partial_\mu \partial_\nu A^\nu_c : \quad g^4_{abc} + g^5_{abc} + g^6_{abc} - g^6_{bac} - g^7_{bac} = 0 \tag{3.2.11}$$

$$u_a \partial^\mu \partial_\nu u_b \partial_\mu A^\nu_c : \quad g^2_{abc} + g^2_{acb} + g^5_{abc} + g^5_{acb} = 0. \tag{3.2.12}$$

First order gauge invariance (3.2.4) now implies linear relations between the f's and g's:

$$\partial^\mu u_a \partial_\mu A_{\nu b}A^\nu_c : \quad -f^1_{cab} = 2g^1_{abc} + g^2_{acb} \tag{3.2.13}$$

$$\partial^\mu u_a A_{\mu b}\partial_\nu A^\nu_c : \quad -2f^2_{abc} - f^5_{bac} = g^3_{abc} + g^4_{acb} \tag{3.2.14}$$

$$u_a \partial_\nu A^\nu_b \partial_\mu A^\mu_c : \quad -f^4_{bac} - f^4_{cab} = g^4_{abc} + g^4_{acb} \tag{3.2.15}$$

$$\partial_\mu u_a u_b \partial^\mu \tilde{u}_c : \quad -f^3_{abc} = 2g^6_{abc} + g^7_{abc} \tag{3.2.16}$$

$$\partial_\mu u_a \partial^\mu u_b \tilde{u}_c : \quad -f^5_{abc} + f^5_{bac} = g^7_{abc} - g^7_{bac} \tag{3.2.17}$$

$$\partial_\mu \partial_\nu u_u A_b^\nu A_c^\mu : \quad -f^1_{cba} - f^1_{bca} = g^3_{abo} + g^3_{acb} \qquad (3.2.18)$$

$$\partial_\mu u_a \partial_\nu A_b^\mu A_c^\nu : \quad -f^1_{acb} = g^3_{abc} + g^5_{acb} \qquad (3.2.19)$$

$$u_a \partial_\nu A_b^\nu A_c^\mu : \quad g^2_{abc} = -g^2_{acb} \qquad (3.2.20)$$

$$u_a \partial_\mu \partial_\nu A_b^\mu A_c^\nu : \quad -f^3_{cab} = g^4_{acb} + g^5_{abc} \qquad (3.2.21)$$

$$u_a \partial_\nu A_b^\mu \partial_\mu A_c^\nu : \quad g^5_{abc} = -g^5_{acb}. \qquad (3.2.22)$$

All information comes out of this linear system. Since the elimination process is somewhat tedious, we give all details to save the readers time. Let us interchange b and c in (3.2.19)

$$-f^1_{abc} = g^3_{acb} + g^5_{acb} \qquad (3.2.23)$$

and add this to (3.2.19)

$$-f^1_{abc} - f^1_{acb} = g^3_{abc} + g^3_{acb} + g^5_{abc} + g^5_{acb}. \qquad (3.2.24)$$

By (3.2.22) g^5 drops out and by (3.2.18) the right side is equal to

$$= -f^1_{cba} - f^1_{bca}. \qquad (3.2.25)$$

This implies

$$f^1_{abc} - f^1_{cba} = f^1_{bca} - f^1_{acb}. \qquad (3.2.26)$$

Let us now decompose f^1_{abc} into symmetric and antisymmetric parts in the first and third indices:

$$f^1_{abc} = d_{abc} + f_{abc}, \quad d_{abc} = d_{cba}, \quad f_{abc} = -f_{cba}, \qquad (3.2.27)$$

then (3.2.26) implies

$$f_{abc} = -f_{acb} = -f_{cba} = f_{cab} = f_{bca} = -f_{bac}. \qquad (3.2.28)$$

So we arrive at the important result that f_{abc} is totally antisymmetric. In the next section we obtain the Jacobi identity from second order gauge invariance, hence, f_{abc} can be regarded as structure constants of a *real* Lie algebra.

The total antisymmetry of f_{abc} implies the total symmetry of d_{abc}. Next we use the representation (3.2.27) in (3.2.13):

$$-f_{cab} - d_{cab} = 2g^1_{abc} + g^2_{abc}. \qquad (3.2.30)$$

Here g^2_{abc} is antisymmetric in b, c (3.2.20) so that g^1_{abc} must be symmetric, hence

$$g^1_{abc} = -\tfrac{1}{2} d_{cab} = -\tfrac{1}{2} d_{abc} \qquad (3.2.31)$$

$$g^2_{abc} = f_{cab} = f_{abc}. \qquad (3.2.32)$$

Now we can write the equations (3.2.18-19) in the form

$$g^3_{abc} + g^3_{acb} = -2d_{cba} = -2d_{abc} \qquad (3.2.33)$$

$$-f_{acb} - d_{acb} - g^3_{abc} + g^5_{acb}. \tag{3.2.34}$$

Since g^5_{abc} is antisymmetric in b, c (3.2.22), the symmetric part of this equation agrees with (3.2.33) and the antisymmetric part is given by

$$-f_{acb} = \tfrac{1}{2}(g^3_{abc} - g^3_{acb}) + g^5_{acb}. \tag{3.2.35}$$

Hence, we find

$$g^5_{abc} = g^3_{abc} - f_{abc} + d_{abc}, \tag{3.2.36}$$

where (3.2.33) has been taken into account.

Now we turn to (3.2.11) and substitute g^5 from (3.2.36)

$$g^4_{abc} = g^7_{cab} - 2g^6_{acb} - (g^3_{acb} - f_{acb} + d_{abc}). \tag{3.2.37}$$

Using this in (3.2.10) we see that g^3 and g^7 cancel out so that finally

$$g^6_{abc} = \tfrac{1}{2}f_{abc}. \tag{3.2.38}$$

Then (3.2.37) can be simplified to

$$g^4_{abc} = g^7_{cab} - g^3_{acb} - d_{abc}. \tag{3.2.39}$$

Substituting this into (3.2.14) gives

$$f^2_{abc} = -\tfrac{1}{2}(f^5_{bac} + g^7_{bac} - d_{abc}). \tag{3.2.40}$$

f^3 follows from (3.2.16):

$$f^3_{abc} = -2g^6_{abc} - g^7_{abc} = -g^7_{abc} - f_{abc}. \tag{3.2.41}$$

On the other hand, from (3.2.21) we get a different result

$$f^3_{abc} = -g^7_{cba} + f_{abc}, \tag{3.2.42}$$

which implies

$$g^7_{abc} = g^7_{cba} - 2f_{abc}. \tag{3.2.43}$$

Finally, from (3.2.15) we conclude

$$f^4_{bac} + g^7_{bac} = -f^4_{cab} - g^7_{cab} \tag{3.2.44}$$

and (3.2.17) gives another symmetry relation

$$f^5_{abc} + g^7_{abc} = f^5_{bac} + g^7_{bac}. \tag{3.2.45}$$

It is easily checked that with the results just obtained all equations are identically satisfied.

Summing up we have obtained the following form of the trilinear coupling

$$T_1 = ig\Big\{(f_{abc} + d_{abc})A_{\mu a}A_{\nu b}\partial^\nu A^\mu_c - \tfrac{1}{2}(f^5_{bac} + g^7_{bac} - d_{abc})A_{\mu a}A^\mu_b\partial_\nu A^\nu_c -$$

$$- (g^7_{abc} + f_{abc})A_{\mu a}u_b\partial^\mu \tilde{u}_c + f^4_{abc}(\partial^\mu A_{\mu a})u_b\tilde{u}_c +$$

$$+f^5_{abc}A_{\mu a}(\partial^\mu u_b)\tilde{u}_c\Big\}. \tag{3.2.46}$$

The terms proportional to d_{abc} give a divergence

$$d_{abc}(A_{\mu a}A_{\nu b}\partial^\nu A^\mu_c + \tfrac{1}{2}A_{\mu a}A^\mu_b\partial_\nu A^\nu_c) = \tfrac{1}{2}d_{abc}\partial^\nu(A_{\mu a}A^\mu_b A_{\nu c}). \tag{3.2.47}$$

It is important to remember the relation (3.2.44) which shows the antisymmetry with respect to b and c. Therefore, we have

$$C_1 \stackrel{\text{def}}{=} ig(f^4_{abc} + g^7_{abc})(\partial^\mu A_{\mu a})u_b\tilde{u}_c = -\tfrac{1}{2}g(f^4_{abc} + g^7_{abc})d_Q(\tilde{u}_a u_b \tilde{u}_c). \tag{3.2.48}$$

Such a term which is d_Q of something is called a co-boundary (see (1.4.32)); this terminology comes from cohomology theory (see W.S.Massey, *Homology and cohomology theory, Dekker 1978*). Using (3.2.48) in (3.2.47) we have to add the term with g^7 which is taken into account as follows

$$-g^7_{abc}(A_{\mu a}u_b\partial^\mu\tilde{u}_c + \partial^\mu A_{\mu a}u_b\tilde{u}_c) =$$

$$= -g^7_{abc}\partial^\mu(A_{\mu a}u_b\tilde{u}_c) + g^7_{abc}A_{\mu a}\partial^\mu u_b\tilde{u}_c). \tag{3.2.49}$$

Now T_1 assumes the following form

$$T_1 = ig\Big\{f_{abc}A_{\mu a}A_{\nu b}\partial^\nu A^\mu_c - \tfrac{1}{2}(f^5_{abc} + g^7_{abc})A_{\mu a}A^\mu_b\partial^\nu A^\nu_c -$$

$$-f_{abc}A_{\mu a}u_b\partial^\mu\tilde{u}_c + (f^5_{abc} + g^7_{abc})A_{\mu a}\partial^\mu u_b\tilde{u}_c\Big\} +$$

$$+\frac{i}{2}gd_{abc}\partial^\nu(A_{\mu a}A^\mu_b A_{\nu c}) - igg^7_{abc}\partial^\mu(A_{\mu a}u_b\tilde{u}_c) + C_1. \tag{3.2.50}$$

Due to (3.2.45) the second and fourth term together give a second co-boundary

$$C_2 = ig(f^5_{abc} + g^7_{abc})(A_{\mu a}\partial^\mu u_b\tilde{u}_c - \tfrac{1}{2}A_{\mu a}A^\mu_b\partial_\nu A^\nu_c) =$$

$$= \frac{i}{2}g(f^5_{abc} + g^7_{abc})d_Q(A_{\mu a}A^\mu_b\tilde{u}_c), \tag{3.2.51}$$

so that we arrive at the final result

$$T_1 = igf_{abc}(A_{\mu a}A_{\nu b}\partial^\nu A^\mu_c - A_{\mu a}u_b\partial^\mu\tilde{u}_c) +$$

$$+ig\partial^\nu\Big[\tfrac{1}{2}d_{abc}A_{\mu a}A^\mu_b A_{\nu c} - g^7_{abc}A_{\nu a}u_b\tilde{u}_c\Big] + C_1 + C_2. \tag{3.2.52}$$

The first two terms $\sim f_{abc}$ form the so-called Yang-Mills plus ghost coupling to lowest order. Hence, the self-coupling of spin-1 gauge fields is of Yang-Mills form plus divergence plus co-boundary couplings. In the next section we will see that the divergence and co-boundary couplings are uninteresting because they have no influence on the physical observables.

We have still to discuss the Q-vertex $T^\mu_{1/1}$. Substituting the previous results for the coupling coefficients g^j_{abc} we have

$$iT^\mu_{1/1} = g\Big\{-\tfrac{1}{2}d_{abc}\partial^\mu u_a A_{\nu b}A^\nu_c + f_{abc}u_a A_{\nu b}\partial^\mu A^\nu_c+$$

$$+g^3_{abc}\partial_\nu u_a A^\nu_b A^\mu_c + (g^7_{cab} - g^3_{acb} - d_{abc})u_a\partial_\nu A^\nu_b A^\mu_c+$$

$$+g^5_{abc}u_a A^\nu_b\partial_\nu A^\mu_c + \tfrac{1}{2}f_{abc}u_a u_b\partial^\mu\tilde{u}_c + g^7_{abc}\partial^\mu u_a u_b\tilde{u}_c\Big\}. \tag{3.2.53}$$

If we subtract the divergence (square bracket) in (3.2.52), then we must subtract in (3.2.53) the following terms proportional to g^7_{abc}

$$d_Q(A_{\nu a}u_b\tilde{u}_c) = i\partial_\nu u_a u_b\tilde{u}_c + iu_b A_{\nu a}\partial_\mu A^\mu_c \tag{3.2.54}$$

and terms proportional to $\tfrac{1}{2}d_{abc}$

$$d_Q(A_{\mu a}A^\mu_b A_{\nu c}) = 2i\partial_\mu u_a A^\mu_b A_{\nu c} + i\partial_\nu u_c A_{\mu a}A^\mu_b. \tag{3.2.55}$$

The resulting new Q-vertex is equal to

$$iT'^\mu_{1/1} = gf_{abc}\Big\{u_a A_{\nu b}(\partial^\mu A^\nu_c - \partial^\nu A^\mu_c) + \tfrac{1}{2}u_a u_b\partial^\mu\tilde{u}_c\Big\}$$

$$+g(g^3_{abc} + d_{abc})\partial_\nu(u_a A^\nu_b A^\mu_c). \tag{3.2.56}$$

The coefficient $(g^3_{abc} + d_{abc})$ in front of the last divergence is antisymmetric in b and c due to (3.2.33). Consequently, we get zero if the derivative ∂_μ is applied. Therefore this term can be dropped without effecting first order gauge invariance. The remaining Q-vertex is then unique. We omit the prime and interchange b and c:

$$iT^\mu_{1/1} = gf_{abc}\Big[u_a(\partial^\nu A^\mu_b - \partial^\mu A^\nu_b)A_{\nu c} + \tfrac{1}{2}u_a u_b\partial^\mu\tilde{u}_c\Big]. \tag{3.2.57}$$

This form will be used in the following.

3.3 Divergence and co-boundary couplings

First we want to investigate the physical equivalence of two S-matrices $S(g)$ and $S'(g)$. Let P be the projection on the physical subspace $\mathcal{F}_{\text{phys}}$ which is assumed to be the same for the two theories. Then we call the two S-matrices physically equivalent if all matrix elements between physical states agree in the adiabatic limit, i.e.

$$\lim_{g\to 1}(\Phi, PS(g)P\Psi) = \lim_{g\to 1}(\Phi, PS'(g)P\Psi), \tag{3.3.1}$$

for arbitrary Φ and Ψ in Fock space. However, this definition cannot be used if the gauge fields are massless, because the adiabatic limit does not exist in this case. It exists only for suitably defined inclusive cross sections. This has been extensively studied for QED where this infrared problem is rather well understood (FQED, Sect.3.11-13). We circumvent this problem by using the following perturbative version of (3.3.1)

$$PT_n P = PT'_n P + \text{div}, \qquad (3.3.2)$$

where T_n and T'_n are the n-point distributions corresponding to $S(g)$ and $S'(g)$, respectively. Obviously, (3.3.2) implies (3.3.1) by partial integration, if the adiabatic limit exists. We will use the definition (3.3.2) also in the massless case where the adiabatic limit of S-matrix elements does not exist, because it is quite plausible that the divergence-terms have no influence on the measurable quantities also in this case. But this has still to be shown by an analysis of the infrared problem, which is an open problem for non-abelian theories.

Specializing (3.3.2) to first order $n = 1$, we immediately see that two couplings T_1 and T'_1 which differ by a divergence are physically equivalent to first order. The same is true if they differ by a co-boundary $d_Q X$ because

$$P d_Q X P = P Q X P - P X Q P = 0. \qquad (3.3.3)$$

Here we have used the property

$$PQ = 0 = QP.$$

This is a consequence of the form (1.4.23)

$$\mathcal{F}_{\text{phys}} = \operatorname{Ker} Q \cap \operatorname{Ker} Q^+$$

of the physical subspace, because $P\Phi$ is in the kernel of Q and Q^+, so that $(PQ\Phi, \Psi) = (\Phi, Q^+ P \Psi) = 0$ for all Ψ. It remains to investigate whether the physical equivalence holds true in higher orders. Regarding divergence couplings this is contained in the following theorem due to M.Dütsch (J.Phys. A 29, 7597 (1996)).

Theorem 3.3.1. Let $T_1(x)$ be of the form

$$T_1(x) = T_1^0(x) + T_1^2(x), \quad \text{where}$$

$$T_1^2(x) = \partial_\nu T_1^{4\nu} \qquad (3.3.4)$$

is a divergence coupling. Assume first order gauge invariance

$$d_Q T_1^0 = i \partial_\mu T_1^{1\mu} \qquad (3.3.5)$$

which implies

$$d_Q T_1 = i \partial_\mu (T_1^{1\mu} + T_1^{3\mu}), \qquad (3.3.6)$$

because

$$d_Q T_1^2 = \partial_\nu d_Q T_1^{4\nu} \stackrel{\text{def}}{=} i \partial_\nu T_1^{3\nu}. \qquad (3.3.7)$$

Then the n-th order is of the following form

$$T_n(x_1,\ldots,x_n) = \sum_{i_1,\ldots i_n \in \{0,2\}} T_n^{i_1\ldots i_n}(x_1,\ldots,x_n) \quad (3.3.8)$$

$$= T_n^{0,0\ldots 0}(x_1,\ldots,x_n) + \text{div}. \quad (3.3.9)$$

Here the upper index i_j specifies the vertex T_1^j at x_j. Furthermore, if $T_n^{0,\ldots 0}$ is gauge invariant, then the same is true for all $T_n^{i_1\ldots i_n}$:

$$d_Q T_n^{i_1\ldots i_n} = i \sum_{l=1}^n \partial_\mu^l T_n^{i_1\ldots i_l+1,\ldots i_n \mu}. \quad (3.3.10)$$

Proof. First we show that the divergence structure of the coupling T_1^2 (3.3.4) can be maintained in the inductive construction of the n-th order (3.3.9), this is the desired physical equivalence. Let us consider a term in (3.3.8) containing r divergence vertices T_1^2 and $n - r$ ordinary ones T_1^0, then we want to prove that

$$T_n^{2\ldots 20\ldots 0} = \partial_{\mu_1}^1 \ldots \partial_{\mu_r}^r T_n^{4\ldots 40\ldots 0} \quad (3.3.11)$$

holds true. As usual, in the induction we assume that (3.3.11) is valid to all orders $\leq n-1$ for arbitrary upper indices $i_j = 0, 2$. Then this is also true for following the sum of direct products (not normally ordered)

$$A_n'^{i_1\ldots i_n}(x_1,\ldots,x_n) = \sum_{P_2} \tilde{T}_{n_1}^{i_1\ldots i_{n_1}}(X) T_{n-n_1}^{i_{n_1+1}\ldots i_n}(Y,x_n) \quad (3.3.12)$$

where $n_2 = n - n_1$, and similarly for R_n' and D_n. This means that the causal distribution D_n has the following form

$$D_n^{2\ldots 20\ldots 0} = \partial_{\mu_1}^1 \ldots \partial_{\mu_r}^r D_n^{4\ldots 40\ldots 0}. \quad (3.3.13)$$

The subtle point is again the splitting of this causal distribution into retarded and advanced parts, but here life is simple. We choose an arbitrary splitting solution $R_n^{4\ldots 40\ldots 0}$ of $D_n^{4\ldots 40\ldots 0}$. Then we get a splitting solution of (3.3.13) by

$$R_n^{2\ldots 20\ldots 0} \stackrel{\text{def}}{=} \partial_{\mu_1}^1 \ldots \partial_{\mu_r}^r R_n^{4\ldots 40\ldots 0}. \quad (3.3.14)$$

because the derivatives do not change the causal support. Furthermore, this splitting solution R_n has the desired divergence structure and the same is then true for

$$T_n' = R_n - R_n'. \quad (3.3.15)$$

The symmetrization with respect to x_n gives T_n and the divergence structure is preserved in this process.

Now we turn to n-th order gauge invariance. We take the gauge variation of (3.3.12)

$$d_Q A_n'^{i_1...i_n} = \sum_{P_2} \Big\{ \Big[d_Q \tilde{T}_{n_1}^{i_1...i_{n_1}}(X) \Big] T_{n-n_1}^{i_{n_1+1}...i_n}(Y, x_n)$$

$$+ \tilde{T}_{n_1}^{i_1...i_{n_1}}(X) d_Q T_{n-n_1}^{i_{n_1+1}...i_n}(Y, x_n) \Big\}. \qquad (3.3.16)$$

For the lower orders we assume gauge invariance as induction hypothesis, hence

$$= i \sum_{l=1}^{n} \partial_\mu^l \Big\{ \sum_{P_2, x_l \in X} \tilde{T}_{n_1}^{i_1...i_l+1,...i_{n_1}\mu}(X) T_{n-n_1}^{i_{n_1+1}...i_n}(Y, x_n)$$

$$+ \sum_{P_2, x_l \notin X} \tilde{T}_{n_1}^{i_1...i_{n_1}}(X) T_{n-n_1}^{i_{n_1+1}...i_l+1,...i_n\mu}(Y, x_n) \Big\}$$

$$= i \sum_{l=1}^{n} \partial_\mu^l A_n'^{i_1...i_l+1,...i_n\mu}, \qquad (3.3.17)$$

and the same is true for R'_n and D_n.

For the distribution splitting we consider the special case of r divergence vertices T_1^2 again

$$d_Q D_n^{2...20...0} = i \Big[\sum_{l=1}^{r} \partial_\mu^l D_n^{2...3...20...0\mu} + \sum_{l=r+1}^{n} \partial_\mu^l D_n^{2...20...10...0\mu} \Big], \qquad (3.3.18)$$

where the upper indices 3 and 1 sit at the position l. All D's herein have divergence structure (3.3.13) with respect to the upper indices 2. For the left-hand side in (3.3.18) we have to compute

$$d_Q \partial_{\mu_1}^1 ... \partial_{\mu_r}^r D_n^{4...40...0\mu_1...\mu_r} = i \Big[\sum_{l=1}^{r} \partial_{\mu_1}^1 ... \partial_{\mu_l}^l ... \partial_{\mu_r}^r D_n^{4...3...40...0\mu_1...\mu_r} +$$

$$+ \sum_{l=r+1}^{n} \partial_{\mu_1}^1 ... \partial_{\mu_r}^r \partial_{\mu_l}^l D_n^{4...40...10...0\mu_1...\mu_r\mu} \Big]. \qquad (3.3.19)$$

Since the derivatives $\partial_{\mu_1}^1 ... \partial_{\mu_r}^r$ appear everywhere, we can forget them in the splitting. Then we perform the splitting in such a way that the identities

$$d_Q R_n^{4...40...0\mu_1...\mu_r} = i \Big[R_n^{34...40...0\mu_1...\mu_r} ... + R_n^{4...430...0\mu_1...\mu_r}$$

$$+ \partial_\mu^{r+1} R_n^{4...410...0\mu_1...\mu_r\mu} + ... + \partial_\mu^n R_n^{4...40...01\mu_1...\mu_r\mu} \Big] \qquad (3.3.20)$$

are satisfied. This is possible in the following way: we choose arbitrary splitting solutions for all retarded distributions except $R_n^{34...}$, the latter is then defined by (3.3.20). Applying the derivatives $\partial_{\mu_1}^1 ... \partial_{\mu_r}^r$ we get

$$d_Q R_n^{2...20...0} = i \Big[\sum_{l=1}^{r} \partial_\mu^l R_n^{2...3...20...0\mu} + \sum_{l=r+1}^{n} \partial_\mu^l R_n^{2...20...10...0\mu} \Big]. \qquad (3.3.21)$$

The same structure is preserved in the remaining operations (3.3.15) and the symmetrization. This completes the proof of (3.3.10). □

Now we turn to co-boundary couplings where we prove the following theorem:

Theorem 3.3.2. Let $T_1(x)$ be of the form

$$T_1(x) = T_1^0(x) + T_1^4(x), \quad \text{where}$$

$$T_1^4 = d_Q T_1^5 \qquad (3.3.22)$$

is a co-boundary coupling. If $T_n^{0,\ldots 0}$ is gauge invariant

$$d_Q T_n^{0,\ldots 0} = \text{div} \qquad (3.3.23)$$

for all n then

$$T_n(x_1, \ldots, x_n) = T_n^{0,\ldots 0}(x_1, \ldots, x_n) + \text{div} + \text{co} - \text{boundary}. \qquad (3.3.24)$$

Proof. We show that all $T_n^{4,\ldots 0 \ldots}$ are of the form

$$T_n^{4,\ldots 0 \ldots} = d_Q X + \partial Y \qquad (3.3.25)$$

if at least one upper index 4 is present. We shall use the short notation ∂Y to denote divergences in the following.

Let us start with one upper index 4, i.e. one co-boundary vertex at x_1. According to the inductive construction we must consider

$$A_n^{\prime 40 \ldots 0}(x_1, \ldots, x_n) = \sum_{P_2} \Big[\tilde{T}_{n_1}^{40,\ldots 0}(x_1, X) T_{n-n_1}^{0,\ldots 0}(Y, x_n) +$$

$$+ \tilde{T}_{n_1}^{0,\ldots}(X') T_{n-n_1}^{40,\ldots}(x_1, Y', x_n) \Big]. \qquad (3.3.26)$$

On the r.h.s. we use the induction hypothesis that (3.3.25) is true for all orders $m < n$. Then we must only focus upon the co-boundary terms because the divergence contributions can be lumped into a total divergence

$$A_n^{\prime 40 \ldots 0} = \sum_{P_2} \Big[(d_Q \tilde{Z}_{n_1}) T_{n-n_1}^{0,\ldots} + \tilde{T}_{n_1}^{0,\ldots} d_Q Z_{n-n_1} \Big] + \text{div}$$

$$= \sum_{P_2} \Big[d_Q (\tilde{Z}_{n_1} T_{n-n_1}^{0,\ldots}) + \tilde{Z}_{n_1} d_Q T_{n-n_1}^{0,\ldots} +$$

$$+ d_Q (\tilde{T}_{n_1}^{0,\ldots} Z_{n-n_1}) - (d_Q \tilde{T}_{n_1}^{0,\ldots}) Z_{n-n_1} \Big] + \text{div}. \qquad (3.3.27)$$

The different signs are due to the fact that the Z's have odd ghost number. By gauge invariance (3.3.23) we conclude

$$A_n^{\prime 40 \ldots 0} = d_Q Z_3 + \partial Z_4. \qquad (3.3.28)$$

The same follows for $R_n'^{40...0}$ and

$$D_n^{40...0} = d_Q Z_5 + \partial Z_6. \tag{3.3.29}$$

Furthermore, a glance to (3.3.26) and the corresponding formula for $R_n'^{40...0}$ shows that Z_5 and Z_6 are sums of commutators of certain T's and, hence, have causal support. Then, in the splitting of (3.3.29) this form can be preserved by simply splitting Z_5 and Z_6 first and then defining the splitting of $D_n^{40...0}$ by (3.3.29), so that we get the desired result

$$T_n^{40,...} = d_Q Z_7 + \partial Z_8. \tag{3.3.30}$$

We now make an induction in the number of 4-vertices. For example, consider two co-boundary vertices. Proceeding in the same way as above we notice that in A' we have T's with two, one and no 4-vertex which all can be written in the form (3.3.25) by the induction assumption. But taking out d_Q we also need

$$d_Q T_m^{40...0} = \partial d_Q Y$$

which follows from (3.3.25) since $d_Q^2 = 0$. All further steps go through as above. In the same way we successively treat all cases with more co-boundary vertices. This completes the proof. □

These two theorems show that the addition of divergence- and co-boundary couplings leads to physically equivalent theories. Note that in virtue of (3.3.3), the co-boundary in (3.3.24) gives no contribution if one restricts to the physical subspace. For this reason we will disregard such couplings in the following.

3.4 Yang-Mills theory to second order

In this section we further investigate pure Yang-Mills theory. By an explicit second order calculation we will understand the important interplay between causal perturbation theory and gauge invariance. We start from T_1 given by (3.2.52) where we neglect the divergence and co-boundary couplings which are physically irrelevant

$$T_1 = ig f_{abc}(\underbrace{A_{\mu a} A_{\nu b} \partial^\nu A_c^\mu}_{(a)} - \underbrace{A_{\mu a} u_b \partial^\mu \tilde{u}_c}_{(b)}). \tag{3.4.1}$$

The corresponding Q-vertex is given by (3.2.57)

$$iT_{1/1}^\mu = g f_{abc}\left[u_a(\underbrace{\partial^\nu A_b^\mu}_{(1)} - \underbrace{\partial^\mu A_b^\nu}_{(2)})A_{\nu c} + \underbrace{\tfrac{1}{2} u_a u_b \partial^\mu \tilde{u}_c}_{(3)}\right]. \tag{3.4.2}$$

Going from first to second order according to the causal construction we first have to compute the causal distribution

$$D_2(x,y) = T_1(x)T_1(y) - T_1(y)T_1(x). \tag{3.4.3}$$

Taking the gauge variation d_Q and using first order gauge invariance we get

$$d_Q D_2(x,y) = [d_Q T_1(x), T_1(y)] + [T_1(x), d_Q T_1(y)]$$
$$= i\partial_\mu^x [T_{1/1}^\mu(x), T_1(y)] + i\partial_\mu^y [T_1(x), T_{1/1}^\mu(y)]. \tag{3.4.4}$$

This means that D_2 is gauge invariant and the same is obviously true for

$$R_2'(x,y) = -T_1(x)T_1(y). \tag{3.4.5}$$

Next we have to split D_2 and the commutators in (3.4.4) *without the derivatives* giving $R_{2/1}^\mu, R_{2/2}^\mu$ and we must look whether the retarded parts still satisfy the equation of gauge invariance

$$d_Q R_2(x,y) = i\partial_\mu^x R_{2/1}^\mu + i\partial_\mu^y R_{2/2}^\mu. \tag{3.4.6}$$

For all $(x-y)^2 \geq 0$ but $x \neq y$ we have $R_2 = D_2$ and similarly for $R_{2/1}^\mu$. Consequently, gauge invariance (3.4.6) can only be violated at $x=y$, i.e. by local terms $\sim D^a \delta^4(x-y)$. We will always call local terms that spoil gauge invariance "anomalies" to distinguish them from other local terms. The usual axial anomalies are precisely of this kind (see sect.4.9). Such anomalies appear in second order pure Yang-Mills theory for tree graphs only.

Concentrating on the tree-graph contributions to the commutators in (3.4.4) we notice that we can interchange normal ordering with the commutator

$$[:ABC:,:DEF:] =: [ABC, DEF]: + \text{loops}. \tag{3.4.7}$$

This is due to the fact that there is only one contraction in the second order tree graphs. We now calculate the first commutator in (3.4.4). The various contributions can be grouped according to the external field operators.

Sector $uA\tilde{u}u$: There is a contribution from the second term in (3.4.2) and the second one in (3.4.1):

$$[(2),(b)] = ig^2 f_{abc} f_{def} u_a A_{\nu c} [\partial^\mu A_b^\nu(x), A_{\lambda d}(y)] u_e \partial^\lambda \tilde{u}_f. \tag{3.4.8}$$

The commutator on the right side is proportional to $\partial_x^\mu D_0(x-y)$. It can trivially be split giving a retarded part $\sim \partial_x^\mu D_0^{\text{ret}}(x-y)$

$$[(2),(b)]_{\text{ret}} = -g^2 f_{abc} f_{bef} u_a A_{\nu c}(x) u_e(y) \partial^\nu \tilde{u}_f \partial_x^\mu D_0^{\text{ret}}(x-y). \tag{3.4.9}$$

Now comes the crucial point: Applying the derivative ∂_μ^x from (3.4.6), we get a local term because

$$\Box_x D_0^{\text{ret}}(x-y) = \delta(x-y) \tag{3.4.10}$$

which gives rise to an anomaly. Before the splitting there was no anomaly because $\Box D_0 = 0$. The anomaly produced by the first term on the r.h.s. of (3.4.6) is equal to

$$A_1 = -g^2 f_{abc} f_{bef} : A_{\nu c} u_a u_e \partial^\nu \tilde{u}_f : \delta(x-y). \tag{3.4.11}$$

The second term with x and y interchanged gives the same result, so that A_1 must be multiplied by a factor 2.

Another contribution to this sector comes from the third term (3) in (3.4.2) commuted with the second one (b) in (3.4.1):

$$[(3),(b)] = -\frac{i}{2} g^2 f_{abc} f_{def} u_a u_b [\partial^\mu \tilde{u}_c, u_e(y) \partial^\lambda \tilde{u}_f] A_{\lambda d}(y). \tag{3.4.12}$$

The commutator herein is calculated by means of the anti-commutation relation (3.1.5)

$$= \{\partial^\mu \tilde{u}_c(x), u_e(y)\} \partial^\lambda \tilde{u}_f = i\delta_{ce} \partial_x^\mu D_0(x-y). \tag{3.4.13}$$

After distribution splitting and application of ∂_μ^x we get the anomaly

$$A_2 = \frac{g^2}{2} f_{abc} f_{dcf} : u_a u_b \partial^\nu \tilde{u}_f A_{\nu d} : \delta(x-y) \tag{3.4.14}$$

in the same way as above (3.4.10-11). Again A_2 gets multiplied by 2 from the second term with x and y interchanged. In the result we replace the summation indices $b \to e$, $c \to b$ and $d \to c$

$$2A_2 = g^2 f_{aeb} f_{cbf} : u_a u_e \partial^\nu \tilde{u}_f A_{\nu c} : \delta(x-y). \tag{3.4.15}$$

There are no other anomalies in this sector, in particular, the l.h.s. of (3.4.6) does not produce a local term. Hence, the theory is gauge invariant if and only if $2A_2$ cancels against $2A_1$. We rewrite the latter by interchanging in one A_1 the indices a and e:

$$2A_1 = g^2 : A_{\nu c} u_a u_e \partial^\nu u_f : (f_{acb} f_{bef} - f_{ecb} f_{baf}) \delta. \tag{3.4.16}$$

The minus sign is due to the anticommutation of u_a and u_e. Using the antisymmetry of the f's the sum $2A_1 + 2A_2$ is zero if and only if

$$f_{abc} f_{bef} + f_{eab} f_{bcf} + f_{ceb} f_{baf} = 0. \tag{3.4.17}$$

This is the Jacobi identity. It is a necessary condition for second order gauge invariance.

Sector $uAAA$: We have learned from the last sector that the anomalies are produced by those terms in (3.4.2) which have a derivative ∂^μ. Therefore, there remains one commutator only to be considered

$$[(2),(a)] = -ig^2 f_{abc} f_{def} u_a A_{\nu c} [\partial^\mu A_b^\nu(x), A_{\varrho d}(y) A_{\sigma e} \partial^\sigma A_f^\varrho]$$

$$= g^2 f_{abc} u_a A_{\nu c}(x) \Big\{ f_{bef} A_{\sigma e} \partial^\sigma A_f^\nu \partial^\mu D_0(x-y)$$

$$+ f_{dbf} A_{\varrho d} \partial^\nu A_f^\varrho \partial^\mu D_0(x-y) + f_{deb} A_d^\nu A_{\sigma e} \partial_x^\mu \partial_y^\sigma D_0(x-y) \Big\}. \tag{3.4.18}$$

The fields in the curly bracket all have arguments y. In the last term herein we have a new situation: $D_0(x - y)$ with two derivatives has singular order $\omega = 0$, consequently, the retarded part is not unique but equal to

$$\partial_x^\mu \partial_y^\sigma D_0^{\text{ret}} + \alpha_1 g^{\mu\sigma} \delta(x - y), \qquad (3.4.19)$$

where α_1 is a free constant. After applying ∂_μ the resulting anomaly is of the form

$$: G(x) F(y) : \partial_y^\sigma \delta(x - y) + \alpha_1 \partial_y^\sigma [: G(x) F(y) : \delta(x - y)]. \qquad (3.4.20)$$

To (3.4.20) we must still add the terms with x and y interchanged. The result can be transformed by the following two identities

$$G(x) F(y) \partial_y^\sigma \delta(x - y) + G(y) F(x) \partial_x^\sigma \delta(x - y) =$$
$$= (\partial^\sigma G)(x) F(x) \delta(x - y) - G(x) (\partial^\sigma F)(x) \delta(x - y) \qquad (3.4.21)$$
$$\partial_y^\sigma [G(x) F(y) \delta(x - y)] + \partial_x^\sigma [G(y) F(x) \delta(x - y)] =$$
$$= (\partial^\sigma G)(x) F(x) \delta(x - y) + G(x) (\partial^\sigma F)(x) \delta(x - y). \qquad (3.4.22)$$

These relations are easily proved by smearing with test functions. They enable us to remove the derivatives from δ in (3.4.20). Summing up we have found the following short rule for the calculation of local terms:

$$G(x) F(y) \partial_x^\mu \partial_y^\sigma D_0(x - y) \Longrightarrow \left[(\alpha_1 + 1) \partial^\sigma G F + (\alpha_1 - 1) G \partial^\sigma F \right] \delta. \qquad (3.4.23)$$

Now we return to (3.4.18). Using the rule (3.4.23) in the last term we obtain the following anomaly

$$A_3 = g^2 f_{abc} u_a A_{\nu c} [f_{bef} A_{\sigma e} \partial^\sigma A_f^\nu + f_{dbf} A_{\varrho d} \partial^\nu A_f^\varrho] 2\delta +$$
$$+ g^2 f_{abc} f_{deb} [(\alpha_1 + 1)(\partial^\sigma u_a A_{\nu c} + u_a \partial^\sigma A_{\nu c}) A_d^\nu A_{\sigma e} +$$
$$+ (\alpha_1 - 1) u_a A_{\nu c} (\partial^\sigma A_d^\nu A_{\sigma e} + A_d^\nu \partial^\sigma A_{\sigma e})] \delta(x - y), \qquad (3.4.24)$$

where again the double dots are omitted. There are no further anomalies so that A_3 must cancel. From the term $\sim u_a A_{\nu c} A_d^\nu \partial^\sigma A_{\sigma e}$ we conclude

$$\alpha_1 = 1. \qquad (3.4.25)$$

Then the contribution $\sim u_a A_{\nu c} A_{\sigma e} \partial^\sigma A_f^\nu$ vanishes

$$2(f_{abc} f_{bef} + f_{abe} f_{cbf}) + 2 f_{abf} f_{ceb} =$$
$$= 2(f_{cab} f_{bef} + f_{aeb} f_{bcf} + f_{ecb} f_{baf}) = 0,$$

due to the Jacobi identity (3.4.17). But there remains a term

$$B_3 = 2 g^2 f_{abc} f_{deb} \partial^\sigma u_a A_{\nu c} A_d^\nu A_{\sigma e} \delta(x - y). \qquad (3.4.26)$$

How do we get rid of this ?

It is our goal to establish gauge invariance of the retarded distributions in the form (3.4.6). If these retarded distributions have singular order $\omega \geq 0$ there is some freedom of normalization

$$d_Q(R_2 + N_2) = i\partial_\mu^x(R_{2/1}^\mu + N_{2/1}^\mu) + i\partial_\mu^y(R_{2/2}^\mu + N_{2/2}^\mu) \qquad (3.4.27)$$

where the N's are local normalization terms $\sim D^a \delta(x-y)$. This offers the possibility to compensate the last anomaly B_3 (3.4.26) by suitable normalization. Indeed, B_3 comes from a normalization term N_2 with four external A's. Let us consider the tree graph with two vertices (a) in (3.4.1) with the two derivatives on the inner line. Its causal distribution is given by

$$D_2 = -f_{abc}f_{def}A_{\mu a}A_{\nu b}[\partial^\nu A_c^\mu(x), \partial^\alpha A_f^\lambda(y)]A_{\lambda d}A_{\alpha e}.$$

The commutator is equal to

$$[\ldots] = i\delta_{cf}g^{\mu\lambda}\partial_x^\nu \partial_y^\alpha D_0$$

and has $\omega = 0$. Consequently, there is a free normalization term $\sim i\beta \delta_{cf} g^{\mu\lambda} g^{\nu\alpha} \delta$ possible:

$$N_2 = -i\beta f_{abc}f_{dec}A_{\mu a}A_d^\mu A_{\nu b}A_e^\nu \delta(x-y), \qquad (3.4.28)$$

where β is an arbitrary constant. Now we calculate the gauge variation

$$d_Q N_2 = 2\beta f_{abc}f_{dec}[\partial_\mu u_a A_d^\mu A_{\nu b} A_e^\nu + \partial_\mu u_d A_a^\mu A_{\nu b} A_e^\nu + \partial_\nu u_b A_e^\nu A_{\mu a} A_d^\mu +$$
$$+ \partial_\nu u_e A_b^\nu A_{\mu a} A_d^\mu]. \qquad (3.4.29)$$

By renaming the summation indices this is equal to

$$d_Q N_2 = 8\beta f_{abc}f_{dec}\partial_\mu u_a A_d^\mu A_{\nu b} A_e^\nu \qquad (3.4.30)$$

and this cancels B_3 (3.4.26) if we choose

$$\beta = -\frac{g^2}{4}. \qquad (3.4.31)$$

This is the four-gluon coupling term. This term is usually obtained from a classical Lagrangian, here we got it from quantum gauge invariance to second order. The last steps in the inductive construction, namely,

$$T_2 = R_2 + N_2 - R_2'$$

and symmetrization cannot violate gauge invariance. Therefore, we have succeeded in preserving gauge invariance in second order tree graphs by proper normalization. Note that the local term $\sim \alpha_1$ in (3.4.19) is actually a normalization term $N_{2/1}^\mu$ in (3.4.27). We will see later that second order loop contributions preserve gauge invariance as well.

3.5 Reductive Lie algebras

The real coupling parameters f_{abc} which specify a Yang-Mills theory have been found to be totally antisymmetric and satisfy the Jacobi identity (3.4.17). This is not quite the same situation as in usual Lie algebra theory. Here one considers a (finite dimensional) algebra generated by N generators $L_a, a = 1, \ldots N$ with an antisymmetric Lie-product defined by

$$[L_a, L_b] = f_{ab}^c L_c \tag{3.5.1}$$

where

$$f_{ab}^c = -f_{ba}^c. \tag{3.5.2}$$

The f's are the so-called structure constants which we assume to be real for our purposes. In addition, the bracket (3.5.1) fulfills the Jacobi identity

$$\Big[[L_a, L_b], L_c\Big] + \Big[[L_c, L_a], L_b\Big] + \Big[[L_b, L_c], L_a\Big] = 0, \tag{3.5.3}$$

which is equivalent to

$$\sum_d \left(f_{ab}^d f_{dc}^e + f_{ca}^d f_{db}^e + f_{bc}^d f_{da}^e \right) = 0. \tag{3.5.4}$$

Note that f_{ab}^c is only antisymmetric with respect to the two lower indices a, b (3.5.2) not totally antisymmetric. In Yang-Mills theory we therefore deal with a special class of Lie algebras. It is our aim now to find out which class that is.

To get information about a given Lie algebra it is often convenient to study representations, in particular the so-called adjoint representation. Here the generators are represented by matrices

$$(L_a)_{bc} = f_{ac}^b \tag{3.5.5}$$

and the bracket is represented by the commutator of the matrices. To verify that this is indeed a representation of (3.5.1) we compute

$$([L_a, L_b])_{cd} = (L_a L_b)_{cd} - (L_b L_a)_{cd}$$
$$= (L_a)_{ce}(L_b)_{ed} - (L_b)_{ce}(L_a)_{ed} =$$
$$= f_{ae}^c f_{bd}^e - f_{be}^c f_{ad}^e = -f_{bd}^e f_{ea}^c - f_{da}^e f_{eb}^c \tag{3.5.6}$$

where we have used antisymmetry (3.5.2) and always sum over double indices. By the Jacobi identity (3.5.4) this is equal to

$$= f_{ab}^e f_{ed}^c = f_{ab}^e (L_e)_{cd}, \tag{3.5.7}$$

which shows the desired property (3.5.1). If for some generator L_a

$$f_{ac}^b = 0 \quad \forall b, c, \tag{3.5.8}$$

so that the corresponding matrix (3.5.5) is zero, we omit this generator. In this way the Abelian part is eliminated. We call the resulting matrix algebra the adjoint representation $\mathcal{A}$.

In the adjoint representation one can define a real symmetric bilinear form by taking the trace over the product of matrices

$$(L_a, L_d) \stackrel{\text{def}}{=} \operatorname{Tr}(L_a L_d) = \sum_{bc} (L_a)_{bc} (L_d)_{cb}$$

$$= \sum_{bc} f_{ac}^b f_{db}^c. \qquad (3.5.9)$$

This is the well-known Killing form. It is very useful for investigating the structure of the Lie algebra. Let us now assume that the f's are totally antisymmetric in all three indices. Then it follows from (3.5.9) that

$$(L_a, L_a) = -\sum_{bc} (f_{ac}^b)^2. \qquad (3.5.10)$$

Since the r.h.s. is < 0 for all $L_a \in \mathcal{A}$ because we have eliminated possible zero matrices (3.5.8), we see that the Killing form is nonsingular. Then, according to the well-known Cartan criterion, the adjoint representation $\mathcal{A}$ is semi-simple (see V.S. Varadarajan, *Lie groups, Lie algebras and their representations*, Prentice-Hall, Inc. *1974*, theorem 3.9.2.). A semisimple Lie algebra is isomorphic to a direct sum of simple Lie algebras, a simple Lie algebra is not Abelian and has no proper ideal.

The fact just observed that the adjoint representation $\mathcal{A}$ is semisimple enables us to answer the question which Lie algebras are relevant for Yang-Mills theories. The result is contained in the following theorem (see Varadarajan, *loc. cit.*, theorem 3.16.3)

Theorem 3.5.1. Let g be a Lie algebra over $\mathbb{R}$ or $\mathbb{C}$. Then the following statements are equivalent:

(1) g is reductive, i.e. it is isomorphic to a direct sum of Abelian and simple Lie algebras.
(2) The adjoint representation of g is semisimple.
(3) The radical of g coincides with its center.

We can still say more because the Killing form is negative definite due to (3.5.10). Then the second Cartan criterion implies that the simple subalgebras must be compact. Summing up, the Lie algebra g which defines the coupling in Yang-Mills theory is isomorphic to a direct sum of abelian and simple compact Lie algebras. In the standard model of combined electroweak and strong interactions this Lie algebra is

$$g = u(1) + su(2) + su(3). \qquad (3.5.11)$$

It is not semisimple because it contains the abelian algebra $u(1)$ for electromagnetism. For aesthetic reasons one likes simple algebras and, therefore, has tried to embed (3.5.11) into a bigger simple algebra like $su(5)$ or $so(10)$. This is the idea of "grand unification" (see the textbooks in the notes). From the point of view of gauge theory this idea is not so great because here we have to deal with reductive Lie algebras which are not necessarily simple.

In classical Yang-Mills theory one usually starts with geometric objects like Lie groups or fibre bundles. In our approach to quantum gauge theory we now have come close to this. There always exists a Lie group (not necessarily unique) corresponding to the reductive Lie algebra g just obtained, and the quantities f_{abc}, A_a^μ, u_b etc. in the Yang-Mills coupling are tensors with respect to this group. But this structure involving only free quantum fields is quite different from the geometry in classical Yang-Mills theory where the basic objects are classical interacting fields. Although geometry is very fascinating for aesthetic reasons, in quantum theory it is less rich and less powerful. The same feature can already be noticed when one compares classical and quantum mechanics: classical mechanics is symplectic geometry, but quantum mechanics contains not much of it.

3.6 Coupling to matter fields

In nature the gauge fields are not only coupled between themselves but also to other fields. These other fields have zero gauge variation and we collect them here under the notion "matter fields". To avoid misunderstanding we remind the reader again that we work exclusively with the free asymptotic fields, interacting matter fields would have non-vanishing gauge variation. Guided by QED we couple the gauge fields A_a^μ to a matter current j_a^μ so that the total coupling is of the following form

$$T_1 = ig f_{abc} (\underbrace{A_{\mu a} A_{\nu b} \partial^\nu A_c^\mu}_{(a)} \underbrace{- A_{\mu a} u_b \partial^\mu \tilde{u}_c}_{(b)}) + \underbrace{ig' j_{\mu a} A_a^\mu}_{(c)}. \quad (3.6.1)$$

The gauge variation

$$d_Q T_1 = i\partial_\mu T_{1/1}^\mu - g' j_{\mu a} \partial^\mu u_a \quad (3.6.2)$$

is a divergence

$$= i\partial_\mu (T_{1/1}^\mu + g' j_a^\mu u_a), \quad (3.6.3)$$

if and only if the matter current is conserved

$$\partial_\mu j_a^\mu = 0. \quad (3.6.4)$$

This will always be assumed in the case of massless gauge fields.

120   Spin-1 gauge theories: massless gauge fields

The most important matter fields are spinor fields describing the leptons and quarks. We therefore consider a collection of spinor fields $\psi_n, n = 1, \ldots d$ satisfying the Dirac equations

$$i\gamma_\nu \partial^\nu \psi_n(x) = M_{nm}\psi_m(x). \tag{3.6.5}$$

Here M is a hermitian mass matrix, $M^+ = M$, in order to have a unitary time evolution for the free fields. The corresponding conserved current is given by

$$j_a^\mu = \tfrac{1}{2}\overline{\psi}_n \gamma^\mu (\lambda_a)_{nm}\psi_m, \tag{3.6.6}$$

where $\lambda_a, a = 1, \ldots N$ are hermitian $d \times d$ matrices in order to have j_a^μ hermitian. All products are normally ordered. This current is conserved if and only if

$$\lambda_a M = M^+ \lambda_a = M\lambda_a \tag{3.6.7}$$

for all $a = 1, \ldots d$. We will see below from second order gauge invariance that the λ_a realize a representation of the Lie algebra g defined by the f_{abc}. If this representation is irreducible, it follows from Schur's lemma that the mass matrix is a multiple of the unit matrix

$$M_{nm} = M\delta_{nm}. \tag{3.6.8}$$

The standard example of a gauge theory with massless gauge fields is the theory of strong interactions called quantum chromodynamics QCD with the Lie algebra $g = su(3)$. The gauge fields are the $N = 8$ gluon fields which interact between themselves and with the quark fields ψ_n. Due to (3.6.8) the colored quarks are degenerate in mass. But the quarks also interact weakly and, therefore, have a second quantum number called flavor. The quarks with different flavor have different masses. As we will see in the following chapter this is due to the fact that the weak interactions are mediated by *massive* gauge fields.

¿From (3.6.3) and (3.4.2) we have the following total Q-vertex

$$iT'^\mu_{1/1} = gf_{abc}\Big[\underbrace{u_a(\partial^\nu A_b^\mu}_{(1)} - \underbrace{\partial^\mu A_b^\nu)}_{(2)} A_{\nu c} + \underbrace{\tfrac{1}{2} u_a u_b \partial^\mu \tilde{u}_c}_{(3)}\Big]$$

$$+ \underbrace{g'\tfrac{1}{2}\overline{\psi}\gamma^\mu \lambda_a \psi u_a}_{(4)}. \tag{3.6.9}$$

We now want to study second order gauge invariance. The sectors $uAAA$ and $uA\tilde{u}u$ can be treated in the same way as for pure Yang-Mills theory in sect. 3.4. The interesting new sector is $uA\overline{\psi}\psi$. Here a first contribution giving rise to an anomaly comes from the commutator

$$[(2), (c)] = \frac{i}{2}gg' f_{abc}u_a A_{\nu c}[\partial^\mu A_b^\nu(x), A_{\varrho d}(y)]\overline{\psi}\gamma^\varrho \lambda_d \psi. \tag{3.6.10}$$

After splitting the commutator $\sim \partial_x^\mu D_0(x-y)$ we get the following retarded part

$$[(2), (c)]_{\text{ret}} = -\tfrac{1}{2}gg' f_{abc} u_a A_{\nu c}(x)\overline{\psi}(y)\gamma^\nu \lambda_b \psi \partial^\mu D_0^{\text{ret}}(x-y). \tag{3.6.11}$$

With the additional derivative ∂_μ (see (3.4.10)) this leads to an anomaly

$$A_1 = -\tfrac{1}{2}gg' f_{abc} u_a A_{\nu c}\overline{\psi}\gamma^\nu \lambda_b \psi \delta(x-y). \tag{3.6.12}$$

The other anomalies in this sector come from the commutator

$$[(4), (c)] = \frac{i}{4}g'^2 u_a [\overline{\psi}\gamma_\mu \lambda_a \psi(x), \overline{\psi}(y)\gamma_\nu \lambda_b \psi] A_b^\nu. \tag{3.6.13}$$

Using the anticommutation rule (1.8.16) twice we obtain

$$= \frac{1}{4}g'^2 u_a \Big(\overline{\psi}(x)\gamma_\mu S(x-y)\gamma_\nu \lambda_a \lambda_b \psi(y) +$$

$$+\overline{\psi}(y)\gamma_\nu S(y-x)\gamma_\mu \lambda_b \lambda_a \psi(x)\Big) A_b^\nu(y). \tag{3.6.14}$$

Applying the external derivative ∂_x^μ after splitting the causal distributions S, the anomalies come from the relations

$$i\gamma_\mu \partial_x^\mu S_{\text{ret}}(x-y) = -\delta(x-y)$$
$$i\partial_x^\mu S_{\text{ret}}(y-x)\gamma^\mu = \delta(x-y). \tag{3.6.15}$$

The δ-distribution makes the arguments equal, so that we get

$$A_2 = \frac{i}{4}g'^2 u_a \overline{\psi}\gamma_\nu (\lambda_a \lambda_b - \lambda_b \lambda_a)\psi A_b^\nu \delta(x-y). \tag{3.6.16}$$

Since there are no other anomalies in this sector and a normalization term $\sim u_a$ is impossible, A_1 and A_2 must cancel each other

$$A_1 + A_2 = \frac{i}{4}g' u_a \overline{\psi}\gamma_\nu \Big[g'(\lambda_a \lambda_b - \lambda_b \lambda_a) - 2ig f_{abc}\lambda_c\Big]$$

$$\times \psi A_b^\nu \delta(x-y). \tag{3.6.17}$$

This gives the following necessary and sufficient condition for second order gauge invariance

$$\lambda_a \lambda_b - \lambda_b \lambda_a = \frac{g}{g'} 2i f_{abc} \lambda_c. \tag{3.6.18}$$

Introducing new λ-matrices by

$$\lambda_a = 2i\frac{g}{g'}\tilde{\lambda}_a, \tag{3.6.19}$$

the condition

$$[\tilde{\lambda}_a, \tilde{\lambda}_b] = f_{abc}\tilde{\lambda}_c \tag{3.6.20}$$

means that the $\tilde{\lambda}$-matrices realize a representation of the Lie algebra g defined by the structure constants f_{abc}. In the standard theory this representation is the smallest non-trivial irreducible one, the so-called fundamental representation (see problem 3.8).

3.7 Gauge invariance to all orders

Gauge invariance to all orders means that the T-products can be normalized in such a way that the relations

$$d_Q T_n(x_1,\ldots,x_n) - i \sum_{l=1}^{n} \partial_\mu^l T_{n/l}^\mu(x_1,\ldots,x_n) = 0 \qquad (3.7.1)$$

are satisfied. In the causal construction, the T-products are obtained in normally ordered form, hence, the l.h.s. in (3.7.1) is a sum of Wick monomials

$$\sum_j t_j(x_1 - x_n, \ldots, x_{n-1} - x_n) : O_j(x_1,\ldots,x_n) := 0. \qquad (3.7.2)$$

Here the t_j are numerical distributions which are translation invariant and $: O_j :$ are the Wick monomials. The latter are products of the basic fields $A_a^\mu, u_b, \tilde{u}_c$ and derivatives thereof. Writing the first order couplings (3.4.1-2) in the form

$$T_1 = ig f_{abc}\left(\tfrac{1}{2} A_{\mu a} A_{\nu b} F_c^{\nu\mu} - A_{\mu a} u_b \partial^\mu \tilde{u}_c\right) \qquad (3.7.3)$$

$$T_{1/1}^\mu = ig f_{abc} u_a \left(F_b^{\mu\nu} A_{\nu c} - \tfrac{1}{2} u_b \partial^\mu \tilde{u}_c\right), \qquad (3.7.4)$$

with

$$F_a^{\mu\nu} = \partial^\mu A^\nu - \partial^\nu A^\mu, \qquad (3.7.5)$$

we realize that $\tilde{u}(x)$ always appears with a derivative and $\partial^\mu A_a^\nu(x)$ in the form of the field strength tensor (3.7.5), only.

Gauge invariance (3.7.2) can now be expressed by a list of identities for numerical (C-number) distributions

$$t_j(x_1,\ldots,x_n) = 0, \qquad (3.7.6)$$

which we call Cg-identities for brevity (C-number identities for gauge invariance). We will work with the original coordinates, in order to have a simpler formulation of the symmetry properties of the t's. Now we want to derive the Cg-identities. For 2-legged distributions, i.e. distributions with two external field operators, we have the operator decomposition

$$T_n(x_1,\ldots,x_n)|_{2-\text{legs}} = \sum_{i<j} t_{AA}^{\mu\nu}(x_i,x_j,x_1,\ldots,\check{x}_i,\check{x}_j\ldots x_n) : A_\mu(x_i) A_\nu(x_j) :$$

$$+ \sum_{i\neq j} t_{AF}^{\alpha\mu\nu}(x_i,x_j,x_1\ldots\check{x}_i,\check{x}_j\ldots x_n) : A_\alpha(x_i) F_{\mu\nu}(x_j) : +$$

$$+ \sum_{i<j} t_{FF}^{\alpha\beta\mu\nu}(x_i,x_j,x_1\ldots) : F_{\alpha\beta}(x_i) F_{\mu\nu}(x_j) : +$$

$$+ \sum_{i\neq j} t_{u\tilde{u}}^\mu(x_i,x_j,\ldots) : u(x_i) \partial_\mu \tilde{u}(x_j) : \qquad (3.7.7)$$

for the main theory (the coordinates with check must be omitted). The numerical 2-legged distributions with one Q-vertex are

$$t_{uA}^{\alpha l \nu}, \quad t_{uF}^{\alpha l \nu \mu}, \tag{3.7.8}$$

which is a set of different distributions depending on the position of the Q-vertex l with respect to the external legs. The following notations are used: The subscripts denote the external field operators ($\tilde{u}$ for $\partial \tilde{u}$). The operator indicated by the first (resp. second) lower index is attached to the first (second) argument of the distribution (see (3.7.7)). The upper Greek indices are the Lorentz indices which are contracted with the field operators, in the same order as the lower indices. For distributions with one Q-vertex we have before these Lorentz indices another Lorentz index, say α, and a number l. This is just the Lorentz index α of $T_{n/l}^{\alpha}$ and the number l denotes the position where the Q-vertex is (in the argument of the numerical distribution). The Q-vertex can be an outer or inner vertex. There are no colour indices, since 2-legged distributions are diagonal in the colour indices.

We now list the Cg-identities corresponding to the 6 different Wick monomials with two factors:

$$: \partial^\mu u_a(x_1) A_a^\nu(x_2) : \quad t_{AA}^{\mu\nu} = t_{uA}^{\mu 1 \nu} \tag{3.7.9}$$

$$: \partial^\mu u_a(x_1) F_a^{\mu\nu}(x_2) : \quad t_{AF}^{\alpha\mu\nu} = t_{uF}^{\alpha 1 \mu \nu} \tag{3.7.10}$$

$$: u_a(x_1) \partial^\alpha F_a^{\mu\nu}(x_2) : \quad \tfrac{1}{2}(t_{u\tilde{u}}^\mu g^{\alpha\nu} - t_{u\tilde{u}}^\nu g^{\alpha\mu}) = t_{uF}^{\alpha 2 \mu\nu}. \tag{3.7.11}$$

Here the left-hand sides come from $d_Q T_n$ and the right-hand sides involving the Q-vertex from $\partial_\mu^l T_{n/l}^\mu, l = 1, 2$. In the remaining three identities there is no contribution from $d_Q T_n$:

$$: u_a(x_1) \tfrac{1}{2} [\partial^\alpha A_a^\nu(x_2) + \partial^\nu A_a^\alpha(x_2)] : \quad 0 = t_{uA}^{\alpha 2 \nu} + t_{uA}^{\nu 2 \alpha} \tag{3.7.12}$$

$$: u_a(x_1) A_a^\nu(x_2) : \quad 0 = \sum_{l=1}^{n} \partial_\alpha^l t_{uA}^{\alpha l \nu} \tag{3.7.13}$$

$$: u_a(x_1) F_a^{\mu\nu}(x_2) : \quad 0 = \sum_{l=1}^{n} \partial_\alpha^l t_{uF}^{\alpha l \mu\nu} + \tfrac{1}{4}(t_{uA}^{\mu 2 \nu} - t_{uA}^{\nu 2 \mu}). \tag{3.7.14}$$

As always in causal perturbation theory, the proof of the Cg-identities is by induction on the order n. The only place where the identities can be violated in the inductive step from $n-1$ to n is in the process of distribution splitting. Here the possible violation is by local terms and we must try to absorb those by appropriate normalization of the t's. It is important to note that every t-distribution has its own freedom of normalization, in particular, the distributions with Q-vertex are independent from the normal t's corresponding to T_n. Then we may regard the identities (3.7.9-11) just as defining equations for the distributions with Q-vertex on the r.h.s, or rather as the fixing of their normalization. The identity (3.7.12) expresses the antisymmetry in the Lorentz indices and can easily be respected in the splitting. The only non-trivial identities are (3.7.13) and (3.7.14).

We substitute (3.7.9) into (3.7.13) and use the antisymmetry (3.7.12)

$$\partial_\alpha^1 t_{AA}^{\alpha\nu} + \tfrac{1}{2}\partial_\alpha^2 [t_{uA}^{\alpha 2\nu} - t_{uA}^{\nu 2\alpha}] + \sum_{l=3}^{n} \partial_\alpha^l t_{uA}^{\alpha l\nu} = 0. \qquad (3.7.15)$$

Similarly we substitute (3.7.10) into (3.7.14) and use (3.7.11)

$$\partial_\alpha^1 t_{AF}^{\alpha\mu\nu} - \tfrac{1}{2}[\partial_2^\nu t_{u\tilde{u}}^\mu - \partial_2^\mu t_{u\tilde{u}}^\nu] + \tfrac{1}{4}[t_{uA}^{\mu 2\nu} - t_{uA}^{\nu 2\mu}] + \sum_{l=3}^{n} \partial_\alpha^l t_{uF}^{\alpha l\mu\nu} = 0. \qquad (3.7.16)$$

To analyze the possible violation of these two identities we define the anomalies

$$a_1^\nu \stackrel{\text{def}}{=} \partial_\alpha^1 t_{AA}^{\alpha\nu} + \tfrac{1}{2}\partial_\alpha^2 [t_{uA}^{\alpha 2\nu} - t_{uA}^{\nu 2\alpha}] + \sum_{l=3}^{n} \partial_\alpha^l t_{uA}^{\alpha l\nu}, \qquad (3.7.17)$$

$$a_2^{\mu\nu} \stackrel{\text{def}}{=} \partial_\alpha^1 t_{AF}^{\alpha\mu\nu} + \tfrac{1}{2}[\partial_2^\nu t_{u\tilde{u}}^\mu - \partial_2^\mu t_{u\tilde{u}}^\nu] + \tfrac{1}{4}[t_{uA}^{\mu 2\nu} - t_{uA}^{\nu 2\mu}] + \sum_{l=3}^{n} \partial_\alpha^l t_{uF}^{\alpha l\mu\nu}. \qquad (3.7.18)$$

Taking covariance and the singular order of the distributions on the right side into account, we write down the most general form of a_1, a_2

$$a_1^\nu(x_1,...,x_n) = [\sum_{i,j,k=1}^{n} C_{ijk}\partial_i^\gamma \partial_{j\gamma}\partial_k^\nu + \sum_{k=1}^{n} D_k \partial_k^\nu]\delta(x_1-x_n,...,x_{n-1}-x_n), \qquad (3.7.19)$$

$$a_2^{\mu\nu}(x_1,...,x_n) = [\sum_{i,j=1}^{n} K_{ij}\partial_i^\mu \partial_j^\nu + Lg^{\mu\nu}]\delta(x_1-x_n,...,x_{n-1}-x_n). \qquad (3.7.20)$$

We first concentrate on $a_2^{\mu\nu}$. The antisymmetry $a_2^{\mu\nu} = -a_2^{\nu\mu}$ implies for the constants in (3.7.20)

$$L=0, \quad K_{ii}=0, \quad K_{ij}=-K_{ji}. \qquad (3.7.21)$$

Taking the permutation symmetry of $a_2^{\mu\nu}(x_1,x_2,x_3,...,x_n)$ in $x_3,x_4,...,x_n$ into account, we conclude

$$K_{ij}=0, \quad \forall 3 \leq i,j \leq n$$

$$K_1 \stackrel{\text{def}}{=} K_{1l}, \quad K_2 \stackrel{\text{def}}{=} K_{2l}, \quad \forall l=3,...,n \qquad (3.7.22)$$

and (3.7.20) becomes

$$a_2^{\mu\nu} = [K_{12}(\partial_1^\mu \partial_2^\nu - \partial_2^\mu \partial_1^\nu) + K_1 \sum_{l=3}^{n}(\partial_1^\mu \partial_l^\nu - \partial_l^\mu \partial_1^\nu)+$$

$$+K_2\sum_{l=3}^{n}(\partial_2^\mu\partial_l^\nu - \partial_l^\mu\partial_2^\nu)]\delta^{(4(n-1))} = (K_{12} - K_1 + K_2)(\partial_1^\mu\partial_2^\nu - \partial_2^\mu\partial_1^\nu)\delta^{(4(n-1))},$$

(3.7.23)

where

$$\sum_{k=1}^{n}\partial_k^\nu\delta(x_1 - x_n,...,x_{n-1} - x_n) = 0,$$

(3.7.24)

has been used. Hence the anomaly $a_2^{\mu\nu}$ assumes the simple form

$$a_2^{\mu\nu}(x_1,...,x_n) = K(\partial_1^\mu\partial_2^\nu - \partial_2^\mu\partial_1^\nu)\delta(x_1 - x_n,...,x_{n-1} - x_n).$$

(3.7.25)

Let us now turn to the anomaly a_1^ν (3.7.19). In the terms with only one derivative, the permutation symmetry of $a_1^\nu(x_1,x_2,x_3,...,x_n)$ in all inner vertices $x_3, x_4, ..., x_n$ (abbreviated by p.i.v. in the following) implies $D_j = D'$, $\forall j = 3,...,n$. The term $D_2\partial_2^\nu\delta^{(4(n-1))}$ can be eliminated by means of (3.7.24). Note that in the terms with three derivatives we can choose $C_{ijk} = 0$, $\forall i > j$, because interchanging i and j gives a term of the same form. We claim that a_1^ν can be simplified to

$$a_1^\nu(x_1,...,x_n) = [C'_{111}\Box_1\partial_1^\nu + C'_{122}\partial_1^\gamma\partial_{2\gamma}\partial_2^\nu + C'_{221}\Box_2\partial_1^\nu + C'_{121}\partial_1^\gamma\partial_{2\gamma}\partial_1^\nu +$$

$$+C'_{112}\Box_1\partial_2^\nu + C'_{222}\Box_2\partial_2^\nu + C_1\sum_{i=3}^{n}\Box_i\partial_1^\nu + C_2\sum_{i=3}^{n}\Box_i\partial_2^\nu + C_3\sum_{i=3}^{n}\partial_1^\gamma\partial_{i\gamma}\partial_i^\nu +$$

$$+C_4\sum_{i=3}^{n}\partial_2^\gamma\partial_{i\gamma}\partial_i^\nu + C_5\sum_{i=3}^{n}\Box_i\partial_i^\nu + D'_1\partial_1^\nu + D\sum_{i=3}^{n}\partial_i^\nu]\delta(x_1 - x_n,...,x_{n-1} - x_n),$$

(3.7.26)

In fact, in the first line all terms with $1 \leq i,j,k \leq 2$, $i \leq j$ (case 1) are written down. Due to p.i.v. and (3.7.24), the terms with two indices in $\{1,2\}$ and one index in $\{3,...,n\}$ (case 2) can be written as case 1 terms, similar to (3.7.22-23). Case 3 is $i \in \{1,2\}$ and $j,k \in \{3,...,n\}$. Let us consider $i = 1$ for example. By means of p.i.v. these terms have the form

$$[C\sum_{j=3}^{n}\partial_1^\gamma\partial_{j\gamma}\partial_j^\nu + C'\sum_{j\neq k}\partial_1^\gamma\partial_{j\gamma}\partial_k^\nu]\delta^{(4(n-1))}.$$

(3.7.27)

With (3.7.24) the second sum is equal to

$$C'\sum_{j=3}^{n}\partial_1^\gamma\partial_{j\gamma}(-\partial_1 - \partial_2 - \partial_j)^\nu\delta^{(4(n-1))}$$

(3.7.28)

and gives no new terms. The case 4: $k \in \{1,2\}$ and $i,j \in \{3,...,n\}$ is similar. Therefore, the cases 3 and 4 give only the terms written in the second line of (3.7.26). There remains case 5 with $i,j,k \in \{3,...,n\}$. Because of p.i.v. these terms have the form

$$[C'_1 \sum_i \Box_i \partial_i^\nu + C'_2 \sum_{i\neq j} \partial_i^\gamma \partial_{j\gamma} \partial_j^\nu + C'_3 \sum_{i\neq k} \Box_i \partial_k^\nu + C'_4 \sum_{i<j\neq k\neq i} \partial_i^\gamma \partial_{j\gamma} \partial_k^\nu]\delta^{(4(n-1))}.$$
(3.7.29)

Analogously to (3.7.28) for example, the last sum is equal to

$$C'_4 \sum_{3\le i<j\le n} \partial_i^\gamma \partial_{j\gamma}(-\partial_1 - \partial_2 - \partial_i - \partial_j)^\nu \delta^{(4(n-1))}. \tag{3.7.30}$$

Using several times the reasoning of (3.7.28), we see that only one of the four sums in (3.7.29) gives new terms. We have chosen the first one in (3.7.26).

Now comes the final step. Performing the *finite* renormalizations

$$t_{AA}^{\alpha\nu} \to t_{AA}^{\alpha\nu} - [D'_1 g^{\alpha\nu} + C'_{122}(\partial_2^\alpha \partial_2^\nu + \partial_1^\alpha \partial_1^\nu) + (C'_{221} - C'_{222})(\Box_1 + \Box_2)g^{\alpha\nu} +$$
$$+ C'_{211}\partial_2^\alpha \partial_1^\nu + (C'_{112} - C'_{111} + C'_{122} + C'_{221} - C'_{222})\partial_1^\alpha \partial_2^\nu]\delta^{(4(n-1))}, \tag{3.7.31}$$

$$t_{uA}^{\alpha l\nu} \to t_{uA}^{\alpha l\nu} - [Dg^{\alpha\nu} - C'_{222}\Box_2 g^{\alpha\nu} - (C'_{111} - C'_{122} - C'_{221} + C'_{222})\Box_1 g^{\alpha\nu} +$$
$$+C_1\partial_l^\alpha\partial_1^\nu + C_2\partial_l^\alpha\partial_2^\nu + C_3\partial_1^\alpha\partial_l^\nu + C_4\partial_2^\alpha\partial_l^\nu + C_5\partial_l^\alpha\partial_l^\nu]\delta^{(4(n-1))}, \quad l\ge 3, \tag{3.7.32}$$

$$t_{AF}^{\alpha\mu\nu} \to t_{AF}^{\alpha\mu\nu} - K(g^{\alpha\mu}\partial_2^\nu - g^{\alpha\nu}\partial_2^\mu)\delta^{(4(n-1))}, \tag{3.7.33}$$

$$t_{u\tilde u}^\nu \to t_{u\tilde u}^\nu, \tag{3.7.34}$$

$$t_{uF}^{\alpha l\mu\nu} \to t_{uF}^{\alpha l\mu\nu}, \quad l\ge 3, \tag{3.7.35}$$

$$t_{uA}^{\alpha 2\nu} \to \frac{1}{2}[t_{uA}^{\alpha 2\nu} - t_{uA}^{\nu 2\alpha}], \tag{3.7.36}$$

the anomalies (3.7.25-26) disappear. We recall that the distributions with one Q-vertex have for $\omega \ge 0$ their own freedom of normalization which is, as far as we do not care about gauge invariance, independent of the normalization of the distributions of the main theory. This gives us the legitimation for the renormalizations (3.7.32), (3.7.36). The latter is necessary to fulfill the Cg-identity (3.7.12). This antisymmetrization is in fact a renormalization, i.e. a change by a local term only. The reason is that D_n is gauge invariant by the induction hypothesis and, therefore, especially $d_{uA}^{\alpha 2\nu}$ is antisymmetric in $\alpha\nu$. This antisymmetry can be destroyed in the distribution splitting by local terms only.

Obviously, the renormalizations (3.7.31-36) preserve covariance, the permutation symmetry in the inner vertices, the symmetry

$$t_{AA}^{\mu\nu}(x_1, x_2, x_3, ...) = t_{AA}^{\nu\mu}(x_2, x_1, x_3, ...), \tag{3.7.37}$$

and the antisymmetry in $\mu\nu$ of $t_{AF}^{\alpha\mu\nu}, t_{uF}^{\alpha l\mu\nu}$. Note that (3.7.31-36) are not the only renormalizations which lead to symmetrical, covariant and gauge invariant $T_n, T_{n/l}$. The normalization of the t-distributions is not uniquely determined by gauge invariance. This finishes the proof of the Cg-identities with two external legs.

For comparison with the Slavnov-Taylor identities in the literature (see the bibliographical notes) we would like to eliminate the distributions with

Q-vertex from the Cg-identities. For this purpose we apply the derivative ∂^2_μ to (3.7.16) and eliminate $t^{\alpha 2\nu}_{uA}$ in (3.7.15). In momentum space we then find the identity

$$p_{1\alpha}\hat{t}^{\alpha\nu}_{AA} + 2ip_{1\alpha}p_{2\mu}\hat{t}^{\alpha\mu\nu}_{AF} + i[p_{2\mu}p_2^\nu\hat{t}^\mu_{u\tilde{u}} - p_2^2\hat{t}^\nu_{u\tilde{u}}] - \sum_{l=3}^n [ip_{l\alpha}\hat{t}^{\alpha l\nu}_{uA} - 2p_{l\alpha}p_{2\mu}\hat{t}^{\alpha l\mu\nu}_{uF}] = 0,$$
(3.7.38)

where in the sum (for $l = n$), p_n must be $p_n \overset{\text{def}}{=} -(p_1 + ... + p_{n-1})$. By translation invariance the arguments of all distributions are $(p_1, p_2, \ldots p_{n-1})$. This Cg-identity contains only distributions of the main theory and divergences with respect to inner vertices. The latter vanish if one considers the infrared limit

$$p_3 \to 0,\ p_4 \to 0,\ \ldots p_n \to 0.$$
(3.7.39)

But this limit only exists if p_1 is off-shell. The contribution of $\hat{t}^\mu_{u\tilde{u}}$ drops out because

$$t^\mu_{u\tilde{u}}(p, -p, 0, \ldots) = p^\mu t_{u\tilde{u}}(p^2).$$

Then we arrive at the following reduced CG-identity

$$p_\alpha \tilde{t}^{\alpha\nu}_{AA}(p, -p, 0, \ldots) = 0.$$
(3.7.40)

It implies the following Slavnov-Taylor identity for the gluon propagator (P.Pascual, R.Tarrach, *QCD: renormalization for the practitioner*, Springer-Verlag 1984)

$$p_\mu D^{\mu\nu}_{ab}(p) = -(2\pi)^{-2}\delta_{ab}\frac{p^\nu}{p^2}.$$

¿From the Cg-identities with more than two legs only those with 3,4 and one with five legs can be violated in the splitting process and need a proof. They can be proved by the same method as above. But the 3- and 4-legged identities require considerably more work, therefore we refer to the literature (M.Dütsch, T.Hurth, G.Scharf, *Nuov.Cim. 108 A, 737 (1995)*). In the existing proof an additional weak infrared assumption is required, but it is not clear whether this is really essential.

3.8 Unitarity

Until now it was not necessary to specify a representation of the various field operators. In this section we have to introduce such a representation as we have already done in Sect.1.3 and 1.4. We avoid any use of indefinite metric and realize all operators on a positive definite Fock-Hilbert space $\mathcal{F}$ as described in chap.1. To be consistent with Lorentz covariance we must then pay a price. The zeroth component of the gauge potentials must be skew-hermitian, in contrast to the hermitian spatial components (see (1.3.12)):

$$A^0(x) = (2\pi)^{-3/2} \int \frac{d^3k}{\sqrt{2\omega}} \left(a^0(\boldsymbol{k}) e^{-ikx} - a^0(\boldsymbol{k})^+ e^{ikx} \right) \tag{3.8.1}$$

$$A^j(x) = (2\pi)^{-3/2} \int \frac{d^3k}{\sqrt{2\omega}} \left(a^j(\boldsymbol{k}) e^{-ikx} + a^j(\boldsymbol{k})^+ e^{ikx} \right), \tag{3.8.2}$$

where $\omega = |\boldsymbol{k}| = k^0$ and the $a(\boldsymbol{k})$, $a(\boldsymbol{k})^+$ are the absorption and emission operators satisfying the usual bosonic commutation relations. We have omitted the colour indices which play no role in this section. A similar asymmetry occurs in the ghost sector (see (1.2.1-2))

$$\tilde{u}(x) = (2\pi)^{-3/2} \int \frac{d^3p}{\sqrt{2E}} \left(-c_1(\boldsymbol{p}) e^{-ipx} + c_2(\boldsymbol{p})^+ e^{ipx} \right) \tag{3.8.3}$$

$$u(x) = (2\pi)^{-3/2} \int \frac{d^3p}{\sqrt{2E}} \left(c_2(\boldsymbol{p}) e^{-ipx} + c_1(\boldsymbol{p})^+ e^{ipx} \right). \tag{3.8.4}$$

Here $E = |\boldsymbol{p}| = p^0$ and the creation and annihilation operators satisfy the anticommutation relations

$$\{c_i(\boldsymbol{p}), c_j(\boldsymbol{q})^+\} = \delta_{ij} \delta(\boldsymbol{p} - \boldsymbol{q}) \tag{3.8.5}$$

and all other anticommutators vanish.

Besides the adjoint we have introduced another conjugation K, so that the vector fields are self-conjugated

$$A^\mu(x) = A^\mu(x)^K. \tag{3.8.6}$$

It can be realized on the boson sector as follows (1.5.20)

$$A(x)^K = \eta A^+(x) \eta, \quad \eta = (-)^{N_0} \tag{3.8.7}$$

where N_0 is the particle number operator for scalar ($\mu = 0$) bosons. In fact, since the field operator $A^0(x)$ changes the number of scalar bosons by 1,

$$A^0(x)^+ = -A^0(x) \quad \text{implies} \quad A^0(x)^K = A^0(x)$$

and

$$A^\mu(x) = (2\pi)^{-3/2} \int \frac{d^3k}{\sqrt{2\omega}} \left(a^\mu(\boldsymbol{k}) e^{-ikx} + a^\mu(\boldsymbol{k})^K e^{ikx} \right). \tag{3.8.8}$$

We extend this conjugation to the ghost sector, in such a way that

$$u(x)^K = u(x), \quad \tilde{u}(x)^K = -\tilde{u}(x). \tag{3.8.9}$$

In view of (3.8.3) and (3.8.4) this means

$$c_2(\boldsymbol{p})^K = c_1(\boldsymbol{p})^+, \quad c_1(\boldsymbol{p})^K = c_2(\boldsymbol{p})^+, \tag{3.8.10}$$

and the ghost fields assume the symmetric form

$$\tilde{u}(x) = (2\pi)^{-3/2} \int \frac{d^3p}{\sqrt{2E}} \left(-c_1(\boldsymbol{p})e^{-ipx} + c_1(\boldsymbol{p})^K e^{ipx} \right) \tag{3.8.11}$$

$$u(x) = (2\pi)^{-3/2} \int \frac{d^3p}{\sqrt{2E}} \left(c_2(\boldsymbol{p})e^{-ipx} + c_2(\boldsymbol{p})^K e^{ipx} \right). \tag{3.8.12}$$

Then the 1-point distribution (3.8.3) is skew-conjugate

$$T_1(x)^K = -T_1(x) = \tilde{T}_1(x),$$

and this holds for all n-point distributions $T_n(X)$ by induction:

$$T_n(X)^K = \tilde{T}_n(X), \tag{3.8.13}$$

if the normalization constants in the distribution splitting are chosen appropriately. Here $\tilde{T}_n$ are the n-point distributions of the inverse S-matrix (2.2.5).

We call the property (3.8.13) perturbative pseudo-unitarity. We want to prove a similar property with the ordinary adjoint $+$ instead of K on the physical subspace. As discussed in Sect.1.4 the latter is defined by means of the gauge charge Q (1.4.21):

$$\mathcal{F}_{\text{phys}} = \text{Ker } K, \tag{3.8.14}$$

where

$$K = \{Q^+, Q\}. \tag{3.8.15}$$

We also recall the direct decomposition (1.4.25)

$$\mathcal{F} = \overline{\text{Ran } Q^+} \oplus \overline{\text{Ran } Q} \oplus \mathcal{F}_{\text{phys}}. \tag{3.8.16}$$

Let P_{Q^+}, P_Q and P be the projections on the three subspaces herein, P is the projection onto the physical subspace, then we can write (3.8.16) as follows:

$$\mathbb{1} = P_{Q^+} + P_Q + P. \tag{3.8.17}$$

K (3.8.15) is a positive self-adjoint operator on the orthogonal complement $\mathcal{F}_{\text{phys}}^\perp$ of $\mathcal{F}_{\text{phys}}$, this follows from the explicit representation in momentum space (1.4.18). Then K has an inverse on $\mathcal{F}_{\text{phys}}^\perp$

$$KK^{-1} = P_{Q^+} + P_Q = K^{-1}K, \tag{3.8.18}$$

$$K^{-1}P = 0. \tag{3.8.19}$$

This allows to rewrite (3.8.17) in the following form

$$\mathbb{1} = P + KK^{-1} = P + Q^+QK^{-1} + QQ^+K^{-1}. \tag{3.8.20}$$

Now it is easy to show that physical unitarity is a consequence of gauge invariance. Let us consider a product of T's between two projections P on $\mathcal{F}_{\text{phys}}$:

130   Spin-1 gauge theories: massless gauge fields

$$PT(X_1)T(X_2)P = PT(X_1)\Big(P + Q^+QK^{-1} + QQ^+K^{-1}\Big)T(X_2)P$$

$$= PT(X_1)PT(X_2)P + PT(X_1)Q^+QK^{-1}T(X_2)P+$$
$$+PT(X_1)QQ^+K^{-1}T(X_2)P, \qquad (3.8.21)$$

where (3.8.20) has been inserted. In the last term we can write

$$PT(X_1)Q = P[T(X_1), Q],$$

because $PQ = 0$. Hence, by gauge invariance this term is a divergence. Since

$$[K, Q] = Q^+QQ + QQ^+Q - QQ^+A - QQQ^+ = 0,$$

K^{-1} also commutes with Q:

$$[K^{-1}, Q] = 0. \qquad (3.8.22)$$

Then the second term in (3.8.21) is a divergence, too, because

$$QK^{-1}T(X_2)P = K^{-1}QT(X_2)P = K^{-1}[Q, T(X_2)]P = \text{div}, \qquad (3.8.23)$$

by gauge invariance. Therefore, we conclude from (3.8.21) that

$$PT(X_1)T(X_2)P = PT(X_1)PT(X_2)P + \text{div}. \qquad (3.8.24)$$

For physical unitarity we have to consider the inverse of the physical S-matrix

$$(PS(g)P)^{-1} = \sum_n \frac{1}{n!} \int d^4x_1 \ldots d^4x_n \, \tilde{T}_n^P(x_1, \ldots, x_n) g(x_1) \ldots g(x_n), \qquad (3.8.25)$$

where the n-point distributions are equal to the following sum over subsets of $X = \{x_1, \ldots, x_n\}$

$$\tilde{T}_n^P(X) = \sum_{r=1}^n (-)^r \sum_{P_r} PT_{n_1}(X_1)P \ldots PT_{n_r}(X_r)P. \qquad (3.8.26)$$

As in (2.1.24) this is the inverse in the sense of formal power series. Using (3.8.24), all internal projection operators can be removed

$$\tilde{T}_n^P(X) = \sum_{r=1}^n (-)^r \sum_{P_r} PT_{n_1}(X_1) \ldots T_{n_r}(X_r)P + \text{div}$$

$$= P\tilde{T}_n(X)P + \text{div}. \qquad (3.8.27)$$

By means of pseudo-unitarity (3.8.13) we finally arrive at the desired perturbative unitarity

$$\tilde{T}_n^P(X) = PT_n(X)^K P + \text{div} = PT_n(X)^+ P + \text{div}. \qquad (3.8.28)$$

In the last step we have used the fact that the conjugation K agrees with the adjoint on $\mathcal{F}_{\text{phys}}$ (1.3.25).

The divergence in (3.8.28) has no physical consequence if the adiabatic limit exists. Roughly speaking we have $SS^+ = \mathbb{1}$ on $\mathcal{F}_{\text{phys}}$. This is the basis of the statistical interpretation of the S-matrix. The above proof is completely general and applies to any gauge theory. It shows that the very basis of unitarity is gauge invariance. Even if the adiabatic limit does not exist in S-matrix elements, it is quite plausible that the divergence in (3.8.28) does not contain any physics. However, it is not known how the real physical scattering states look like in massless gauge theories, so that the physical content of (3.8.28) is not clear in this case.

3.9 Other gauges

Classical gauge independence means that the gauge can be changed without changing the physics. In quantum theory a change of gauge has drastic consequences: the free Fock representation must be altered, the S-matrix is defined in a new Fock space. But at the end the observable quantities like scattering cross-sections are the same. We want to demonstrate this for a special class of gauges, the so-called λ-gauges.

We shall use asymptotic gauge fields satisfying the modified wave equation

$$\Box A_\mu^{(\lambda)} = (1-\lambda)\partial_\mu \partial^\nu A_\nu^{(\lambda)}. \tag{3.9.1}$$

Here λ is a real gauge parameter, $\lambda = 1$ corresponds to the Feynman gauge. We have omitted colour indices etc. which are unimportant in this section, only the Lorentz structure matters here. The upper index (λ) indicates that the field corresponds to the gauge parameter λ, and we are going to consider the fields with different λ simultaneously.

First we want to solve the Cauchy problem for (3.9.1) with Cauchy data specified at time $t = 0$ in the whole $\mathbb{R}^3$. For this reason we isolate the highest time derivatives in (3.9.1)

$$\lambda \partial_0^2 A_0^{(\lambda)} = \triangle A_0^{(\lambda)} + (1-\lambda)\partial_0 \partial^j A_j^{(\lambda)} \tag{3.9.2}$$

$$\partial_0^2 A_j^{(\lambda)} = \triangle A_j^{(\lambda)} + (1-\lambda)\partial_j(\partial^0 A_0^{(\lambda)} + \partial^l A_l^{(\lambda)}). \tag{3.9.3}$$

Consequently, in agreement with the ordinary wave equation, the Cauchy data are given by $A_\mu^{(\lambda)}(0,\boldsymbol{x})$ and $(\partial_0 A_\mu^{(\lambda)})(0,\boldsymbol{x})$. Taking the divergence ∂^μ of (3.9.1) we get for $\lambda \neq 0$

$$\Box \partial^\mu A_\mu^{(\lambda)} = 0, \tag{3.9.4}$$

so that $A_\mu^{(\lambda)}$ satisfies the iterated wave equation

$$\Box^2 A_\mu^{(\lambda)} = 0. \tag{3.9.5}$$

The Cauchy problem for this equation is considered in Appendix A (Sect.3.11) at the end of this chapter. The solution can be written in terms of the Lorentz invariant distributions $D(x)$ and $E(x)$ which are defined in Appendix A:

$$A_\mu^{(\lambda)}(x) = \int_{y_0=0} d^3y\, D(x-y) \overset{\leftrightarrow}{\partial}_0^y A_\mu^{(\lambda)}(y) + \int_{y_0=0} d^3y\, E(x-y) \overset{\leftrightarrow}{\partial}_0^y \Box A_\mu^{(\lambda)}(y). \tag{3.9.6}$$

Here, all second and third order time derivatives under the last integral must be expressed by spatial derivatives of the Cauchy data by means of (3.9.2) and (3.9.3).

It is very important to notice that the decomposition (3.9.6) is Lorentz covariant. Indeed, instead of selecting the plane $y_0 = 0$ we may consider a smooth space-like surface σ with a surface measure $d\sigma_\nu(y)$. Then, with help of Gauss' theorem, the integrals in (3.9.6) can be written in invariant form

$$\int_{y_0=0} d^3y \ldots \overset{\leftrightarrow}{\partial}_0 \to \int_\sigma d\sigma_\nu(y) \ldots \overset{\leftrightarrow}{\partial}^\nu,$$

showing that each term on the r.h.s. of (3.9.6) is a Lorentz four-vector.

Let us denote the first term in (3.9.6) which satisfies the ordinary wave equation by $A_\mu^w(x)$ (w for wave). The second term denoted by B_μ is equal to

$$B_\mu(x) = (1-\lambda) \int d^3y\, E(x-y) \overset{\leftrightarrow}{\partial}_0 \partial_\mu \partial A^{(\lambda)}(y), \tag{3.9.7}$$

where (3.9.1) has been inserted. For $\mu = j = 1, 2, 3$ the derivative can be taken out by partial integration, so that

$$B_j(x) = \partial_j \chi(x) \tag{3.9.8}$$

is a spatial gradient with

$$\chi(x) = (1-\lambda) \int_{y_0=0} d^3y\, E(x-y) \overset{\leftrightarrow}{\partial}_0 \partial A^{(\lambda)}(y). \tag{3.9.9}$$

But this is impossible for the zeroth component

$$B_0(x) = \partial_0 \chi(x) + B(x). \tag{3.9.10}$$

The difference $B(x)$ can be transformed as follows

$$B(x) = (1-\lambda) \int d^3y\, [E(x-y) \overset{\leftrightarrow}{\partial}_0 \partial_0^y \partial A^{(\lambda)}(y) - \partial_0^x E(x-y) \overset{\leftrightarrow}{\partial}_0 \partial A^{(\lambda)}(y)]$$

$$= (1-\lambda) \int d^3y\, \partial_0^y [E(x-y) \overset{\leftrightarrow}{\partial}_0 \partial A^{(\lambda)}(y)]$$

$$= (1-\lambda) \int d^3y [\partial_0^y E \partial_0 \partial A^{(\lambda)} + E \partial_0^2 \partial A^{(\lambda)} - \partial_{0y}^2 E \partial A^{(\lambda)} - \partial_0^y E \partial_0 \partial A^{(\lambda)}]$$

$$= -(1-\lambda) \int d^3y \, \Box E(x-y) \partial A^{(\lambda)}(y) = -(1-\lambda) \int d^3y \, D(x-y) \partial^\nu A_\nu^{(\lambda)}(y). \tag{3.9.11}$$

This shows that the field $B(x)$ also fulfills the wave equation $\Box B = 0$. Therefore, it is tempting to combine it with $A_0^w(x)$. The resulting four-component field

$$A_\mu^L = (A_0^w + B, A_1^w, A_2^w, A_3^w) \tag{3.9.12}$$

satisfies the wave equation and we have the simple decomposition

$$A_\mu^{(\lambda)} = A_\mu^L + \partial_\mu \chi. \tag{3.9.13}$$

However, this decomposition has the serious defect of not being covariant (see (3.9.12)). Therefore, we must make a sharp distinction between the field A_μ^L and the *covariant* field A_μ^F in the Feynman gauge $\lambda = 1$, although both fields satisfy the wave equation and the same commutation relation, as we shall see.

Next we want to quantise the $A^{(\lambda)}$-field. It follows from (3.9.6) that the commutation relations for arbitrary times must involve the distributions D and E. Then, Poincaré covariance and the singular order $\omega = -2$ of the resulting distribution suggest the following form

$$[A_\mu^{(\lambda)}(x), A_\nu^{(\lambda)}(y)] = ig_{\mu\nu} D(x-y) + i\alpha \partial_\mu \partial_\nu E(x-y), \tag{3.9.14}$$

where a common factor $(h/2\pi)$ has been set $=1$. When operating with $\Box g^{\kappa\mu} - (1-\lambda)\partial^\kappa \partial^\mu$ on the variable x, we must get zero. This determines the parameter α. Using $\Box E(x) = D(x)$ from App.A (A.6), we find

$$[A_\mu^{(\lambda)}(x), A_\nu^{(\lambda)}(y)] = ig_{\mu\nu} D(x-y) + i\frac{1-\lambda}{\lambda} \partial_\mu \partial_\nu E(x-y). \tag{3.9.15}$$

The corresponding commutation relations for the positive and negative frequency parts read

$$[A_\mu^{(\lambda)(-)}(x), A_\nu^{(\lambda)(+)}(y)] = ig_{\mu\nu} D^{(+)}(x-y) + i\frac{1-\lambda}{\lambda} (\partial_\mu \partial_\nu E)^{(+)}(x-y). \tag{3.9.16}$$

Note that the positive frequency part of the derivative $(\partial_\nu E)$ is well defined, in contrast to $E^{(+)}$ (see App.A).

¿From the initial values of the D- and E-distributions

$$D(0,\boldsymbol{x}) = 0, \quad (\partial_0 D)(0,\boldsymbol{x}) = \delta^3(\boldsymbol{x}) \tag{3.9.17}$$

$$(\partial_0^n E)(0,\boldsymbol{x}) = 0, \quad n=0,1,2, \quad (\partial_0^3 E)(0,\boldsymbol{x}) = \delta^3(\boldsymbol{x}) \tag{3.9.18}$$

we obtain the equal-time commutation relations

$$[\partial_0 A_\mu^{(\lambda)}(x), A_\nu^{(\lambda)}(y)]_0 = ig_{\mu\nu}\left(1 + g_{\mu 0}\frac{1-\lambda}{\lambda}\right)\delta(\boldsymbol{x}-\boldsymbol{y}) \tag{3.9.19}$$

$$[\partial_0 A_0^{(\lambda)}(x), \partial_0 A_j^{(\lambda)}(y)]_0 = i\frac{\lambda - 1}{\lambda}\partial_j \delta(\boldsymbol{x} - \boldsymbol{y}), \qquad (3.9.20)$$

where the subscript 0 means $x_0 = y_0$. All other commutators are zero. It follows from (3.9.19) and (3.9.20) that 3-dimensional smearing with a space dependent test function $f(\boldsymbol{x})$ is sufficient to get a well-defined operator in Fock space.

¿From the fundamental commutation relations (3.9.15) the commutators of all other fields can be calculated because they are all expressed by $A_\mu^{(\lambda)}$. We find

$$[\chi(x), \chi(y)] = 0 \qquad (3.9.21)$$

$$[A_j^w(x), A_k^w(y)] = ig_{jk}D(x-y) \qquad (3.9.22)$$

$$[A_0^w(x), A_0^w(y)] = \frac{i}{\lambda}D(x-y) \qquad (3.9.23)$$

$$[A_\mu^L(x), A_\nu^L(y)] = ig_{\mu\nu}D(x-y). \qquad (3.9.24)$$

Now, $A_j^w(x), j = 1, 2, 3$ are the spatial components of a covariant vector field satisfying the wave equation. The commutation relations (3.9.22) are the same as for the Feynman field $A_j^F(x)$. Nevertheless, we cannot identify the two as we shall see now by constructing a concrete representation of the field operators.

Most authors who consider the λ-gauges leave the construction of a concrete representation to the reader. We try to be more polite to our readers. Since three-dimensional smearing is enough to render $A_\mu^{(\lambda)}(x)$ well defined, we will construct all fields as three-dimensional Fourier integrals, leaving aside manifest Lorentz covariance. Our strategy will be to start with a representation of the time-zero fields which satisfies the equal-time commutation relations (3.9.19) (3.9.20) and then calculating the time evolution by the previous formulas (e.g. (3.9.6)). We follow our previous procedure of assuming a Fock space with *positive definite metric* and changing the form of the zeroth component $A_0^{(\lambda)}$ instead. This is very natural in the λ-gauge because the zeroth component plays a special role here, anyway.

We use the usual emission and absorption operators for all four components satisfying

$$[a_\nu^{(\lambda)}(\boldsymbol{p}), a_\mu^{(\lambda)+}(\boldsymbol{q})] = \delta_{\nu\mu}\delta(\boldsymbol{p} - \boldsymbol{q}). \qquad (3.9.25)$$

The adjoint is defined with respect to the positive definite scalar product so that these operators can be represented in the usual way in a Fock space $\mathcal{F}^{(\lambda)}$, if smeared out with test functions $f(\boldsymbol{p}) \in L^2(\mathbf{R}^3)$. In addition we will use the operators for the longitudinal mode

$$a_\parallel^{(\lambda)}(\boldsymbol{p}) = \frac{p^j}{\omega}a_j^{(\lambda)}(\boldsymbol{p}) = -\frac{p_j}{\omega}a_j^{(\lambda)}(\boldsymbol{p}), \qquad (3.9.26)$$

where always $\omega = |\boldsymbol{p}| = p^0$. Introducing the linear combinations

$$b_1^{(\lambda)}(\boldsymbol{p}) = \frac{1}{\sqrt{2}}(a_\parallel^{(\lambda)}(\boldsymbol{p}) + a_0^{(\lambda)}(\boldsymbol{p}))$$

$$b_2^{(\lambda)}(\boldsymbol{p}) = \frac{1}{\sqrt{2}}(a_\parallel^{(\lambda)}(\boldsymbol{p}) - a_0^{(\lambda)}(\boldsymbol{p})), \qquad (3.9.27)$$

we have the following commutators

$$[b_1^{(\lambda)}(\boldsymbol{p}), b_2^{(\lambda)+}(\boldsymbol{q})] = 0$$

$$[a_\nu^{(\lambda)}(\boldsymbol{p}), b_2^{(\lambda)+}(\boldsymbol{q})] = -\frac{1}{\sqrt{2}}\frac{p_\nu}{\omega}\delta(\boldsymbol{p}-\boldsymbol{q})$$

$$[a_\nu^{(\lambda)+}(\boldsymbol{p}), b_1^{(\lambda)}(\boldsymbol{q})] = -\frac{1}{\sqrt{2}}\frac{p^\nu}{\omega}\delta(\boldsymbol{p}-\boldsymbol{q}). \qquad (3.9.28)$$

In addition to the adjoint we have to introduce a second conjugation K which appears in all Lorentz covariant expressions and defines the Krein structure as in Sect.1.5. The conjugation is defined by

$$a_0^{(\lambda)}(\boldsymbol{p})^K = -a_0^{(\lambda)}(\boldsymbol{p})^+, \quad a_j^{(\lambda)}(\boldsymbol{p})^K = a_j^{(\lambda)}(\boldsymbol{p})^+, \quad j=1,2,3. \qquad (3.9.29)$$

Note that $a_\mu^{(\lambda)}, a_\mu^{(\lambda)+}$ are not treated as four-vectors, therefore, we write the indices always downstairs.

The gauge field $A_\mu^{(\lambda)}(x)$ must be self-conjugated $A_\mu^{(\lambda)K} = A_\mu^{(\lambda)}$, in order to get a pseudo-unitary S-matrix. Then, a little experimentation shows that the time-zero fields must be of the following form:

$$A_\mu^{(\lambda)}(0,\boldsymbol{x}) = (2\pi)^{-3/2}\int \frac{d^3p}{\sqrt{2\omega}}\left\{a_\mu^{(\lambda)}(\boldsymbol{p})e^{i\boldsymbol{p}\boldsymbol{x}} + a_\mu^{(\lambda)K}(\boldsymbol{p})e^{-i\boldsymbol{p}\boldsymbol{x}}\right.$$

$$-\frac{1-\lambda}{2\sqrt{2}\lambda}\left[\frac{p_\mu}{\omega}b_1^{(\lambda)}(\boldsymbol{p})e^{i\boldsymbol{p}\boldsymbol{x}} + \frac{p_\mu}{\omega}b_2^{(\lambda)+}(\boldsymbol{p})e^{-i\boldsymbol{p}\boldsymbol{x}}\right.$$

$$\left.\left.-2g_{\mu 0}b_1^{(\lambda)}(\boldsymbol{p})e^{i\boldsymbol{p}\boldsymbol{x}} - 2g_{\mu 0}b_2^{(\lambda)+}(\boldsymbol{p})e^{-i\boldsymbol{p}\boldsymbol{x}}\right]\right\}, \qquad (3.9.30)$$

$$(\partial_0 A_\mu^{(\lambda)})(0,\boldsymbol{x}) = -i(2\pi)^{-3/2}\int \frac{d^3p}{\sqrt{2\omega}}\left\{\omega a_\mu^{(\lambda)}(\boldsymbol{p})e^{i\boldsymbol{p}\boldsymbol{x}} - \omega a_\mu^{(\lambda)K}(\boldsymbol{p})e^{-i\boldsymbol{p}\boldsymbol{x}}\right.$$

$$-\frac{1-\lambda}{2\sqrt{2}\lambda}\left[-p_\mu b_1^{(\lambda)}(\boldsymbol{p})e^{i\boldsymbol{p}\boldsymbol{x}} + p_\mu b_2^{(\lambda)+}(\boldsymbol{p})e^{-i\boldsymbol{p}\boldsymbol{x}}\right.$$

$$\left.\left.-2g_{\mu 0}\omega b_1^{(\lambda)}(\boldsymbol{p})e^{i\boldsymbol{p}\boldsymbol{x}} + 2g_{\mu 0}\omega b_2^{(\lambda)+}(\boldsymbol{p})e^{-i\boldsymbol{p}\boldsymbol{x}}\right]\right\}. \qquad (3.9.31)$$

It is straight-forward to verify the commutation relations (3.9.15), (3.9.19) and (3.9.20).

The fields for arbitrary times can now be found from (3.9.6). For this purpose we need the following three-dimensional Fourier transforms (for $y_0 = 0$)

$$\int d^3y\, D(x-y)e^{ipy} = -\frac{i}{2\omega}\left(e^{i\omega x^0} - e^{-i\omega x^0}\right)e^{ipx} \qquad (3.9.32)$$

$$\int d^3y\, E(x-y)e^{ipy} = \frac{-1}{4\omega^2}\left[e^{i\omega x^0}\left(x^0 + \frac{i}{\omega}\right) + e^{-i\omega x^0}\left(x^0 - \frac{i}{\omega}\right)\right]e^{ipx}. \qquad (3.9.33)$$

We first compute

$$(\partial^\mu A^{(\lambda)}_\mu)(0,\boldsymbol{x}) = \frac{-i}{\lambda}(2\pi)^{-3/2}\int d^3p\,\sqrt{\omega}\left(b_1^{(\lambda)}(\boldsymbol{p})e^{ipx} - b_2^{(\lambda)+}(\boldsymbol{p})e^{-ipx}\right), \qquad (3.9.34)$$

$$(\partial_0\partial^\mu A^{(\lambda)}_\mu)(0,\boldsymbol{x}) = \frac{-\sqrt{2}}{\lambda}(2\pi)^{-3/2}\int \frac{d^3p}{\sqrt{2\omega}}\,\omega^2\left(b_1^{(\lambda)}(\boldsymbol{p})e^{ipx} + b_2^{(\lambda)+}(\boldsymbol{p})e^{-ipx}\right). \qquad (3.9.35)$$

Then we obtain from the first term in (3.9.6)

$$A_j^w(x) = (2\pi)^{-3/2}\int\frac{d^3p}{\sqrt{2\omega}}\left(a_j^{(\lambda)}(\boldsymbol{p})e^{-ipx} + a_j^{(\lambda)+}(\boldsymbol{p})e^{ipx}\right) + \partial_j f(x), \qquad (3.9.36)$$

where

$$f(x) = -i\frac{1-\lambda}{4\lambda(2\pi)^{3/2}}\int\frac{d^3p}{\omega^{3/2}}\left[b_1^{(\lambda)}(-\boldsymbol{p})e^{ipx} - b_2^{(\lambda)+}(-\boldsymbol{p})e^{-ipx}\right].$$

The zeroth component behaves differently

$$A_0^w(x) = (2\pi)^{-3/2}\int\frac{d^3p}{\sqrt{2\omega}}\left[a_0^{(\lambda)}(\boldsymbol{p})e^{-ipx} - a_0^{(\lambda)+}(\boldsymbol{p})e^{ipx}\right.$$

$$\left. + \frac{1-\lambda}{\sqrt{2}\lambda}\left(b_1^{(\lambda)}(\boldsymbol{p})e^{-ipx} + b_2^{(\lambda)+}(\boldsymbol{p})e^{ipx}\right)\right]$$

$$-\frac{1-\lambda}{2\sqrt{2}\lambda}(2\pi)^{-3/2}\int\frac{d^3p}{\sqrt{2\omega}}\left(b_1^{(\lambda)}(-\boldsymbol{p})e^{ipx} + b_2^{(\lambda)+}(-\boldsymbol{p})e^{-ipx}\right). \qquad (3.9.37)$$

Next we calculate $\chi(x)$ from (3.9.9)

$$\chi(x) = \frac{1-\lambda}{\sqrt{2}\lambda}(2\pi)^{-3/2}\int\frac{d^3p}{\sqrt{2\omega}}\left[b_1^{(\lambda)}(\boldsymbol{p})e^{-ipx}\left(x^0 - \frac{i}{2\omega}\right)\right.$$

$$\left. + b_2^{(\lambda)+}(\boldsymbol{p})e^{ipx}\left(x^0 + \frac{i}{2\omega}\right)\right] - f(x) \qquad (3.9.38)$$

and $B(x)$ from (3.9.11)

$$B(x) = \frac{\lambda-1}{\sqrt{2}\lambda}(2\pi)^{-3/2}\int\frac{d^3p}{\sqrt{2\omega}}\left(b_1^{(\lambda)}(\boldsymbol{p})e^{-ipx} + b_2^{(\lambda)+}(\boldsymbol{p})e^{ipx}\right)$$

$$-\frac{\lambda-1}{\sqrt{2}\lambda}(2\pi)^{-3/2}\int\frac{d^3p}{\sqrt{2\omega}}\left(b_1^{(\lambda)}(-\boldsymbol{p})e^{ipx} + b_2^{(\lambda)+}(-\boldsymbol{p})e^{-ipx}\right). \qquad (3.9.39)$$

This cancels against the second line in (3.9.37) so that

$$A_0^{(\lambda)}(x) = (2\pi)^{-3/2} \int \frac{d^3p}{\sqrt{2\omega}} \left[a_0^{(\lambda)}(p) e^{-ipx} - a_0^{(\lambda)+}(p) e^{ipx} \right] + \partial_0(\chi + f). \tag{3.9.40}$$

The first integral in (3.9.40) and (3.9.36) formally agrees with the Feynman field A_μ^F, but the latter is defined by means of different annihilation and creation operators $a_\mu^{(1)}(p), a_\mu^{(1)+}(p)$

$$A_\mu^F(x) = (2\pi)^{-3/2} \int \frac{d^3p}{\sqrt{2\omega}} \left[a_\mu^{(1)}(p) e^{-ipx} + a_\mu^{(1)K}(p) e^{ipx} \right]. \tag{3.9.41}$$

In $A_\mu^{(\lambda)}$ the terms with wrong frequencies $\sim b_1^{(\lambda)}(-p)$ etc. cancel out. Then the resulting decomposition $A_\mu^{(\lambda)} = \tilde{A}_\mu^L + \partial_\mu \tilde{\chi}$ is identical with the one introduced by B.Lautrup (*Mat. Fys. Medd. Dan. Vid. Selsk. 35 (1967)*).

Until now every field $A^{(\lambda)}$ operates in its own Fock space $\mathcal{F}^{(\lambda)}$. But there must exist a λ-independent intersection of these $\mathcal{F}^{(\lambda)}$ where the gauge independent objects live. Indeed, in the foregoing equations the λ-dependence is only through the unphysical scalar and longitudinal modes b_1, b_2 (3.9.27), all equations involving only transverse modes which can be written down contain no λ. Therefore, we can safely identify the transverse emission and absorption operators for different λ. Let $\varepsilon^\mu = (0, \varepsilon)$ and $\eta^\mu = (0, \eta)$ be two transverse polarization vectors

$$p \cdot \varepsilon(p) = 0 = p \cdot \eta(p), \quad \varepsilon^2 = 1 = \eta^2, \quad \varepsilon \cdot \eta = 0. \tag{3.9.42}$$

Then we put

$$\varepsilon^\mu a_\mu^{(\lambda)}(p) = a_\varepsilon(p), \quad \eta^\mu a_\mu^{(\lambda)}(p) = a_\eta(p) \tag{3.9.43}$$

independent of λ. Choosing one unique vacuum Ω for all field operators

$$a_\varepsilon(p)\Omega = 0 = a_\eta(p)\Omega = b_1^{(\lambda)}(p)\Omega = b_2^{(\lambda)}(p)\Omega = 0$$

for all p (or rather after smearing with test functions $f(p)$), then the different Fock spaces $\mathcal{F}^{(\lambda)}$ hang together. Their common intersection is the physical subspace $\mathcal{H}_{\text{phys}}$ which is spanned by the transverse states $(a_\varepsilon^+)^m (a_\eta^+)^n \Omega$.

3.10 Gauge independence

The nilpotent gauge charge Q_λ for the λ-gauge is defined by

$$Q_\lambda = \lambda \int d^3x \partial^\mu A_\mu^{(\lambda)}(x) \overleftrightarrow{\partial}_0 u(x), \tag{3.10.1}$$

where the colour indices are always suppressed if the meaning is clear. The ghost fields $u, \tilde{u}$ are quantized as follows

$$\Box u = 0, \quad \Box \tilde{u} = 0$$

$$\{u_a(x), \tilde{u}_b(y)\} = -i\delta_{ab}D(x-y). \tag{3.10.2}$$

Since there is no λ-dependence here, they can be represented in the usual way

$$u(x) = (2\pi)^{-3/2} \int \frac{d^3p}{\sqrt{2\omega}} \left(c_2(\boldsymbol{p})e^{-ipx} + c_1^+(\boldsymbol{p})e^{ipx} \right) \tag{3.10.3}$$

$$\tilde{u}(x) = (2\pi)^{-3/2} \int \frac{d^3p}{\sqrt{2\omega}} \left(-c_1(\boldsymbol{p})e^{-ipx} + c_2^+(\boldsymbol{p})e^{ipx} \right) \tag{3.10.4}$$

where

$$\{c_i(\boldsymbol{p}), c_j^+(\boldsymbol{q})\} = \delta_{ij}\delta(\boldsymbol{p}-\boldsymbol{q}), \quad i,j=1,2.$$

The conjugation K is extended to the ghost sector by

$$c_2(\boldsymbol{p})^K = c_1(\boldsymbol{p})^+, \quad c_1(\boldsymbol{p})^K = c_2(\boldsymbol{p})^+ \tag{3.10.5}$$

so that $u^K = u$ is K-selfadjoint and $\tilde{u}^K = -\tilde{u}$. Then Q_λ (3.10.1), if densely defined, becomes K-symmetric $Q_\lambda \subset Q_\lambda^K$. It is not necessary for the following to give an explicit description of the domain. According to a general result (A.Galindo, *Comm. Pure Appl. Math.* 15 (1962) 423), it has a K-selfadjoint extension $Q_\lambda^K = Q_\lambda$ which is a closed operator and this is all we need for our purpose.

Using (3.9.34), (3.9.35) and (3.10.3) it is easy to calculate Q_λ in momentum space

$$Q_\lambda = \sqrt{2} \int d^3p\, \omega(\boldsymbol{p})[b_1(\boldsymbol{p})c_1^+(\boldsymbol{p}) + b_2^+(\boldsymbol{p})c_2(\boldsymbol{p})]. \tag{3.10.6}$$

For typographical simplicity we have not written the λ-dependence in b_1, b_2. Q_λ together with its adjoint

$$Q_\lambda^+ = \sqrt{2} \int d^3p\, \omega(\boldsymbol{p})[c_1(\boldsymbol{p})b_1^+(\boldsymbol{p}) + c_2^+(\boldsymbol{p})b_2(\boldsymbol{p})] \tag{3.10.7}$$

are unbounded closed operators; the unboundedness is not only due to the emission and absorption operators but also because of $\omega(\boldsymbol{p}) = |\boldsymbol{p}|$.

Since Q_λ, Q_λ^+ are closed operators, we have the following direct decompositions of the Fock space (see (1.4.24))

$$\mathcal{F}^{(\lambda)} = \overline{\operatorname{Ran} Q_\lambda} \oplus \operatorname{Ker} Q_\lambda^+ = \overline{\operatorname{Ran} Q_\lambda^+} \oplus \operatorname{Ker} Q_\lambda, \tag{3.10.8}$$

where Ran is the range and Ker the kernel of the operator. The overline denotes the closure; note that $\operatorname{Ran} Q_\lambda$ is not closed because 0 is in the essential spectrum of Q_λ. Now, $Q_\lambda^2 = 0$ implies $\operatorname{Ran} Q_\lambda \perp \operatorname{Ran} Q_\lambda^+$, therefore, it follows from (3.10.8) that

$$\mathcal{F}^{(\lambda)} = \overline{\operatorname{Ran} Q_\lambda} \oplus \overline{\operatorname{Ran} Q_\lambda^+} \oplus \left(\operatorname{Ker} Q_\lambda \cap \operatorname{Ker} Q_\lambda^+ \right). \tag{3.10.9}$$

The range of Q_λ and Q_λ^+ certainly consists of unphysical states because (3.10.6) and (3.10.7) only contains emission operators of unphysical particles (scalar and longitudinal "gluons" and ghosts). The physical states must therefore be contained in the last subspace in (3.10.9).

We know from (1.4.21) and (1.4.23) that

$$\text{Ker}\, Q_\lambda \cap \text{Ker}\, Q_\lambda^+ = \text{Ker}\, \{Q_\lambda, Q_\lambda^+\} \qquad (3.10.10)$$

where the curly bracket is the anticommutator. Calculating the anticommutator from (3.10.6) and (3.10.7) we find

$$\{Q_\lambda, Q_\lambda^+\} = 2\int d^3p\, \omega^2(\boldsymbol{p}) \Big[b_1^+(\boldsymbol{p})b_1(\boldsymbol{p}) + b_2^+(\boldsymbol{p})b_2(\boldsymbol{p}) +$$

$$+ c_1^+(\boldsymbol{p})c_1(\boldsymbol{p}) + c_2^+(\boldsymbol{p})c_2(\boldsymbol{p}) \Big]. \qquad (3.10.11)$$

Up to the (positive) factor ω^2 this is just the particle number operator of the unphysical particles. The physical subspace is characterized by the fact that there are no unphysical particles, hence,

$$\mathcal{H}_{\text{phys}} = \text{Ker}\, \{Q_\lambda, Q_\lambda^+\} \qquad (3.10.12)$$

and this is a closed subspace. As discussed above, it is the intersection of all $\mathcal{F}^{(\lambda)}$.

We introduce the projection operator P_λ on $\mathcal{H}_{\text{phys}}$. It is our goal to prove the gauge independence of the physical S-matrix $P_\lambda S^{(\lambda)}(g) P_\lambda$. The perturbative formulation in terms of time-ordered products $T_n^{(\lambda)}$ would be

$$P_\lambda T_n^{(\lambda)} P_\lambda = P_1 T_n^{(1)} P_1 + \text{div}. \qquad (3.10.13)$$

Here div denotes a sum of divergences. As discussed at the end of Sect.3.8 we may regard these divergences as physically irrelevant. In (3.10.13) we have compared the physical n-point functions in the λ-gauge with the Feynman gauge $\lambda = 1$.

Gauge independence (3.10.13) is a direct consequence of gauge invariance

$$[Q_\lambda, T_n^{(\lambda)}] = \text{div}. \qquad (3.10.14)$$

As in the case of the ordinary Feynman gauge (see (3.8.24)), gauge invariance implies the following important relation

$$PT(X_1)PT(X_2)P = PT(X_1)T(X_2)P + \text{div}. \qquad (3.10.15)$$

Here we have omitted indices n and subscripts λ to indicate that (3.10.15) holds for arbitrary λ and arbitrary n-point functions.

The proof of gauge independence is by induction on n. The beginning $n=1$ can be easily verified because $P_\lambda A_\mu^{(\lambda)} P_\lambda = P_1 A_\mu^{(1)} P_1$ and

$$T_1^{(\lambda)} = igf_{abc}\left(\tfrac{1}{2} A_{\mu a}^{(\lambda)} A_{\nu b}^{(\lambda)} F_c^{(\lambda)\nu\mu} - A_{\mu a}^{(\lambda)} u_b \partial^\mu \tilde{u}_c\right)$$

does not depend explicitly on λ. Let us now assume that

$$P_\lambda T_i^{(\lambda)} P_\lambda = P_1 T_i^{(1)} P_1 + \text{div} \qquad (3.10.16)$$

holds for all $i \leq n-1$. Then we consider arbitrary products

$$P_\lambda T^{(\lambda)}(X_1) T^{(\lambda)}(X_2) P_\lambda = P_\lambda T^{(\lambda)}(X_1) P_\lambda T^{(\lambda)}(X_2) P_\lambda + \text{div}_1, \qquad (3.10.17)$$

where we have used (3.10.15). Due to the induction assumption (3.10.16) this is equal to

$$= P_1 T^{(1)}(X_1) P_1 T^{(1)}(X_2) P_1 + \text{div}_2 = P_1 T^{(1)}(X_1) T^{(1)}(X_2) P_1 + \text{div}_3.$$

Here we have used (3.10.15) again. The causal D-distribution of order n in the Epstein-Glaser construction is a sum of such products (3.10.17), hence, it follows that

$$P_\lambda D_n^{(\lambda)} P_\lambda = P_1 D_n^{(1)} P_1 + \text{div}. \qquad (3.10.18)$$

All three terms in here have separately causal support, therefore they can individually be split into retarded and advanced parts. The local normalization terms can be chosen in such a way that

$$P_\lambda R_n^{(\lambda)} P_\lambda = P_1 R_n^{(1)} P_1 + \text{div}, \qquad (3.10.19)$$

where R denotes the retarded distributions. We must check that this way of normalization is not in conflict with the normalization which we adopt to achieve gauge invariance (3.10.14). But this is not the case for the following reason. We decompose

$$T_n = P T_n P + W_n.$$

The condition (3.10.19) concerns the physical part $P T_n P$, only. But the latter is gauge invariant for any normalization

$$Q P T_n P - P T_n P Q = 0$$

because $PQ = 0 = QP$. Therefore, the normalization in the proof of gauge invariance involves only the unphysical part W_n. ¿From the gauge independence of the retarded distributions (3.10.19) we get the same result for the n-point distributions

$$P_\lambda T_n^{(\lambda)} P_\lambda = P_1 T_n^{(1)} P_1 + \text{div} \qquad (3.10.20)$$

in the usual way. This completes the inductive proof. For the proof of gauge invariance (3.10.14) in an arbitrary λ-gauge we refer to the literature (see the bibliographical notes).

3.11 Appendix A: Cauchy problem for the iterated wave equation

First we formulate the Cauchy problem for the equation

$$\Box^2 u \equiv (\partial_0^2 - \partial_1^2 - \partial_2^2 - \partial_3^2)^2 u = 0. \quad (3.11.1)$$

Since (3.11.1) is of fourth order in time $x_0 = t$, a complete set of Cauchy data at $t = 0$ is given by

$$(\partial_0^n u)(0, \boldsymbol{x}) = u_n(\boldsymbol{x}), \quad n = 0, 1, 2, 3. \quad (3.11.2)$$

For simplicity we assume the u_n to be in Schwartz space; then the initial-value problem (3.11.1), (3.11.2) has a unique solution. This solution can be constructed by means of the tempered distributions $D(x)$ and $E(x)$, defined by

$$\Box D = 0, \quad D(0, \boldsymbol{x}) = 0, \quad (\partial_0 D)(0, \boldsymbol{x}) = \delta^3(\boldsymbol{x}), \quad (3.11.3)$$

$$\Box^2 E = 0, \quad (\partial_0^n E)(0, \boldsymbol{x}) = 0, \quad n = 0, 1, 2, \quad (\partial_0^3 E)(0, \boldsymbol{x}) = \delta^3(\boldsymbol{x}), \quad (3.11.4)$$

where D is the Jordan-Pauli distribution and E is sometimes called the dipole distribution and we will soon compute it.

We now claim that the solution of the Cauchy problem (3.11.1), (3.11.2) is given by

$$u(x) = \int d^3y \left[D(x-y) u_1(\boldsymbol{y}) - \partial_0^y D(x-y) u_0(\boldsymbol{y}) \right. \quad (3.11.5)$$
$$\left. + E(x-y)(u_3 - \Delta u_1)(\boldsymbol{y}) - \partial_0^y E(x-y)(u_2 - \Delta u_0)(\boldsymbol{y}) \right],$$

where Δ denotes the three-dimensional Laplace operator. This formula is the same as the covariant equation

$$u(x) = \int_{y_0=0} d^3y\, D(x-y) \overset{\leftrightarrow{y}}{\partial_0} u(y) + \int_{y_0=0} d^3y\, E(x-y) \overset{\leftrightarrow{y}}{\partial_0} \Box u(y),$$

which is an obvious generalization of the solution of the ordinary wave equation. Using (3.11.3) and (3.11.4), it is a simple task to verify (3.11.1) and (3.11.2). Therefore it remains to construct the dipole distribution E.

From (3.11.3) and (3.11.4) we get

$$\Box E(x) = D(x) \quad (3.11.6)$$

and we want to obtain E as solution of this equation. We solve this problem in momentum space. The Fourier transform of D is well known,

$$\hat{D}(p) = \frac{i}{2\pi} \operatorname{sgn} p_0 \delta(p^2), \quad (3.11.7)$$

so that

$$p^2 \hat{E}(p) = \frac{i}{2\pi} \text{sgn}\, p_0 \delta(p^2). \tag{3.11.8}$$

A solution of this equation can immediately be written down by means of the identity

$$p^2 \delta'(p^2) = \frac{d}{dp^2}\left(p^2 \delta(p^2)\right) - \delta(p^2) = -\delta(p^2), \tag{3.11.9}$$

namely

$$\hat{E}(p) = \frac{i}{2\pi} \text{sgn}\, p_0 \delta'(p^2). \tag{3.11.10}$$

By inverse Fourier transform the initial conditions (3.11.4) can be verified and $E(x)$ can be computed:

$$E(x) = \frac{1}{8\pi} \text{sgn}\, (x_0) \Theta(x^2). \tag{3.11.11}$$

Note that the positive-frequency part

$$"\hat{E}^{(+)}(p)" = \frac{i}{2\pi} \Theta(p_0) \delta'(p^2)$$

is ill-defined. This never occurs in rigorous calculations. Only derivatives of E have to be split into positive- and negative-frequency parts and these are well-defined.

3.12 Problems

3.1 Consider the Yang-Mills coupling with *bosonic* ghost fields $v_a(x)$:

$$T_1(x) = ig f_{abc}(A_{\mu a} A_{\nu b} \partial^\nu A_c^\mu + \tfrac{1}{2} A_{\mu a} v_b \partial^\mu v_c), \tag{3.12.1}$$

where

$$[v_a(x), v_b(y)] = -i\delta_{ab} D(x-y). \tag{3.12.2}$$

Derive from

$$Q = \int d^3x (\partial_\nu A_a^\nu \overleftrightarrow{\partial_0} v_a) \tag{3.12.3}$$

the gauge variations and show that the theory is gauge invariant to first order.

3.2 Verify by the methods of Sect.3.4 that second order gauge invariance cannot be established for the theory of problem 3.1.

3.3 Prove the identities (3.4.21-22).

3.4 Calculate the finite gauge transformation

$$A'^\mu(x) = e^{-i\lambda Q} A^\mu(x) e^{i\lambda Q} \tag{3.12.4}$$

and similarly for the ghost and anti-ghost fields.

3.5 Show that the sum of all second order vacuum polarization graphs

$$T_2(x,y) =: A_{\mu a}(x) t^{\mu\nu}(x-y) A_{\nu a}(y) :$$

in Yang-Mills theory is gauge invariant.

Hint: Show that the tensor structure is given by

$$t^{\mu\nu}(x) = (\partial^\mu \partial^\nu - g^{\mu\nu}\Box) t(x) \tag{3.12.5}$$

as a consequence of first order gauge invariance.

3.6 Prove gauge invariance of the 1-loop graphs with external field operators $A_{\mu a}$, $F_a^{\mu\nu} = \partial^\mu A_a^\nu - \partial^\nu A_a^\mu$, u_a, $\partial_\mu \tilde{u}_a$.

3.7 Compute the loop graphs of problem 3.6
Hint: Use the methods of Sect.2.6.

3.8 Verify that for second order tree graphs in Yang-Mills theory it is well possible to write $d_Q R_2(x,y)$ as a divergence *without introducing any normalization term*. This does not mean gauge invariance because the divergence does not have the required form (3.4.6).

Hint: Perform in (3.4.4) first the derivative ∂_μ and then the splitting. Note that in order to get the form (3.4.6) one has to do this the other way around.

3.9 In the fundamental representation of $SU(N)$ the $SU(N)$ generators are represented by $N \times N$ hermitian and traceless matrices $-(i/2)\lambda_a$, $a = 1,\ldots N^2 - 1$, satisfying

$$[\lambda_a, \lambda_b] = 2i f_{abc} \lambda_c \tag{3.12.6}$$

$$\lambda_a \lambda_b = \frac{2}{N} \delta_{ab} \mathbb{1} + d_{abc} \lambda_c + i f_{abc} \lambda_c, \tag{3.12.7}$$

where d_{abc} is real and totally symmetric. Calculate the traces $\text{Tr}(\lambda_a \lambda_b)$, $\text{Tr}(\lambda_a \lambda_b \lambda_c)$.

3.10 Prove

$$d_{abc} = \frac{1}{4}\Big(\text{Tr}(\lambda_a \lambda_b \lambda_c) + \text{Tr}(\lambda_b \lambda_a \lambda_c)\Big) \tag{3.12.8}$$

$$f_{abc} = \frac{1}{4i}\Big(\text{Tr}(\lambda_a \lambda_b \lambda_c) - \text{Tr}(\lambda_b \lambda_a \lambda_c)\Big) \tag{3.12.9}$$

$$(\lambda_a)_{jk}(\lambda_a)_{lm} = 2(\delta_{jm}\delta_{kl} - \frac{1}{N}\delta_{jk}\delta_{lm}) \tag{3.12.10}$$

by using problem 3.9. These relations can be used for calculating traces of products of f's and d's.

3.12 The adjoint representation of $SU(N)$ is defined by

$$(f_a)_{bc} = f_{abc} \tag{3.12.11}$$

$$[f_a, f_b] = f_{abc} f_c$$

$$(d_a)_{bc} = d_{abc}. \qquad (3.12.12)$$

Use problem 3.10 to prove

$$\text{Tr}\,(f_a f_b) = -N\delta_{ab} \qquad (3.12.13)$$

$$\text{Tr}\,(d_a d_b) = \left(N - \frac{4}{N}\right)\delta_{ab} \qquad (3.12.14)$$

$$\text{Tr}\,(f_a d_b) = 0. \qquad (3.12.15)$$

3.12 Prove:

$$\text{Tr}\,(d_a d_b d_c) = \frac{1}{2}\left(N - \frac{12}{N}\right)d_{abc} \qquad (3.12.16)$$

$$\text{Tr}\,(f_a d_b d_c) = \frac{1}{2}\left(\frac{4}{N} - N\right)f_{abc} \qquad (3.12.17)$$

$$\text{Tr}\,(f_a d_b f_c) = -\frac{N}{2}d_{abc} \qquad (3.12.18)$$

$$\text{Tr}\,(f_a f_b f_c) = \frac{N}{2}f_{abc}. \qquad (3.12.19)$$

4. Spin-1 gauge theories: massive gauge fields

For a long time massive gauge fields were a mystery. For this reason it was not possible to formulate a consistent theory of weak interactions which are mediated by the massive $W^\pm$- and Z vector bosons. Even today there remains some mystery if one says that the masses are generated by the so-called Higgs mechanism. The latter requires hypothetical scalar particles. Since these scalar fields have asymmetric self-couplings, one physical scalar (the Higgs field) gets a symmetry-breaking vacuum expectation value. Then the gauge symmetry is spontaneously broken and the gauge fields can acquire mass.

Instead of entering this mysterious scene we simply ask the question whether it is possible to proceed with massive vector fields introduced in Sect.1.5 in the same way as we did with massless ones in the last chapter. The answer is a clear yes! We have already seen in Sect.1.5 that we need scalar fields for the construction of the gauge charge Q. These scalar modes do not contribute to the physical subspace. If the massive gauge fields are self-interacting then second order gauge invariance requires the introduction of an additional *physical* scalar particle, the Higgs boson. All couplings are determined by causal gauge invariance, spontaneous symmetry breaking plays no role. We will study this phenomenon first in two simple theories and then turn to the general theory. As an application of it, we finally obtain the electroweak theory.

The discovery of the Higgs boson in 2012 at CERN is really a triumph of the theory, i.e. the gauge principle - and experiment. The current value of the mass is

$$m_{\rm H} = 125\,{\rm GeV}/c^2. \qquad (4.0.1)$$

It cannot be predicted by gauge invariance alone. We show in Sect. 4.5 that more Higgs fields are possible. On the other hand with one Higgs boson all couplings are fixed by gauge invariance (Sect. 4.6). One always says that the Higgs "generates" the masses of the gauge bosons, because one starts from massless gauge fields. We start from massive gauge fields and then the Higgs is required in order to satisfy gauge invariance.

4.1 Massive QED and Abelian Higgs model

To see how causal gauge invariance can be extended to massive gauge fields, we first discuss the simple case of quantum electrodynamics with massive photons. The theory is defined by the coupling (3.1.12)

$$T_1(x) = ie : \overline{\psi}(x)\gamma^\mu \psi(x) : A_\mu(x). \tag{4.1.1}$$

Here $\psi, \overline{\psi}$ are free Dirac fields but A_μ is a massive vector field

$$(\Box + m^2)A_\mu(x) = 0 \tag{4.1.2}$$

quantized according to

$$[A^\mu(x), A^\nu(y)] = g^{\mu\nu} i D_m(x-y). \tag{4.1.3}$$

As discussed in sect.1.5, we need a scalar field Φ with the same mass m in order to define a nilpotent gauge charge Q (1.5.31)

$$Q = \int d^3x \, (\partial_\nu A^\nu + m\Phi) \overleftrightarrow{\partial}_0 u. \tag{4.1.4}$$

$u(x)$ is the massive fermionic ghost field

$$(\Box + m^2)u = 0 \tag{4.1.5}$$

$$\{u(x), \tilde{u}(y)\} = -iD_m(x-y). \tag{4.1.6}$$

The scalar field satisfies

$$(\Box + m^2)\Phi = 0 \tag{4.1.7}$$

$$[\Phi(x), \Phi(y)] = -iD_m(x-y). \tag{4.1.8}$$

Next we determine the gauge variations (cf. (1.4.7))

$$d_Q A^\mu(x) = [Q, A^\mu(x)] = i\partial^\mu u(x) \tag{4.1.9}$$

$$d_Q \Phi(x) = [Q, \Phi(x)] = imu(x) \tag{4.1.10}$$

$$d_Q u(x) = \{Q, u(x)\} = 0 \tag{4.1.11}$$

$$d_Q \tilde{u}(x) = \{Q, \tilde{u}(x)\} = -i(\partial_\mu A^\mu + m\Phi(x)) \tag{4.1.12}$$

$$d_Q \psi(x) = 0 = d_Q \overline{\psi}(x). \tag{4.1.13}$$

Now it is easily seen that massive QED is gauge invariant to first order:

$$d_Q T_1 = -e : \overline{\psi}\gamma^\mu\psi : \partial_\mu u$$

$$= i\partial_\mu (ie : \overline{\psi}\gamma^\mu\psi : u) \stackrel{\text{def}}{=} i\partial_\mu T^\mu_{1/1}, \tag{4.1.14}$$

because the free Dirac current is conserved.

We now investigate n-th order gauge invariance. From the inductive construction we obtain T_n in normally ordered form by means of Wick's theorem. In this form we collect all terms with the external field operator $A_\mu(x_l)$

$$T_n(x_1,\ldots,x_n) =: T_l^\mu(x_1,\ldots,x_n)A_\mu(x_l):+\ldots, \quad (4.1.15)$$

where the dots represent terms without $A_\mu(x_l)$. There is a certain ambiguity with terms $\sim \delta(x_k-x_l)A_\mu(x_k)$, such a term is not included in (4.1.15). Now gauge invariance requires

$$T_l^\mu(x_1,\ldots,x_n)\partial_\mu u(x) = \partial_\mu^l[T_l^\mu(x_1,\ldots,x_n)u(x)]$$

or

$$\partial_\mu^l T_l^\mu(x_1,\ldots,x_n) = 0. \quad (4.1.16)$$

This is the same Ward identity as in the massless case (3.1.22). If the theory is constructed, i.e. suitably normalized, in such a way that (4.1.16) holds, then we have massive QED perturbatively gauge invariant. It is interesting to note that neither the ghost- nor physical or unphysical scalar fields enter in the coupling. This is due to the fact that the gauge field does not couple with itself.

We next consider a simple theory which does have self-coupling of the gauge field A_μ. To form a trilinear coupling which is a Lorentz scalar we introduce an additional physical scalar field φ with arbitrary mass m_H (H stands for "Higgs"-field) which allows a self-coupling term $\sim A_\mu A^\mu \varphi$. Then a general ansatz leading to a power-counting normalizable theory is

$$T_1(x) = igm[A_\mu A^\mu \varphi + aA_\mu A^\mu \Phi + b_1 u\tilde{u}\varphi + b_2 u\tilde{u}\Phi + b_3 A_\mu u\partial^\mu \tilde{u}+$$
$$+cA_\mu(\varphi\partial^\mu\Phi - \Phi\partial^\mu\varphi) + d_1\Phi^3 + d_2\Phi^2\varphi + d_3\Phi\varphi^2 + d_4\varphi^3], \quad (4.1.17)$$

where all products are normally ordered. We calculate $d_Q T_1$ and obtain

$$d_Q T_1 = -gm\Big[2\partial_\mu\Big(u(A^\mu\varphi + aA^\mu\Phi)\Big) + c\partial_\mu\Big(u(\varphi\partial^\mu\Phi - \Phi\partial^\mu\varphi)\Big)+$$
$$+cm\partial_\mu(uA^\mu\varphi) - 2u\partial_\mu A^\mu\varphi - 2uA^\mu\partial_\mu\varphi - 2au\partial_\mu A^\mu\Phi - 2auA^\mu\partial_\mu\Phi+$$
$$+amuA_\mu A^\mu + b_1 u\partial_\mu A^\mu\varphi + b_1 mu\Phi\varphi + b_2 u\partial_\mu A^\mu\Phi + b_2 mu\Phi^2+$$
$$+b_3\Big(\partial^\mu u u\partial_\mu \tilde{u} + A^\mu u\partial_\mu(\partial_\nu A^\nu + m\Phi)\Big) + cm^2 u\varphi\Phi - cm_H^2 u\varphi\Phi-$$
$$-cmu\partial_\mu A^\mu\varphi - 2cmuA_\mu\partial^\mu\varphi + 3d_1 mu\Phi^2 + 2d_2 mu\Phi\varphi + d_3 mu\varphi^2\Big]. \quad (4.1.18)$$

Here we have taken out the derivatives of the ghost fields. Since for a gauge theory $d_Q T_1$ has to be a pure divergence, the terms which are not of this form must cancel. This fixes most of the free parameters. We then get

$$T_1 = img\Big[A^\mu A_\mu\varphi + u\tilde{u}\varphi - \frac{1}{m}A_\mu(\varphi\partial^\mu\Phi - \Phi\partial^\mu\varphi) - \frac{m_H^2}{2m^2}\varphi\Phi^2 + d_4\varphi^3\Big], \quad (4.1.19)$$

and

$$dQT_1 = -gm\partial_\mu\left[uA^\mu\varphi - \frac{1}{m}u(\varphi\partial^\mu\Phi - \Phi\partial^\mu\varphi)\right] \stackrel{\text{def}}{=} i\partial_\mu T^\mu_{1/1}. \quad (4.1.20)$$

Further results come from second and third order gauge invariance.

Following the inductive construction, we have first to calculate the causal distribution
$$D_2(x,y) = T_1(x)T_1(y) - T_1(y)T_1(x),$$
the products herein are ordinary products, only the individual factors are normally ordered. D_2 is obviously gauge invariant
$$d_Q D_2(x,y) = i\partial^x_\mu [T^\mu_{1/1}(x), T_1(y)] + i\partial^y_\mu [T_1(x), T^\mu_{1/1}(y)]. \quad (4.1.21)$$

The main problem is whether gauge invariance can be preserved in the distribution splitting. As discussed in sect.3.4, gauge invariance of the retarded distribution R_2 can only be violated by local terms $\sim D^a\delta(x-y)$. But such local terms are precisely the freedom of normalization in the distribution splitting. If the normalization terms $N_2, N^\mu_{2/1}, N^\mu_{2/2}$ can be chosen in such a way that
$$d_Q(R_2 + N_2) = i\partial^x_\mu(R_{2/1} + N_{2/1}) + i\partial^x_\mu(R_{2/2} + N_{2/2}) \quad (4.1.22)$$

holds, then the theory is gauge invariant to second order. The local terms on the r.h.s. of (4.1.22), which come from the causal splitting, are the anomalies according to our previous terminology (see after (3.4.6)), because these local terms would violate gauge invariance if they do not cancel.

We consider the following example: in the commutator $[T^\mu_{1/1}(x), T_1(y)]$ appears the term
$$-g^2 m u(x)\Phi(x)[\partial^\mu \varphi(x), \varphi(y)]A_\nu(y)A^\nu(y) =$$
$$= ig^2 m u(x)\Phi(x)A_\nu(y)A^\nu(y)\partial^\mu D_{m_H}(x-y). \quad (4.1.23)$$
After splitting the Jordan-Pauli distribution D_{m_H} is replaced by the retarded distribution $D^{\text{ret}}_{m_H}$. If we now calculate for (4.1.21) the divergence of (4.1.23), we obtain an anomaly
$$\tfrac{1}{2}A_1 = ig^2 m u \Phi A_\nu A^\nu \delta(x-y) \quad (4.1.24)$$
because
$$\partial^x_\mu \partial^\mu_x D^{\text{ret}}_m(x-y) = -m^2 D^{\text{ret}}_m(x-y) + \delta(x-y).$$
The terms with x and y interchanged lead to the same contribution. But in the causal distribution $D_2 = [T_1(x), T_1(y)]$ the term
$$-g^2 A_\mu(x)\Phi(x)[\partial^\mu\varphi(x), \partial^\nu\varphi(y)]A_\nu(y)\Phi(y) =$$
$$= -ig^2 A_\mu(x)A_\nu(y)\Phi(x)\Phi(y)\partial^\mu_x \partial^\nu_x D_{m_H}(x-y) \quad (4.1.25)$$
appears, which has singular order $\omega = 0$ and therefore allows a normalization term in the split distribution

$$\partial_x^\nu \partial_x^\mu D^{\text{ret}}(x-y) \longrightarrow \partial_x^\nu \partial_x^\mu D^{\text{ret}}(x-y) + Cg^{\mu\nu}\delta(x-y). \qquad (4.1.26)$$

Since

$$d_Q(\Phi^2 A_\mu A^\mu C \delta(x-y)) = 2iCmu\Phi A_\mu A^\mu \delta(x-y) + \ldots \qquad (4.1.27)$$

we can compensate the anomaly (4.1.24) by choosing $C = ig^2$. In this way we obtain the quadrilinear couplings of the theory as normalization terms in second and third order.

We give here the complete list of all normalization terms for tree diagrams in second order:

$$N_1 = ig^2 A_\mu A^\mu \Phi^2 \delta(x-y)$$

$$N_2 = ig^2 A_\mu A^\mu \varphi^2 \delta(x-y)$$

$$N_3 = -ig^2 \frac{m_H^2}{4m^2} \Phi^4 \delta(x-y)$$

$$N_4 = ig^2 \left(\frac{m_H^2}{m^2} + 3d_4\right) \varphi^2 \Phi^2 \delta(x-y)$$

$$N_5 = ig^2 e \varphi^4 \delta(x-y), \quad e \text{ still free.} \qquad (4.1.28)$$

The remaining free parameters d_4 and e can be determined by considering the anomalies $\sim \delta(x-z)\delta(y-z)$ of tree graphs in third order. They arise in the splitting of terms

$$D_{3/1}^\mu(x,y,z) = [T_{1/1}^\mu(x), T_2(y,z)] + \ldots \qquad (4.1.29)$$

where $T_{1/1}^\mu$ (4.1.20) gets contracted with a normalization term N_{1-5} (4.1.28) in T_2. Considering all anomalies $\sim u\Phi\varphi^3$, gauge invariance requires

$$2e = \frac{m_H^2}{m^2} + 3d_4, \qquad (4.1.30)$$

and from the anomalies $\sim u\varphi\Phi^3$ we obtain

$$d_4 = -\frac{m_H^2}{2m^2}. \qquad (4.1.31)$$

For the discussion let us collect all purely scalar couplings

$$V_1 = -ig\frac{m_H^2}{2m}(\varphi\Phi^2 + \varphi^3) \qquad (4.1.32)$$

$$V_2 = -ig^2 \frac{m_H^2}{4m^2}(\varphi^2 + \Phi^2)^2. \qquad (4.1.33)$$

V_2 gets multiplied by $1/2!$ because it comes from second order. Then the total scalar coupling is equal to

$$V = V_1 + \tfrac{1}{2}V_2 = -ig^2 \frac{m_H^2}{8m^2}\left(\varphi^2 + \Phi^2 + \frac{2m}{g}\varphi\right)^2 + i\frac{m_H^2}{2}\varphi^2. \qquad (4.1.34)$$

Apart from the quadratic mass term this is the so-called Higgs potential in its asymmetric form. The symmetric form is obtained if we introduce the shifted Higgs field

$$\tilde{\varphi} = \varphi + \frac{m}{g}. \tag{4.1.35}$$

Then we arrive at the symmetric double-well potential

$$V \sim \left(\tilde{\varphi}^2 + \Phi^2 - \frac{m^2}{g^2}\right)^2. \tag{4.1.36}$$

In the old theory this potential is put in by hand, we got it out as a consequence of causal gauge invariance. Since our original Higgs field $\varphi(x)$ has vacuum expectation value zero, the shifted field has a non-vanishing vacuum expectation value

$$(\Omega, \tilde{\varphi}(x)\Omega) = \frac{m}{g}. \tag{4.1.37}$$

This is in complete agreement with the old scenario of spontaneous symmetry breaking, but now without any mystery. Summing up all coupling terms, we obtain

$$L_{\text{int}} = g\Big\{mA_\mu A^\mu \varphi + mu\tilde{u}\varphi - A_\mu(\varphi \partial^\mu \Phi - \Phi \partial^\mu \varphi) -$$

$$-g\frac{m_H^2}{8m^2}\left(\varphi^2 + \Phi^2 + 2\frac{m}{g}\varphi\right)^2\Big\}. \tag{4.1.38}$$

This is the interaction Lagrangian of the Abelian Higgs model.

4.2 General massive gauge theory

We consider r massive and s massless gauge fields A_a^μ, $a = 1,\ldots,r+s$ together with $(r+s)$ fermionic ghost and anti-ghost fields $u_a, \tilde{u}_a$. These free asymptotic fields are quantized as follows

$$(\Box + m_a^2)A_a^\mu(x) = 0, \quad [A_a^\mu(x), A_b^\nu(y)]_- = i\delta_{ab}g^{\mu\nu}D_{m_a}(x-y), \tag{4.2.1}$$

$$(\Box + m_a^2)u_a(x) = 0 = (\Box + m_a^2)\tilde{u}_a(x) \tag{4.2.2}$$

$$\{u_a(x), \tilde{u}_b(y)\}_+ = -i\delta_{ab}D_{m_a}(x-y), \tag{4.2.3}$$

all other commutators vanish, D_m are the Jordan-Pauli distributions. The masses of a gauge field and the corresponding ghost and anti-ghost fields must be equal, otherwise causal gauge invariance cannot be achieved. We have $m_a = 0$ for $a > r$.

In order to get a gauge charge Q which is nilpotent

$$Q^2 = 0, \tag{4.2.4}$$

we have to introduce for every massive gauge vector field $A_a^\mu(x), a \leq r$, a scalar partner $\Phi_a(x)$ with the same mass m_a. This was already discussed in Sect.1.5 (see (1.5.35)). The scalar fields are quantized according to

$$(\Box + m_a^2)\Phi_a(x) = 0, \quad [\Phi_a(x), \Phi_b(y)] = -i\delta_{ab}D_{m_a}(x-y). \tag{4.2.5}$$

Then the gauge charge Q is defined by

$$Q \stackrel{\text{def}}{=} \int d^3x \, (\partial_\nu A_a^\nu + m_a \Phi_a) \overleftrightarrow{\partial}_0 u_a. \tag{4.2.6}$$

Calculating Q^2 as one half of the anticommutator $\{Q, Q\}$ one easily verifies the nilpotency (4.2.4).

The scalar and ghost fields appearing in Q (4.2.6) are all unphysical because their excitations do not belong to the physical subspace

$$\mathcal{H}_{\text{phys}} = \text{Ker}(Q^+ Q + Q Q^+). \tag{4.2.7}$$

This has been discussed in detail at the end of Sect.1.5. Here we want to to stress the fact that our definition of gauge invariance refers to a structural property independent of representation. We simply call a field unphysical if it appears in Q (4.2.6), otherwise it is physical. For the gauge fields that means $\partial_\nu A^\nu$ is unphysical. Second order gauge invariance will force us to introduce additional *physical* scalar fields $\varphi_p, p = 1, \ldots, t$, called Higgs fields, with arbitrary masses μ_p. We shall use indices $p, q, \ldots = 1, \ldots t$ from the end of the alphabet to number the Higgs fields, letters $h, j, k, l, \ldots = 1, \ldots r$ from the middle denote the other massive scalar fields and $a, b, c, d, e, f, \ldots = 1, \ldots r+s$ is used for the gauge fields and ghosts.

With this field content we are going to analyze the following trilinear couplings:

$$T_1(x) = T_1^0 + T_1^1 + \ldots + T_1^{11} \tag{4.2.8}$$

where

$$T_1^0 = ig f_{abc}(A_{\mu a} A_{\nu b} \partial^\nu A_c^\mu - A_{\mu a} u_b \partial^\mu \tilde{u}_c) \tag{4.2.9}$$

$$T_1^1 = ig f_{ahj}^1 A_a^\mu (\Phi_h \partial_\mu \Phi_j - \Phi_j \partial_\mu \Phi_h), \quad f_{ahj}^1 = -f_{ajh}^1 \tag{4.2.10}$$

$$T_1^2 = ig f_{abh}^2 A_{\mu a} A_b^\mu \Phi_h, \quad f_{abh}^2 = f_{bah}^2 \tag{4.2.11}$$

$$T_1^3 = ig f_{abh}^3 \tilde{u}_a u_b \Phi_h \tag{4.2.12}$$

$$T_1^4 = ig f_{hjk}^4 \Phi_h \Phi_j \Phi_k, \tag{4.2.13}$$

where f_{hjk}^4 is totally symmetric in h, j, k and g is a coupling constant. All f's are real because T_1 must be skew-adjoint. For reasons of economy we assume the pure Yang-Mills coupling f_{abc} in (4.2.9) to be totally antisymmetric. If one starts with the most general ansatz, one must repeat the discussion in sect.3.2 to derive the antisymmetry. The Jacobi identity need not be assumed, it follows again below in second order (Sect.4.4). In T_1^1 we have only considered the antisymmetric combination because the symmetric one can be expressed by a divergence

$$A_a^\mu(\Phi_h\partial_\mu\Phi_j + \Phi_j\partial_\mu\Phi_h) = \partial_\mu(A_a^\mu\Phi_h\Phi_j) - \partial_\mu A_a^\mu\Phi_h\Phi_j.$$

The remaining $\partial_\mu A_a^\mu$ term is a co-boundary $d_Q(\tilde{u}_a\Phi_h\Phi_j)$ plus terms of the form T_1^3, T_1^4. But divergence and co-boundary couplings can always be skipped because they lead to physically equivalent S-matrices (sect.3.3).

The Higgs couplings are obtained by replacing the scalar fields in (4.2.10-13) by Higgs fields:

$$T_1^5 = igf_{ahp}^5 A_a^\mu(\Phi_h\partial_\mu\varphi_p - \varphi_p\partial_\mu\Phi_h) \tag{4.2.14}$$

$$T_1^6 = igf_{apq}^6 A_a^\mu(\varphi_p\partial_\mu\varphi_q - \varphi_q\partial_\mu\varphi_p), \quad f_{apq}^6 = -f_{aqp}^6 \tag{4.2.15}$$

$$T_1^7 = igf_{abp}^7 A_{\mu a} A_b^\mu \varphi_p, \quad f_{abp}^7 = f_{bap}^7 \tag{4.2.16}$$

$$T_1^8 = igf_{abp}^8 \tilde{u}_a u_b \varphi_p \tag{4.2.17}$$

$$T_1^9 = igf_{hjp}^9 \Phi_h\Phi_j\varphi_p, \quad f_{hjp}^9 = f_{jhp}^9 \tag{4.2.18}$$

$$T_1^{10} = igf_{hpq}^{10} \Phi_h\varphi_p\varphi_q, \quad f_{hpq}^{10} = f_{hqp}^{10} \tag{4.2.19}$$

$$T_1^{11} = igf_{pqu}^{11} \varphi_p\varphi_q\varphi_u, \tag{4.2.20}$$

where f^{11} is totally symmetric. *All products of field operators throughout are normally ordered (Wick) products of free fields.*

4.3 First order gauge invariance

As before the gauge charge Q (4.2.6) defines a gauge variation according to

$$d_Q F \stackrel{\text{def}}{=} QF - (-1)^{n_F} FQ, \tag{4.3.1}$$

where n_F is the number of ghost plus anti-ghost fields in the Wick monomial F. We get the following gauge variations of the fundamental fields

$$d_Q A_a^\mu(x) = i\partial^\mu u_a(x), \quad d_Q\Phi_h(x) = im_h u_h(x) \tag{4.3.2}$$

$$d_Q u_a(x) = 0, \quad d_Q \tilde{u}_a(x) = -i(\partial_\mu A_a^\mu(x) + m_a\Phi_a(x)) \tag{4.3.3}$$

$$d_Q\varphi_p = 0. \tag{4.3.4}$$

We now calculate the gauge variation of all terms in T_1 and transform the result to a divergence form

$$d_Q T_1 = i\partial_\mu T_{1/1}^\mu. \tag{4.3.5}$$

The $T_{1/1}^\mu$ appearing here is the Q-vertex. It is not unique, but the possible modification has no influence on gauge invariance of higher orders. The

First order gauge invariance   153

most convenient way to achieve the divergence form (4.3.5) is to take out the derivatives of the ghost fields. In this procedure we always use the field equations. In this way we find:

$$d_Q T_1^0 = g f_{abc} \Big\{ \partial_\mu [A_{\nu a} u_b (\partial^\nu A_c^\mu - \partial^\mu A_c^\nu) + \frac{1}{2} u_a u_b \partial^\mu \tilde{u}_c]$$

$$- m_c^2 A_{\nu a} u_b A_c^\nu + \frac{1}{2} m_c^2 u_a u_b \tilde{u}_c + m_c A_{\nu a} u_b \partial^\nu \Phi_c \Big\} \quad (4.3.6)$$

$$d_Q T_1^1 = -g f_{ahj}^1 \Big\{ \partial^\mu [u_a (\Phi_h \partial_\mu \Phi_j - \Phi_j \partial_\mu \Phi_h)$$

$$+ m_j A_a^\mu \Phi_h u_j - m_h A_a^\mu \Phi_j u_h] + (m_j^2 - m_h^2) u_a \Phi_h \Phi_j$$

$$+ m_h (\partial_\mu A_a^\mu \Phi_j + 2 A_a^\mu \partial_\mu \Phi_j) u_h - m_j (\partial_\mu A_a^\mu \Phi_h + 2 A_a^\mu \partial_\mu \Phi_h) u_j \Big\} \quad (4.3.7)$$

$$d_Q T_1^2 = -g f_{abh}^2 \Big\{ \partial_\mu [(u_a A_b^\mu + A_a^\mu u_b) \Phi_h] - u_a \partial_\mu A_b^\mu \Phi_h - u_a A_b^\mu \partial_\mu \Phi_h$$

$$- u_b \partial_\mu A_a^\mu \Phi_h - u_b A_a^\mu \partial_\mu \Phi_h + m_h A_{\mu a} A_b^\mu u_h \Big\} \quad (4.3.8)$$

$$d_Q T_1^3 = g f_{abh}^3 \Big\{ (\partial_\mu A_a^\mu + m_a \Phi_a) u_b \Phi_h - m_h \tilde{u}_a u_b u_h \Big\} \quad (4.3.9)$$

$$d_Q T_1^4 = -g f_{hjk}^4 \Big\{ m_h u_h \Phi_j \Phi_k + m_j \Phi_h u_j \Phi_k + m_k \Phi_h \Phi_j u_k \Big\} \quad (4.3.10)$$

$$d_Q T_1^5 = -g f_{ahp}^5 \Big\{ \partial^\mu [u_a (\Phi_h \partial_\mu \varphi_p - \varphi_p \partial_\mu \Phi_h) - m_h A_a^\mu \varphi_p u_h]$$

$$- (m_h^2 - \mu_p^2) u_a \Phi_h \varphi_p + 2 m_h A_a^\mu u_h \partial_\mu \varphi_p + m_h \partial_\mu A_a^\mu u_h \varphi_p \Big\} \quad (4.3.11)$$

$$d_Q T_1^6 = -g f_{apq}^6 \Big\{ \partial^\mu [u_a (\varphi_p \partial_\mu \varphi_q - \varphi_q \partial_\mu \varphi_p)] + (\mu_q^2 - \mu_p^2) u_a \varphi_p \varphi_q \Big\} \quad (4.3.12)$$

$$d_Q T_1^7 = -g f_{abp}^7 \Big\{ \partial^\mu [(u_a A_{\mu b} + u_b A_{\mu a}) \varphi_p]$$

$$- (u_a \partial_\mu A_b^\mu + u_b \partial_\mu A_a^\mu) \varphi_p - (u_a A_b^\mu + u_b A_a^\mu) \partial_\mu \varphi_p \Big\} \quad (4.3.13)$$

$$d_q T_1^8 = g f_{abp}^8 (\partial_\mu A_a^\mu + m_a \Phi_a) u_b \varphi_p \quad (4.3.14)$$

$$d_Q T_1^9 = -g f_{hjp}^9 (m_h u_h \Phi_j + m_j u_j \Phi_h) \varphi_p \quad (4.3.15)$$

$$d_Q T_1^{10} = -g f_{hpq}^{10} m_h u_h \varphi_p \varphi_q, \quad d_Q T_1^{11} = 0. \quad (4.3.16)$$

We have given this long list in detail because a lot of information can directly be read off. The divergence terms give the Q-vertex

$$i T_{1/1}^\mu = g f_{abc} \Big[A_{\nu a} u_b (\partial^\nu A_c^\mu - \partial^\mu A_c^\nu) + \frac{1}{2} u_a u_b \partial^\mu \tilde{u}_c \Big] \quad (4.3.17.a)$$

$$- g f_{ahj}^1 \Big[2 u_a \Phi_h \partial^\mu \Phi_j + m_j A_a^\mu \Phi_h u_j - m_h A_a^\mu \Phi_j u_h \Big] \quad (4.3.17.b)$$

$$- g f_{abh}^2 (u_a A_b^\mu + u_b A_a^\mu) \Phi_h \quad (4.3.17.c)$$

$$gf^5_{ahp}\left[u_a(\Phi_h\partial^\mu\varphi_p - \varphi_p\partial^\mu\Phi_h) - m_h A''_a u_h\varphi_p\right] \tag{4.3.17.d}$$

$$-2gf^6_{apq}u_a\varphi_p\partial^\mu\varphi_q \tag{4.3.17.e}$$

$$-gf^7_{abp}(u_a A^\mu_b + u_b A^\mu_a)\varphi_p. \tag{4.3.17.f}$$

The remaining terms must cancel out. Collecting the terms $\sim u_b A_{\mu a} A^\mu_c$ we get the relation

$$2m_b f^2_{acb} = (m_a^2 - m_c^2)f_{abc}. \tag{4.3.18}$$

Hence, if $m_b = 0$ and $f_{abc} \neq 0$ we must have

$$m_a = m_c. \tag{4.3.19}$$

For $m_b, m_h \neq 0$ we find

$$f^2_{abh} = \frac{m_b^2 - m_a^2}{2m_h}f_{abh}. \tag{4.3.20}$$

Then, collecting terms $\sim A_{\mu a} u_h \partial^\mu \Phi_j$ we get

$$f^1_{ahj} = \frac{m_j^2 + m_h^2 - m_a^2}{4m_h m_j}f_{ahj}. \tag{4.3.21}$$

From $u_a u_b \tilde{u}_c$ we obtain

$$m_b f^3_{cab} - m_a f^3_{cba} = m_c^2 f_{abc}.$$

Using all these results in the equation $\sim \partial_\mu A^\mu_a \Phi_h u_j$ we arrive at

$$f^3_{ahj} = \frac{m_j^2 - m_h^2 + m_a^2}{2m_j}f_{ahj}, \tag{4.3.22}$$

and then from $u_h \Phi_j \Phi_k$ we obtain

$$f^4_{hjk} = 0. \tag{4.3.23}$$

We have succeeded in expressing all couplings so far by f_{abc}. With these results all remaining terms without Higgs couplings cancel.

We next turn to the Higgs couplings. From $A^\mu_a u_b \partial_\mu \varphi_p$ we find

$$f^7_{abp} = m_b f^5_{abp}, \quad f^7_{abp} = 0 \quad \text{for} \quad a > r \quad \text{or} \quad b > r, \tag{4.3.24}$$

and from $\partial_\mu A^\mu_a u_h \varphi_p$ we get

$$f^8_{abp} = -m_b f^5_{abp} \tag{4.3.25}$$

and $=0$ for $b > r$. Finally the terms $\sim u_a \Phi_h \varphi_p$ give

$$f^9_{ahp} = -\frac{\mu_p^2}{2m_a}f^5_{ahp}, \quad a \leq r \tag{4.3.26}$$

and zero for $a > r$. The terms $\sim u_a \varphi_p \varphi_q$ lead to

$$f^{10}_{u\nu pq} = \frac{\mu_p^2 - \mu_q^2}{m_a} f^6_{apq}, \quad a \le r \qquad (4.3.27)$$

and zero for $a > r$. We see that the Higgs couplings are not completely fixed by first order gauge invariance. So far the Higgs couplings could be set equal to zero, but then we would find a breakdown of gauge invariance at second order.

4.4 Second order gauge invariance

The method to analyze the gauge invariance of second order tree graphs has already been described in sect.3.4. According to this method we must consider the commutator

$$D^\mu_{2/1} \stackrel{\text{def}}{=} [T^\mu_{1/1}(x), T_1(y)]. \qquad (4.4.1)$$

Here the terms in $T^\mu_{1/1}$ (4.3.17) which contain a derivative ∂^μ give rise to local contributions (anomalies) after distribution splitting. These are the second and third term in (4.3.17.a), the first term in (4.3.17.b), the first two in (4.3.17.d) and the first in (4.3.17.e). We shall abbreviate these terms by 17.a/2... 17.e/1 in the following. Commuting the factors with derivative ∂^μ in these terms with all terms in $T_1(y)$ (4.2.9-20) we get tree-graph contributions with four external legs (sectors) which we now have to examine.

Sector $uA\tilde{u}u$:

These field operators come out if we commute the second term in (4.3.17.a) with the second one in (4.2.9)

$$[(17.a/2), (2.9/2)] = i f_{abc} f_{def} A_{\nu a} u_b [\partial^\mu A^\nu_c(x), A_{\lambda d}(y)] u_e \partial^\lambda \tilde{u}_f, \qquad (4.4.2)$$

where we set the coupling constant $g = 1$ from now on. This gives a result $\sim \partial^\mu_x D(x - y)$. After splitting this causal distribution we get the retarded part $\partial^\mu_x D_{\text{ret}}(x - y)$. If now the derivative ∂^x_μ is applied

$$\partial_\mu \partial^\mu_x D_{\text{ret}}(x - y) = -m^2 D_{\text{ret}} + \delta(x - y) \qquad (4.4.3)$$

we get a local term

$$A_1 = -f_{abc} f_{cef} A_{\nu a} u_b u_e \partial_\nu \tilde{u}_f \delta(x - y) \qquad (4.4.3)$$

which is the anomaly. We emphasize that the splitting of a distribution into a local and a nonlocal part is unique *for tree graphs* (in general it is not unique). There is a second term with x and y interchanged giving the same contribution so that we notice the short rule for the following. Proceeding in the same way with the third term in (4.3.17.a) commuted with the second one in (4.2.9) we get

$$[(17.a/3), (2.9/2)] = f_{abc}f_{dcf}u_a u_b \partial^\nu \tilde{u}_f A_{\nu d}\delta(x-y). \tag{4.4.5}$$

There are no further contributions in this sector so that (4.4.3) must cancel against (4.4.5) in order to have gauge invariance. We interchange the indices of summation b and e in (4.4.3)

$$2A_1 = (-f_{abc}f_{cef} + f_{aec}f_{cbf})u_b u_e \partial^\nu \tilde{u}_f A_{\nu a}$$

and add (4.4.5), then the total anomaly becomes

$$(-f_{abc}f_{cef} + f_{aec}f_{cbf} - f_{ebc}f_{acf})u_b u_e \partial^\nu \tilde{u}_f A_{\nu a}. \tag{4.4.6}$$

Taking the total asymmetry of f_{abc} into account the bracket vanishes if and only if the Jacobi identity is satisfied.

Sector $uAAA$:

As the foregoing one this is a pure Yang-Mills sector. From the commutator between (4.3.17.a/2) and (2.9/1) we get three contributions

$$[(17.a/2), (2.9/1)] = f_{abc}A_{\nu a}u_b(x)\Big\{f_{cef}A_{\alpha e}\partial^\alpha A_f^\nu \partial^\mu D + f_{dcf}A_{\lambda d}\partial^\nu A_f^\lambda \partial^\mu D \tag{4.4.7}$$

$$+ f_{dec}A_d^\nu(y)A_{\alpha e}\partial_x^\mu \partial_y^\alpha D(x-y)\Big\}.$$

Here in the last term we have a new situation because the distribution $\partial^\mu \partial^\alpha D$ has singular order 0. It gives rise to a local term according to the rule (3.4.23)

$$G(x)F(y)\partial_x^\mu \partial_y^\alpha D(x-y) \Longrightarrow \big[(\alpha_1+1)\partial^\alpha GF + (\alpha_1-1)G\partial^\alpha F\big]\delta(x-y), \tag{4.4.8}$$

where α_1 is a free normalization constant. Using this in (4.4.7) we get the following total result for the local terms

$$= f_{abc}A_{\nu a}u_b(x)\Big\{f_{cef}A_{\alpha e}\partial^\alpha A_f^\nu + f_{dcf}A_{\lambda d}\partial^\nu A_f^\lambda\Big\}2\delta +$$

$$+ f_{abc}f_{dec}\Big\{(\alpha_1+1)(\partial_\alpha u_b A_{\nu a} + u_b \partial_\alpha A_{\nu a})A_d^\nu A_e^\alpha$$

$$+ (\alpha_1-1)u_b A_{\nu a}(\partial_\alpha A_d^\nu A_e^\alpha + A_d^\nu \partial_\alpha A_e^\alpha)\Big\}\delta. \tag{4.4.9}$$

¿From the vanishing of the term $\sim u_b A_{\nu a} A_d^\nu \partial_\alpha A_e^\alpha$ we conclude

$$(\alpha_1 - 1)f_{abc}f_{dec} = 0, \tag{4.4.10}$$

which implies $\alpha_1 = 1$. Then the terms $\sim u_b A_e^\alpha \partial_\alpha A_f^\nu A_{\nu a}$ cancel due to the Jacobi identity. But the term $\sim \partial_\alpha u_b A_{\nu a} A_d^\nu A_e^\alpha$ does not vanish

$$(\alpha_1 + 1) f_{abc} f_{dec} = 4\beta_1. \qquad (4.4.11)$$

Here a normalization term N_2 (3.4.27) is necessary. In fact, the 4-boson coupling (3.4.28)

$$N_1 = -i\beta_1 A_{\nu a} A_d^\nu A_{\alpha b} A_e^\alpha \delta(x-y) \qquad (4.4.12)$$

with the gauge variation

$$d_Q N_1 = 4\beta_1 \partial_\alpha u_b A_e^\alpha A_{\nu a} A_d^\nu \delta \qquad (4.4.13)$$

gives just the desired local term. Such a normalization term (4.4.12) is indeed possible because the first term in (4.2.9) commuted with itself gives the following second order tree graph contribution

$$D_2 = -f_{abc} f_{def} A_{\mu a} A_{\nu a} [\partial^\nu A_c^\mu(x), \partial^\alpha A_f^\lambda(y)] A_{\lambda d} A_{\alpha e}.$$

The commutator $\sim \partial^\nu \partial^\alpha D(x-y)$ has singular order 0 again, which allows the normalization term (4.4.12). $\alpha_1 = 1$ in (4.4.11) fixes β_1:

$$N_1 = -\frac{i}{2} f_{abc} f_{dec} A_{\nu a} A_d^\nu A_{\alpha b} A_e^\alpha \delta(x-y). \qquad (4.4.14)$$

This is the mechanism how additional couplings are generated by gauge invariance. Note that in (4.4.10) no normalization term is possible.

For later use we list the form of all possible normalization terms. They come from second order tree graphs with two derivatives on the inner line:

$(2.9) - (2.9) : AAAA, (2.10) - (2.10) : A\Phi A\Phi, (2.10) - (2.14) : A\Phi A\varphi,$

$(2.14) - (2.14) : A\Phi A\Phi, A\varphi A\varphi, (2.14) - (2.15) : A\Phi A\varphi, (2.15) - (2.15) : A\varphi A\varphi.$

In addition we shall need further normalization terms

$$\Phi\Phi\Phi\Phi, \Phi\Phi\varphi\varphi, \Phi\varphi\varphi\varphi, \varphi\varphi\varphi. \qquad (4.4.15)$$

They are produced by fourth order box diagrams.

Sector $uA\Phi\varphi$:

Now we have the tools to discuss all cases of compensation of local terms. For $u_a\varphi_q A_d^\nu \partial_\nu \Phi_k$ we find the relation

$$2(\alpha_2+1)f^1_{akj}f^5_{djq} - 2f_{dac}f^5_{ckq} - 2(\alpha_2+1)f^5_{akp}f^6_{dqp}$$
$$-2(\alpha_3-3)f^5_{ajq}f^1_{djk} - 2(\alpha_3-3)f^6_{aqp}f^5_{dkp} = 0. \qquad (4.4.16)$$

For $u_a \partial_\nu \varphi_q A_d^\nu \Phi_k$ we have

$$2(\alpha_2-3)f^1_{akj}f^5_{djq} + 2f_{dac}f^5_{ckq} - 2(\alpha_2-3)f^5_{akp}f^6_{dqp}$$
$$-2(\alpha_3+1)f^5_{ajq}f^1_{djk} - 2(\alpha_3+1)f^6_{aqp}f^5_{dkp} = 0. \qquad (4.4.17)$$

For $\partial_\nu u_a \varphi_q A_d^\nu \Phi_k$ we find

$$2(\alpha_2+1)f^1_{akj}f^5_{djq} - 2(\alpha_2+1)f^5_{akp}f^6_{dqp} - 2(\alpha_3+1)f^5_{ajq}f^1_{djk}$$
$$-2(\alpha_3+1)f^6_{aqp}f^5_{dkp} = \beta_2. \qquad (4.4.18)$$

Finally, for $u_a \varphi_q \partial_\nu A_d^\nu \Phi_k$ we get

$$2(\alpha_2-1)f^1_{akj}f^5_{djq} - 2(\alpha_2-1)f^5_{akp}f^6_{dqp} - 2(\alpha_3-1)f^5_{ajq}f^1_{djk} -$$
$$-2(\alpha_3-1)f^6_{aqp}f^5_{dkp} = 0. \qquad (4.4.19)$$

Note that we have to collect all terms with the same Wick monomial before splitting, otherwise we would get too many normalization constants α_j from (4.4.8), although this has no influence on the final result. β_2 belongs to the normalization term

$$N_2 = -\frac{i}{2}\beta_2(a,d,k,q)A_{\nu a}A_d^\nu \Phi_k \varphi_q \delta \qquad (4.4.20)$$

with

$$d_Q N_2 = \beta_2 \partial_\nu u_a A_d^\nu \Phi_k \varphi_q \delta + \beta_2 \frac{m_k}{2} u_k A_{\nu a} A_d^\nu \varphi_q \delta. \qquad (4.4.21)$$

The last term herein couples this sector to the sector $uAA\varphi$.

The three equations without β_2 are linearly dependent. Subtracting (4.4.17) from (4.4.16) and taking the antisymmetry of f^1_{ajk} in j,k (4.3.21) into account, we obtain

$$2(f^1_{akj}f^5_{djq} - f^1_{dkj}f^5_{ajq}) + 2(f^6_{aqp}f^5_{dkp} - f^6_{dqp}f^5_{akp}) = f_{dac}f^5_{ckp}. \qquad (4.4.22)$$

To get rid of the many indices we use a matrix notation. Let

$$(F_a^1)_{kj} = f^1_{akj}, \quad (F_a^5)_{jq} = f^5_{ajq}, \quad (F_a^6)_{pq} = f^6_{apq}, \qquad (4.4.23)$$

then (4.4.22) can be written in compact form

$$2[F_a^1, F_b^5] + 2[F_a^5, F_b^6] = -f_{abc}F_c^5. \qquad (4.4.24)$$

Next subtracting (4.4.16) from (4.4.18) we find the quartic coupling (4.4.20)

$$\beta_2(a,d,k,q) = 2f_{dac}f^5_{ckq} - 8(f^5_{ajq}f^1_{djk} + f^6_{aqp}f^5_{dkp}). \qquad (4.4.25)$$

Second order gauge invariance   159

Sector $uu\tilde{u}\varphi$:

In this sector we have only one combination of external legs, namely $u_a u_b \tilde{u}_d \varphi_p$. The corresponding relation is

$$f_{abc}f^8_{dcp} - f^5_{ajp}f^3_{dbj} + f^5_{bjp}f^3_{daj} + 2(f^6_{apq}f^8_{dbq} - f^6_{bpq}f^8_{daq}) = 0. \qquad (4.4.26)$$

The origin of the terms is clear from the upper indices. We have antisymmetrized the second and fourth terms in a, b corresponding to the antisymmetry of $u_a u_b$. Again we us matrix notation

$$(F^3_a)_{bj} = f^3_{baj}, \quad (F^8_a)_{bq} = f^8_{baq}. \qquad (4.4.27)$$

Then we have the following constraint

$$[F^3_b, F^5_a] + 2[F^8_a, F^6_b] = f_{abc}F^8_c. \qquad (4.4.28)$$

If we have one Higgs field only ($p = q = 1$) then f^6_{apq} vanishes by antisymmetry (4.2.15). In this case using (4.3.25) we can simplify (4.4.26):

$$f^3_{dbj}f^5_{aj1} - f^3_{daj}f^5_{bj1} + m_j f_{abj}f^5_{dj1} = 0, \qquad (4.4.29)$$

We specialize to $d = a$ and insert (4.3.22):

$$\sum_{j=1}^{r} \frac{3m_j^2 - m_b^2 + m_a^2}{2m_j} f_{abj}f^5_{aj1} = 0. \qquad (4.4.30)$$

If we write a summation symbol then only the indicated index is summed over. For a, b, j all different, f_{abj} defines a non-singular matrix and the mass-dependent factor does not alter that. Consequently f^5 vanishes for different indices, only $f^5_{jj1}, j = 1, \ldots r$ are different from 0. That means the Higgs couplings is diagonal in this case, in contrast to the couplings of the unphysical scalars which are non-diagonal.

Now (4.4.23) can be further simplified

$$f^3_{dba}f^5_{aa1} - f^3_{dab}f^5_{bb1} + m_d f_{abd}f^5_{dd1} = 0 \qquad (4.4.31)$$

where no summation takes place. Interchanging a with d and b with d, we get a homogeneous linear system for f^5

$$m_a f_{dba}f^5_{aa1} - f^3_{adb}f^5_{bb1} + f^3_{abd}f^5_{dd1} = 0$$

$$f^3_{bda}f^5_{aa1} + m_b f_{adb}f^5_{bb1} - f^3_{bad}f^5_{dd1} = 0. \qquad (4.4.32)$$

Using (4.3.22) it is easy to check that the 3×3 determinant vanishes so that we get a non-trivial solution. The latter is very simple

$$f^5_{aa1} = \frac{m_a}{m_d} f^5_{dd1}, \qquad (4.4.33)$$

in particular, $f^5_{aap} = 0$ for $a > r$. We have still to show that necessarily $f^5 \neq 0$. This follows from the following sector.

Sector $uA\Phi\Phi$:

From $u_a\partial_\nu\Phi_j\Phi_h A_d^\nu$ we get

$$4f_{dac}f^1_{chj} - 4(\alpha_4 - 3)f^1_{ahk}f^1_{djk} - 4(\alpha_4 + 1)f^1_{ajk}f^1_{dhk}$$
$$-(\alpha_4 + 1)f^5_{ajp}f^5_{dhp} - (\alpha_4 - 3)f^5_{ahp}f^5_{djp} = 0, \quad (4.4.34)$$

and, assuming $j \neq h$, $u_a\Phi_j\Phi_h\partial_\nu A_d^\nu$ gives

$$-(\alpha_4-1)(f^5_{ajp}f^5_{dhp}+f^5_{ahp}f^5_{djp})-4(\alpha_4-1)(f^1_{ajk}f^1_{dhk}+f^1_{ahk}f^1_{djk}) = 0. \quad (4.4.35)$$

Finally $\partial_\nu u_a\Phi_j\Phi_h A_d^\nu$ gives

$$-(\alpha_4+1)(f^5_{ajp}f^5_{dhp}+f^5_{ahp}f^5_{djp})-4(\alpha_4+1)(f^1_{ajk}f^1_{dhk}+f^1_{ahk}f^1_{djk}) = 2\beta_3, \quad (4.4.36)$$

with

$$N_3 = -\frac{i}{2}\beta_3(a,d,j,h)A_{\nu a}A_d^\nu\Phi_j\Phi_h\delta \quad (4.4.37)$$

$$\partial_Q N_3 = \beta_3\partial_\nu u_a A_d^\nu\Phi_j\Phi_h\delta + \beta_3 m_j u_j A_{\nu a}A_d^\nu\Phi_h\delta. \quad (4.4.38)$$

Subtracting (4.4.34) from (4.4.36) we find

$$\beta_3(a,d,j,h) = -2f_{dac}f^1_{chj} - 8f^1_{ahk}f^1_{djk} - 2f^5_{ahp}f^5_{djp} \quad (4.4.39)$$

where the first term does not contribute to (4.4.37). The result (4.4.39) remains valid for $j = h$.

Subtracting now (4.4.34) and (4.4.35) it follows

$$4(f^1_{ajk}f^1_{dhk} - f^1_{ahk}f^1_{djk}) + f^5_{ajp}f^5_{dhp} - f^5_{ahp}f^5_{djp} = 2f_{dac}f^1_{chj}. \quad (4.4.40)$$

In matrix notation this constraint reads

$$4[F^1_b, F^1_a] + [F^5_a, F^{5T}_b] = 2f_{abc}F^1_c \quad (4.4.41)$$

where T denotes the transposed matrix. Using previous results (4.4.40) gives for $j \neq h$

$$f^5_{ajp}f^5_{dhp} - f^5_{ahp}f^5_{djp} = \frac{m_j^2 + m_h^2 - m_c^2}{2m_h m_j}f_{dac}f_{chj}$$
$$-\frac{m_k^2 + m_j^2 - m_a^2}{m_j m_k}f_{ajk}\frac{m_k^2 + m_h^2 - m_d^2}{4m_h m_k}f_{dhk}$$
$$+\frac{m_k^2 + m_h^2 - m_a^2}{m_h m_k}f_{ahk}\frac{m_k^2 + m_j^2 - m_d^2}{4m_j m_k}f_{djk}. \quad (4.4.42)$$

In the special case $a = j$ and $d = h$ ($j \neq h$) we have

$$\sum_{p=1}^t f^5_{jjp}f^5_{jjp} = \frac{1}{2m_h^2}\left\{\sum_{c=1}^{r+s}(m_j^2 + m_h^2 - m_c^2)f_{jhc}f_{jhc}\right.$$

$$-\sum_{k=1}^{r} \frac{m_k^4 - (m_j^2 - m_h^2)^2}{2m_k^2} f_{jhk} f_{jhk} \Bigg\}. \qquad (4.4.43)$$

The r.h.s. is known from the basic Lie algebra and generally different from 0, consequently f^5 must be also different from 0. In case of only one Higgs field $t = 1$, the Higgs coupling f^5 can be calculated from (4.4.43) as a square root. For fixed j equation (4.4.43) holds for all $h \neq j$ and gives the same value on the l.h.s. This implies relations between the masses and the Yang-Mills couplings (see Sect.4.5).

Sector $uAA\Phi$:

In this sector there is only one Wick monomial $u_a A_{\nu b} A_c^{\nu} \Phi_h$ which gives the relation

$$4(f_{bad} f_{dch}^2 + f_{cad} f_{dbh}^2) - 4(f_{ahj}^1 f_{bcj}^2 + f_{ahj}^1 f_{cbj}^2)$$

$$-2(f_{ahp}^5 f_{bcp}^7 + f_{ahp}^5 f_{cbp}^7) = m_a \Big(\beta_3(b,c,a,h) + \beta_3(c,b,a,h)\Big), \qquad (4.4.44)$$

where (4.4.38) has been taken into account. This constraint does not have a simple matrix structure because of the two terms with f_{abc}. Substituting (4.4.39) and previous results we obtain

$$2m_a(f_{bhp}^5 f_{cap}^5 + f_{chp}^5 f_{bap}^5) - 2m_b f_{ahp}^5 f_{cbp}^5 - 2m_c f_{ahp}^5 f_{bcp}^5 =$$

$$= \frac{2}{m_h}\Big[(m_d^2 - m_c^2) f_{bad} f_{chd} + (m_d^2 - m_b^2) f_{cad} f_{bhd}\Big]$$

$$+ \frac{m_j^2 + m_h^2 - m_a^2}{m_h m_j^2}(m_c^2 - m_b^2) f_{ahj} f_{bcj} - \frac{m_j^2 + m_h^2 - m_b^2}{2m_h m_j^2}(m_j^2 + m_a^2 - m_c^2) f_{bhj} f_{caj}$$

$$- \frac{m_j^2 + m_h^2 - m_c^2}{2m_h m_j^2}(m_j^2 + m_a^2 - m_b^2) f_{chj} f_{baj}. \qquad (4.4.45)$$

In the case $h = b \neq c$ this leads to

$$m_a m_b f_{aap}^5 f_{bbp}^5 \delta_{ac} = -m_c^2 \sum_{d>r} f_{abd} f_{bcd}$$

$$+ \sum_{j=1}^{r} \frac{f_{abj} f_{bcj}}{4m_j^2}\Big[(m_j^2 - m_b^2)(3m_j^2 - m_a^2 + m_b^2) - m_c^2(m_j^2 + m_a^2 - m_b^2)\Big]. \qquad (4.4.46)$$

In the special case $c = a$ we get the simple relation

$$(m_a^2 - m_b^2) \sum_{d>r}(f_{abd})^2 = 0 \qquad (4.4.47)$$

where (4.4.43) has been used. As a cross-check this relation follows directly from (4.3.18). From (4.4.45) we find

$$\sum_{d>r}(m_c^2 f_{bad}f_{chd} + m_b^2 f_{cad}f_{bhd}) - m_h(m_b + m_c)f_{ahp}^5 f_{bcp}^5 =$$

$$= \sum_{j=1}^{r} \frac{1}{4m_j^2}\Big\{ f_{bhj}f_{caj}[(m_j^2 - m_b^2)(3m_j^2 - m_a^2 + m_c^2) - m_h^2(m_j^2 + m_a^2 - m_c^2)]$$

$$+ f_{baj}f_{chj}[(m_j^2 - m_c^2)(3m_j^2 - m_a^2 + m_b^2) - m_h^2(m_j^2 + m_a^2 - m_b^2)]$$

$$- 2f_{bcj}f_{ahj}(m_j^2 + m_h^2 - m_a^2)(m_b^2 - m_c^2)\Big\}. \quad (4.4.48)$$

In the special case $b = c$ equation (4.4.45) becomes

$$2m_a f_{bap}^5 f_{bhp}^5 - 2m_b f_{ahp}^5 f_{bbp}^5 =$$

$$= -\frac{(m_k^2 + m_a^2 - m_b^2)(m_k^2 + m_h^2 - m_b^2)}{2m_h m_k^2} f_{bak}f_{bhk} - 2\frac{m_b^2 - m_d^2}{m_h}f_{bad}f_{dbh}. \quad (4.4.49)$$

For $a = h \neq b = c$ this gives

$$m_b f_{aap}^5 f_{bbp}^5 = \frac{m_b^2}{m_a}\sum_{d>r}(f_{bad})^2$$

$$+ \sum_k \frac{(f_{bak})^2}{4m_a m_k^2}\Big[(m_k^2 - m_b^2)(-3m_k^2 - m_b^2 + 2m_a^2) + m_a^4\Big] \quad (4.4.50)$$

and for $h \neq a \neq b = c \neq h$ we get

$$m_b^2 \sum_{d>r} f_{bad}f_{bhd} = -\sum_{k=1}^{r} f_{bak}f_{bhk}\frac{1}{4m_k^2}$$

$$\times \Big[(m_k^2 - m_b^2)(-3m_k^2 + m_a^2 - m_b^2 + m_h^2) + m_a^2 m_h^2\Big]. \quad (4.4.51)$$

Sector $uA\varphi\varphi$:

From $u_a\varphi_p\partial_\nu\varphi_q A_b^\nu$ we get

$$4f_{bac}f_{cpq}^6 - (\alpha_5 + 1)f_{ajq}^5 f_{bjp}^5 - (\alpha_5 - 3)f_{ajp}^5 f_{bjq}^5$$

$$- 4(\alpha_5 + 1)f_{aqv}^6 f_{bpv}^6 - 4(\alpha_5 - 3)f_{apv}^6 f_{bqv}^6 = 0, \quad (4.4.52)$$

and, assuming $p \neq q$, $u_a\varphi_p\varphi_q\partial_\nu A_b^\nu$ gives

$$-(\alpha_5 - 1)(f_{ajp}^5 f_{bjq}^5 + f_{ajq}^5 f_{bjp}^5) - 4(\alpha_5 - 1)(f_{aqv}^6 f_{bpv}^6 + f_{apv}^6 f_{bqv}^6) = 0. \quad (4.4.53)$$

Finally $\partial_\nu u_a\varphi_p\varphi_q A_b^\nu$ gives

$$-(\alpha_5 + 1)(f_{ajp}^5 f_{bjq}^5 + f_{ajq}^5 f_{bjp}^5) - 4(\alpha_5 + 1)(f_{aqv}^6 f_{bpv}^6 + f_{apv}^6 f_{bqv}^6) = 2\beta_4, \quad (4.4.54)$$

with

$$N_4 = -\frac{i}{2}\beta_4(a,b,p,q)\Lambda_{\nu a}A_b^\nu \varphi_p \varphi_q \delta. \qquad (4.4.55)$$

Subtracting (4.4.53) from (4.4.52) we have

$$f_{ajp}^5 f_{bjq}^5 - f_{ajq}^5 f_{bjp}^5 + 4(f_{aqv}^6 f_{bvp}^6 - f_{apv}^6 f_{bvq}^6) = 2f_{abc}f_{cpq}^6 \qquad (4.4.56)$$

or in matrix form :

$$[F_a^{5T}, F_b^5] + 4[F_b^6, F_a^6] = 2f_{abc}F_c^6. \qquad (4.4.57)$$

Adding (4.4.47) and (4.4.48) and using previous results we get

$$\beta_4(a,b,p,q) = -f_{ajp}^5 f_{bjq}^5 - f_{ajq}^5 f_{bjp}^5 - 4(f_{aqv}^6 f_{bpv}^6 + f_{apv}^6 f_{bqv}^6). \qquad (4.4.57)$$

The same result remains valid for $p = q$.

Sector $uAA\varphi$:

This sector is similar to $uAA\Phi$. We obtain the constraint

$$4(f_{bad}f_{dcp}^7 + f_{cad}f_{dbp}^7) - 4(f_{apq}^6 f_{bcq}^7 + f_{apq}^6 f_{cbq}^7) +$$

$$+2(f_{ajp}^5 f_{bcj}^2 + f_{ajp}^5 f_{cbj}^2) = \frac{m_a}{2}[\beta_2(a,b,c,p) + \beta_2(a,c,b,p)] \qquad (4.4.58)$$

which again does not have a simple matrix structure.

Sector $uu\tilde{u}\Phi$:

This sector is similar to $uu\tilde{u}\varphi$. In analogy to (4.4.26) we get the constraint

$$f_{abc}f_{dcj}^3 - f_{ajp}^5 f_{dbq}^8 + f_{bjp}^5 f_{dap}^8 + 2(f_{ajk}^1 f_{dbk}^3 - f_{bjk}^1 f_{dak}^3) = 0. \qquad (4.4.59)$$

Note that according to (4.4.15) no normalization term is possible here. In matrix notation this constraint reads

$$2[F_b^3, F_a^1] + [F_b^{5T}, F_a^8] = f_{abc}F_c^3. \qquad (4.4.60)$$

Sector $u\Phi\Phi\varphi$:

From the monomial $u_a\Phi_h\Phi_j\varphi_p$ we obtain the relation

$$f_{ajq}^5 f_{hpq}^{10} + f_{ahq}^5 f_{jpq}^{10} + f_{apq}^6 (f_{hjq}^9 + f_{apq}^6 f_{jhq}^9) = 0.$$

Again there is no normalization term in this sector. Using previous results we arrive at

$$\sum_q \frac{1}{m_h}\left[(\mu_p^2 - \mu_q^2)f_{ajq}^5 f_{hpq}^6 - \frac{\mu_q^2}{2}f_{hjq}^5 f_{apq}^6\right] + h \leftrightarrow j = 0. \qquad (4.4.61)$$

This relation is interesting because it involves the Higgs masses.

Sector $u\Phi\varphi\varphi$:

From the monomial $u_a\Phi_h\varphi_p\varphi_q$ we get

$$4f^6_{avp}f^{10}_{hvq} - 3f^5_{ahv}f^{11}_{vqp} + 2f^5_{ajp}f^9_{hjq} + p \leftrightarrow q = 2m_a\beta_5(a,h,p,q). \quad (4.4.62)$$

Here a quartic coupling

$$N_5 = -i\beta_5(h,k,p,q)\Phi_h\Phi_k\varphi_p\varphi_q \quad (4.4.63)$$

is obtained.

Sectors $u\Phi\Phi\Phi$ and $u\varphi\varphi\varphi$:

In the first sector we find from $u_a\Phi_h\Phi_j\Phi_k$

$$S_{hjk}(f^5_{ahp}f^9_{jkp}) = -2m_a\beta_6(a,h,j,k) \quad (4.4.64)$$

where S_{hjk} means symmetrization in the indices h,j,k. According to (4.4.15) we find the quartic coupling

$$N_6 = -i\beta_6(h,k,l,m)\Phi_h\Phi_k\Phi_l\Phi_m. \quad (4.4.65)$$

In the last sector we similarly obtain

$$S_{pqv}(6f^6_{aup}f^{11}_{uqv} + f^5_{ahp}f^{10}_{hqv}) = \frac{m_a}{2}\beta_7(a,p,q,v), \quad (4.4.66)$$

with the coupling

$$N_7 = -i\beta_7(h,p,q,v)\Phi_h\varphi_p\varphi_q\varphi_v. \quad ((4.4.67)$$

This completes the long list of constraints.

After these tedious and error-sensitive calculations comes the reward. Following Grigore (J. Phys. A **33** (2000) 8443, eq.(3.3.3)) we introduce the $(r+t) \times (r+t)$ antisymmetric square matrix

$$F_a = \begin{pmatrix} -2F^1_a & -F^5_a \\ F^{5T}_a & -2F^6_a \end{pmatrix} \quad (4.4.68)$$

which involves the scalar degrees of freedom. Then the matrix equations (4.4.24), (4.4.41) and (4.4.57) can be written in the form

$$[F_a, F_b] = f_{abc}F_c, \quad a,b,c = 1,\ldots,r+s. \quad (4.4.69)$$

That means the matrices F_a realize a $(r+t)$-dimensional representation of the basic real Lie algebra defined by the structure constants f_{abc}. If we know the representations of this Lie algebra then (4.4.69) gives strong restrictions of the Higgs coupling constants.

The matrices (4.4.67) can also been used to write the original scalar couplings in more compact form. Following the above authors we collect the scalar fields into a $(r+t)$-dimensional vector

$$\Psi^T = (\Phi_1, \ldots, \Phi_r, \varphi_1, \ldots, \varphi_t).$$

Then the couplings (4.2.10),(4.2.14) and (4.2.15) can be combined as follows

$$T_1^1 + T_1^5 + T_1^6 = \frac{ig}{2} A_a^\mu (\partial_\mu \Psi^T F_a \Psi - \Psi^T F_a \partial_\mu \Psi). \qquad (4.4.70)$$

The right-hand side is the trilinear part of the covariant derivative coupling $(D\Psi)^T \cdot D\Psi$ where

$$D^\mu \Psi = (\partial^\mu + g A_a^\mu F^a) \Psi \qquad (4.4.71)$$

is the covariant derivative. For further details we refer to the original paper.

4.5 Third order gauge invariance

Instead of (4.4.1) we now have to look for local terms $\sim \delta^8(x-z, y-z)$ in

$$D_{3/1}^\mu(x,y,z) = [T_{1/1}^\mu(x), T_2(y,z)] + [T_1(y), T_{2/1}^\mu(x,z)] + [T_1(z), \tilde{T}_{2/1}^\mu(x,y)] \qquad (4.5.1)$$

$$D_{3/2}^\mu(x,y,z) = [T_{1/1}^\mu(y), T_2(x,z)] + [T_1(x), T_{2/1}^\mu(y,z)] + [T_1(z), \tilde{T}_{2/2}^\mu(x,y)] \qquad (4.5.2)$$

$$D_{3/3}^\mu(x,y,z) = [T_1(x), T_{2/2}^\mu(y,z)] + [T_1(y), T_{2/2}^\mu(x,z)] + [T_{1/1}^\mu(z), \tilde{T}_2(x,y)] \qquad (4.5.3)$$

where $\tilde{T}_2$ refers to the inverse S-matrix (cf. (2.2.6)). The first term in (4.5.1) produces a local term if the second term in (4.3.17.a) is commuted with the second order normalization term N_1 (4.4.14). The latter contains $\delta(y-z)$ and the commutator $\sim \partial^\mu D(x-y)$ gives another $\delta(x-y)$ by the usual mechanism (4.4.4). The result is

$$(17.a/2) - (4.4.14) = -2 f_{abc} f_{cb'e} f_{a'd'e} u_b A_{\nu a} A_{a'}^\nu A_{\lambda b'} A_{d'}^\lambda \delta(x-y)\delta(y-z). \qquad (4.5.4)$$

To examine the second term in (4.5.1) we use the fact that $\tilde{T}_{2/1}^\mu = -T_{2/1}^\mu + \ldots$ plus terms which give no local contribution. From (4.4.14) we have

$$\partial_\mu^x T_{2/1}^\mu(x,z)|_{\text{loc}} = 2 f_{abc} f_{dec} \partial_\lambda u_b A_e^\lambda A_{\nu a} A_d^\nu \delta(x-z). \qquad (4.5.5)$$

If this is commuted with the second term in (4.2.9), the anti-ghost - ghost contraction has two derivatives so that the resulting C-number distribution has $\omega = 0$ and, after splitting, allows a normalization term

$$(2.9/2) - (4.5.5) = -2i\alpha f_{a'b'b} f_{abc} f_{dec} u_{b'} A_{\lambda a'} A_e^\lambda A_{\nu a} A_d^\nu \delta(y-x)\delta(x-z). \qquad (4.5.6)$$

After renaming the summation indices this has the same form as (4.5.4). However, the contributions from the second and third member in (4.5.1) cancel each other and similarly in (4.5.2). But in (4.5.3) these normalization terms survive and after suitable choice of α in (4.5.6) compensate the anomaly

(4.5.4). Then the sector $uAAAA$ is gauge invariant. The situation is the same in the other sectors $uAA\Phi\Phi$, $uAA\Phi\varphi$ and $uAA\varphi\varphi$ containing A's. Here, instead of N_1 the normalization terms N_2, N_3 and N_4 come into play.

Next we turn to the sector $u\Phi^3\varphi$ where we get two anomalies

$$(4.3.17.d/1) - N_5 = 4 f^5_{ahp} \beta_5(j,k,p,q) u_a \Phi_h \Phi_j \Phi_k \varphi_q$$

$$(4.3.17.d/2) - N_6 = -8 f^5_{alq} \beta_6(l,k,j,h) u_a \Phi_h \Phi_j \Phi_k \varphi_q.$$

They must cancel each other because no normalization term is possible. This leads to the relation

$$\sum_{p=1}^{t} f^5_{ahp} \beta_5(j,k,p,q) = 2 \sum_{l=1}^{r} f^5_{alq} \beta_6(h,j,k,l). \qquad (4.5.7)$$

In case of one physical scalar ($t = 1$) this allows to determine the pure Higgs coupling f^{11} (see below). Similarly, in the sector $u\Phi^2\varphi^2$ we get the relation

$$8 \sum_{q=1}^{t} f^6_{apq} \beta_5(h,k,q,v) = -3 \sum_{q=1}^{t} f^5_{ahq} \beta_7(k,p,q,v). \qquad (4.5.8)$$

Finally, in the sector $u\Phi\varphi^3$ we find the relation

$$2 \sum_{v=1}^{t} f^5_{akv} \beta_8(p,q,v,u) = \sum_{h=1}^{r} f^5_{ahp} \beta_5(h,k,q,u), \qquad (4.5.9)$$

where the quartic Higgs coupling N_8

$$N_8 = -i \beta_8(p,q,u,v) \varphi_p \varphi_q \varphi_u \varphi_v \qquad (4.5.10)$$

is needed. The sector $u\Phi^4$ does not have an anomaly because there is no normalization term in (4.4.61).

It is instructive to discuss the important special case $t = 1$ of one physical scalar in detail. Then (4.5.7) can be simplified as follows

$$\beta_5(j,k,1,1) = 2\beta_6(j,k).$$

From (4.4.64) we have

$$\beta_6(j,k) = -\frac{1}{m_j} f^9_{jj1} f^5_{kk1} = \frac{\mu_1}{4} \left(\frac{f^5_{jj1}}{m_j} \right)^2 \qquad (4.5.11)$$

where (4.3.26) has been used, μ_1 is the Higgs mass. From (4.5.9) we get

$$\beta_8(1,1,1,1) = \frac{1}{2} \beta_5(j,j,1,1) = \frac{\mu_1^2}{4} \left(\frac{f^5_{jj1}}{m_j} \right)^2, \qquad (4.5.12)$$

which is actually independent of j. Finally from (4.4.62) we obtain

$$f^{11}_{111} - \frac{\mu_1^2}{2m_j} f^5_{jj1}. \qquad (4.5.13)$$

Let us now collect all trilinear purely scalar coupling terms

$$V_1 = i\left(f^9_{hj1}\Phi_h\Phi_j\varphi + f^{11}_{111}\varphi^3\right) =$$

$$= -\frac{i}{2}\frac{\mu_p^2}{m_j} f^5_{jj1}\varphi\left(\sum_k \Phi_k^2 + \varphi^2\right) \qquad (4.5.14)$$

and the quartic terms N_5, N_6 and N_7

$$V_2 = -i\left(\sum_{lk}\beta_6(l,k)\Phi_l^2\Phi_k^2 + \beta_5\varphi^2\sum_k\Phi_k^2 + \beta_8\varphi^4\right) =$$

$$= -\frac{i}{2}\frac{\mu_1^2}{2}\left(\frac{f^5_{jj1}}{m_j}\right)^2\left(\varphi^2 + \sum_k \Phi_k^2\right)^2. \qquad (4.5.15)$$

Introducing the coupling constant g again, we must multiply (4.5.12) by g and (4.5.13) by $g^2/2!$ because this is the second order contribution. Then the total scalar potential is equal to

$$V_\varphi = -ig^2 \frac{\mu_1^2}{8m_j^2}(f^5_{jj1})^2\left[\left(\varphi^2 + \sum_k \Phi_k^2\right)^2 + \frac{4m_j}{gf^5_{jj1}}\varphi\left(\varphi^2 + \sum_k \Phi_k^2\right)\right]. \qquad (4.5.16)$$

Completing the square inside the square bracket just amounts to addition of a mass term for the Higgs field

$$V(\varphi) = V_\varphi - \frac{i}{2}\mu_1^2\varphi^2 =$$

$$= -ig^2 \frac{\mu_1^2}{8m_j^2}(f^5_{jj1})^2\left[\varphi^2 + \sum_k \Phi_k^2 + \frac{2m_j}{gf^5_{jj1}}\varphi\right]^2. \qquad (4.5.17)$$

This is the asymmetric Higgs potential. In fact, introducing the shifted Higgs field

$$\tilde\varphi = \varphi + a, \qquad a = \frac{m_j}{gf^5_{jj1}}, \qquad (4.5.18)$$

the Higgs potential (4.5.15) assumes a symmetric double-well form

$$V \sim \left(\tilde\varphi^2 + \sum_j \Phi^2 - a^2\right)^2.$$

The shifted Higgs field then has a non-vanishing vacuum expectation a (4.5.16), so that we have recovered (i.e. actually deduced) the usual Higgs mechanism.

4.6 Derivation of the electroweak gauge theory

Let us seek all gauge theories with three massive gauge fields $m_1, m_2, m_3 \neq 0$ and one massless photon field $m_4 = 0$. There are many 4-dimensional Lie algebras, but we will see that gauge invariance is strong enough to fix the f_{abc} uniquely if we assume a single Higgs field. Then f^6 in (4.2.15) vanishes and f^5_{jj1} is diagomal. In 2012 a single Higgs particle with a mass of 125 GeV/c^2 was discovered at CERN. Therefore the special case considered in this section is the most interesting.

We put $a = 4, d = 2, j = 1, h = 2$ in (4.4.42)

$$0 = f_{243} f_{321} \frac{m_1^2 + m_2^2 - m_3^2}{2m_1 m_2} + f_{423} f_{213} \frac{m_3^2 + m_2^2}{m_2 m_3} \cdot \frac{m_3^2 + m_1^2 - m_2^2}{4m_1 m_3}$$

$$= \frac{f_{243} f_{321}}{4 m_1 m_2 m_3^2} (m_3^2 m_1^2 + 2 m_3^2 m_2^2 - 3 m_3^4 - m_1^2 m_2^2 + m_2^4). \tag{4.6.1}$$

Since the bracket is different from zero, we must either have $f_{243} = 0$ or $f_{321} = 0$. We shall verify below that the second alternative leads to the trivial solution $f = 0$ so we concentrate on the first case. For $a = 4, d = 1, h = 2, j = 1$ we find from (4.4.42)

$$0 = \frac{f_{143} f_{321}}{4 m_1 m_2 m_3^2} [m_3^2 (m_2^2 + 2m_1^2 - 3m_3^2) + m_1^2 (m_1^2 - m_2^2)], \tag{4.6.2}$$

which implies $f_{143} = 0$. Next we put $j = 1, h = 2$ in (4.4.43) and also $j = 1, h = 3$:

$$\sum_p (f^5_{11p})^2 = \frac{1}{2m_2^2} \Big[(f_{123})^2 (m_1^2 + m_2^2 - m_3^2) + (f_{124})^2 (m_1^2 + m_2^2) -$$

$$- (f_{123})^2 \frac{m_3^4 - (m_1^2 - m_2)^2}{2m_3^2} \Big]$$

$$= \frac{1}{2m_3^2} \Big[(f_{132})^2 (m_1^2 + m_3^2 - m_2^2) - (f_{132})^2 \frac{m_2^4 - (m_1^2 - m_3^2)^2}{2m_2^2} \Big]. \tag{4.6.3}$$

This implies

$$\left(\frac{f_{124}}{f_{123}} \right)^2 = 2 \frac{m_3^2 (m_3^2 - m_1^2) + m_2^2 (m_1^2 - m_2^2)}{m_3^2 (m_1^2 + m_2^2)}. \tag{4.6.4}$$

If the r.h.s. is different from zero we have $f_{124} \neq 0$, otherwise the solution would be trivial. Then it follows from (4.3.19) that $m_1 = m_2$ which is the equal mass of the W-bosons. This simplifies (4.6.4) as follows

$$\left(\frac{f_{124}}{f_{123}} \right)^2 = \frac{m_3^2}{m_1^2} - 1, \tag{4.6.5}$$

which implies $m_3 > m_1$. Defining the weak mixing angle Θ by

$$\frac{m_1}{m_3} = \cos\Theta, \qquad (4.6.6)$$

we have

$$\left(\frac{f_{124}}{f_{123}}\right)^2 = \tan^2\Theta.$$

Since a common factor in the f's can be absorbed in the coupling constant g, we end up with

$$f_{124} = -\sin\Theta, \qquad f_{123} = -\cos\Theta \qquad (4.6.7)$$

in agreement with the Weinberg-Salam model. All other structure constants follow by antisymmetry. The signs in (4.6.7) have been chosen according to standard convention, as well as m_3 for m_Z. Of course any permutation of the indices 1,2,3 is possible, but the solution remains the same.

It remains to discuss the possibility $f_{123} = 0$. Then it follows from (4.6.3) that $m_1 = m_2 = m_3$ and $|f_{124}| = |f_{134}|$. If we now put $a = h = 4$ and $d = j = 1$ in (4.4.42) we arrive at

$$0 = \frac{1}{2}(f_{142}f_{241} + f_{143}f_{341}) - \frac{1}{4}(f_{412}f_{142} + f_{413}f_{143})$$

$$= -\frac{1}{4}\left((f_{142})^2 + (f_{143})^2\right).$$

Hence, all f's vanish in this case.

It is not hard to verify that for the unique non-trivial solution (4.6.7) all conditions for gauge invariance are satisfied. By means of (4.6.7) all couplings can be calculated in the case of one Higgs field. This will be done below. But we could use any number $t \geq 1$ of Higgs fields for a gauge inariant model. Only for $t = 1$ the couplings are completely determined by gauge invariance.

In the same way one can construct the gauge theory with only two massive gauge fields $m_1, m_2 \neq 0$ and one massless field $m_3 = 0$. This is not the $SU(2)$ Higgs-Kibble model often discussed in the literature which has three massive fields (see problem 4.11). It turns out that $m_1 = m_2$ must be equal, this theory is the electroweak theory with mixing angle $\Theta = \pi/2$ (problem 4.2). It is also interesting to consider the case of four massive gauge fields. This would be an electroweak theory with a massive photon. Such a theory does not exist (problem 4.12), which in a way explains why the photon must be massless..

In the electroweak theory we have three massive gauge fields W_1^μ, W_2^μ, Z^μ and the massless photon A^μ. The corresponding ghosts are denoted by u_1, u_2, u_3 and u_4 is the massless ghost corresponding to A^μ, analogously for the antighosts $\tilde{u}_1, \ldots \tilde{u}_4$. The three unphysical scalar fields are Φ_1, Φ_2, Φ_3 and the one Higgs field is denoted by φ. The masses are related by (4.6.6)

$$\frac{m_W}{m_Z} = \cos\Theta. \qquad (4.6.8)$$

Then from (4.3.20-23) we find the following scalar couplings

$$T_1^1 + \ldots + T_1^4 = ig\Big\{\sin\Theta A^\nu(\Phi_2\partial_\nu\Phi_1 - \Phi_1\partial_\nu\Phi_2) \tag{4.6.9a}$$

$$+\left(\cos\Theta - \frac{1}{2\cos\Theta}\right)Z^\nu(\Phi_2\partial_\nu\Phi_1 - \Phi_1\partial_\nu\Phi_2) + \frac{1}{2}W_1^\nu(\Phi_3\partial_\nu\Phi_2 - \Phi_2\partial_\nu\Phi_3) \tag{4.6.9b}$$

$$-\frac{1}{2}W_2^\nu(\Phi_3\partial_\nu\Phi_1 - \Phi_1\partial_\nu\Phi_3) + m_W\sin\Theta A_\nu(W_2^\nu\Phi_1 - W_1^\nu\Phi_2) \tag{4.6.9c}$$

$$+m_W\left(\cos\Theta - \frac{1}{\cos\Theta}\right)(W_{2\nu}Z^\nu\Phi_1 - W_{1\nu}Z^\nu\Phi_2) + m_W\sin\Theta(\tilde{u}_1u_3\Phi_2 - \tilde{u}_2u_3\Phi_1) \tag{4.6.9d}$$

$$+\frac{m_W}{2}(\tilde{u}_2u_1 - \tilde{u}_1u_2)\Phi_3 + \frac{m_W}{2\cos\Theta}\tilde{u}_3(u_2\Phi_1 - u_1\Phi_2) \tag{4.6.9e}$$

$$+m_W\left(\frac{1}{2\cos\Theta} - \cos\Theta\right)(\tilde{u}_2u_3\Phi_1 - \tilde{u}_1u_3\Phi_2)\Big\}. \tag{4.6.9f}$$

The Higgs couplings follow from (4.3.24-27) and (4.4.43). The last equation leaves an ambiguity in sign, which has no observable consequence because it can be removed by a redefinition $\varphi \to -\varphi$. Therefore, we choose the plus sign.

$$T_1^5 + \ldots T_1^{11} = i\frac{g}{2}\Big\{W_1^\nu(\varphi\partial_\nu\Phi_1 - \Phi_1\partial_\nu\varphi) - m_W W_{1\nu}W_1^\nu\varphi \tag{4.6.10a}$$

$$+\frac{m_H^2}{2m_W}\varphi\Phi_1^2 + m_W\tilde{u}_1u_1\varphi\Big] + \tag{4.6.10b}$$

$$+W_2^\nu(\varphi\partial_\nu\Phi_2 - \Phi_2\partial_\nu\varphi) - m_W W_{2\nu}W_2^\nu\varphi \tag{4.6.10c}$$

$$+\frac{m_H^2}{2m_W}\varphi\Phi_2^2 + m_W\tilde{u}_2u_2\varphi\Big] + \tag{4.6.10d}$$

$$+\frac{1}{\cos\Theta}\Big[Z^\nu(\varphi\partial_\nu\Phi_3 - \Phi_3\partial_\nu\varphi) - m_Z Z_\nu Z^\nu\varphi + \frac{m_H^2}{2m_Z}\varphi\Phi_3^2 + \tag{4.6.10e}$$

$$+m_Z\tilde{u}_3u_3\varphi\Big]\Big\} + ig\frac{m_H^2}{4m_W}\varphi(\varphi^2 + \Phi_1^2 + \Phi_2^2 + \Phi_3^2). \tag{4.6.10f}$$

The last term follows from the normalization term (4.5.12). The other normalization terms (4.5.13) give the quadrilinear coupling

$$T_2^N = -ig^2\frac{m_H^2}{16m_W^2}(\varphi^2 + \Phi_1^2 + \Phi_2^2 + \Phi_3^2)^2. \tag{4.6.11}$$

It must be multiplied by $1/2!$ because this is a second order coupling, as can be seen from the power g^2. Together with the last term in (4.6.10f) it represents the Higgs potential.

For completeness we also list the pure Yang-Mills couplings (4.2.8)

$$T_1^0 = T_1^A + T_1^u, \tag{4.6.12}$$

Derivation of the electroweak gauge theory 171

$$T_1^A = ig f_{abc} A_{\mu a} A_{\nu b} \partial^\nu A_c^\mu = \quad (4.6.13)$$

$$= ig\Big\{\sin\Theta(W_1^\nu W_2^\mu - W_1^\mu W_2^\nu)\partial_\nu A_\mu +$$
$$+ \cos\Theta(W_2^\nu Z^\mu - W_2^\mu Z^\nu)\partial_\nu W_{1\mu} - \sin\Theta(A^\nu W_2^\mu - A^\mu W_2^\nu)\partial_\nu W_{1\mu}+$$
$$+ \cos\Theta(Z^\nu W_1^\mu - Z^\mu W_1^\nu)\partial_\nu W_{2\mu} + \sin\Theta(A^\nu W_1^\mu - A^\mu W_1^\nu)\partial_\nu W_{2\mu}+$$
$$+ \cos\Theta(W_1^\nu W_2^\mu - W_1^\mu W_2^\nu)\partial_\nu Z_\mu\Big\}. \quad (4.6.14)$$

The ghost coupling is given by

$$T_1^u = -ig f_{abc} A_{\mu a} u_b \partial^\mu \tilde{u}_c = \quad (4.6.15)$$

$$= ig\Big\{\sin\Theta(W_1^\mu u_2 - u_1 W_2^\mu)\partial_\mu \tilde{u}_4 +$$
$$+ \cos\Theta(W_2^\mu u_3 - u_2 Z^\mu)\partial_\mu \tilde{u}_1 - \sin\Theta(A^\mu u_2 - u_4 W_2^\mu)\partial_\mu \tilde{u}_1 +$$
$$+ \cos\Theta(Z^\mu u_1 - u_3 W_1^\mu)\partial_\mu \tilde{u}_2 + \sin\Theta(A^\mu u_1 - u_4 W_1^\mu)\partial_\mu \tilde{u}_2 +$$
$$+ \cos\Theta(W_1^\mu u_2 - u_1 W_2^\mu)\partial_\mu \tilde{u}_3\Big\}. \quad (4.6.16)$$

The second order normalization terms $N_1 - N_4$ (4.4.14-55) give the following quartic couplings

$$T_2^{N_1} = ig^2\Big\{\cos^2\Theta(W_1^\mu W_1^\nu + W_2^\mu W_2^\nu)Z_\mu Z_\nu + \sin^2\Theta(W_1^\mu W_1^\nu + W_2^\mu W_2^\nu)A_\mu A_\nu -$$
$$- \cos^2\Theta(W_1^\mu W_{1\mu} + W_2^\mu W_{2\mu})Z^\nu Z_\nu - \sin^2\Theta(W_1^\mu W_{1\mu} + W_2^\mu W_{2\mu})A^\nu A_\nu +$$
$$+ \sin 2\Theta(W_1^\mu W_1^\nu + W_2^\mu W_2^\nu)Z_\mu A_\nu - \sin 2\Theta(W_1^\mu W_{1\mu} + W_2^\mu W_{2\mu})Z^\nu A_\nu -$$
$$- W_1^\mu W_{1\mu} W_2^\nu W_{2\nu} + W_1^\mu W_{2\mu} W_1^\nu W_{2\nu}\Big\} \quad (4.6.17)$$

$$T_2^{N_2} = ig^2 \sin\Theta\Big\{(W_1^\mu \Phi_2 - W_2^\mu \Phi_1)\varphi A_\mu - \tan\Theta(W_1^\mu \Phi_2 - W_2^\mu \Phi_1)\varphi Z_\mu\Big\} \quad (4.6.18)$$

$$T_2^{N_3} = ig^2\Big\{\sin^2\Theta(\Phi_1^2 + \Phi_2^2)A_\mu A^\mu + (\sin 2\Theta - \tan\Theta)(\Phi_1^2 + \Phi_2^2)A_\mu Z^\mu +$$
$$+ \Big(\cos\Theta - \frac{1}{2\cos\Theta}\Big)^2 (\Phi_1^2 + \Phi_2^2)Z_\mu Z^\mu - \sin\Theta(W_1^\mu \Phi_1 + W_2^\mu \Phi_2)\Phi_3 A_\mu +$$
$$+ \sin\Theta\tan\Theta(W_1^\mu \Phi_1 + W_2^\mu \Phi_2)\Phi_3 Z_\mu + \frac{1}{4}(W_1^\mu W_{1\mu}\Phi_1^2 + W_2^\mu W_{2\mu}\Phi_2^2)+$$
$$+ \frac{1}{4}(W_1^\mu W_{1\mu}\Phi_2^2 + W_2^\mu W_{2\mu}\Phi_1^2) + \frac{1}{4}(W_1^\mu W_{1\mu} + W_2^\mu W_{2\mu})\Phi_3^2 + \frac{1}{4\cos^2\Theta}\Phi_3^2 Z_\mu Z^\mu\Big\}$$
$$\quad (4.6.19)$$

$$T_2^{N_4} = ig^2\Big\{\frac{1}{4}(W_1^\mu W_{1\mu} + W_2^\mu W_{2\mu})\varphi^2 + \frac{1}{4\cos^2\Theta}\varphi^2 Z^\mu Z_\mu\Big\}. \quad (4.6.20)$$

Again these second order couplings get multiplied by 1/2!. This finishes the complete list of couplings for the electroweak theory, beside the matter couplings which are considered in the following sections. The list was rather long which might be the reason why it is not completely given in most textbooks. But it is not hard to find the relevant coupling terms if one computes a specific process. Note that the structure constants f_{abc} do not satisfy the relations one usually has in massless pure Yang-Mills theory. This is a consequence of our choice of the basic gauge fields as mass eigenstates (see problem 4.13).

4.7 Coupling to leptons

To determine the coupling to leptons by the same method as before, we start from the following ansatz

$$\begin{aligned}
T_1^F = ig\Big\{ & b_1 W_\mu^+ \bar{e}\gamma^\mu \nu + b_1' W_\mu^+ \bar{e}\gamma^\mu \gamma^5 \nu \\
& + b_2 W_\mu^- \bar{\nu}\gamma^\mu e + b_2' W_\mu^- \bar{\nu}\gamma^\mu \gamma^5 e \\
& + b_3 Z_\mu \bar{e}\gamma^\mu e + b_3' Z_\mu \bar{e}\gamma^\mu \gamma^5 e \\
& + b_4 Z_\mu \bar{\nu}\gamma^\mu \nu + b_4' Z_\mu \bar{\nu}\gamma^\mu \gamma^5 \nu \\
& + b_5 A_\mu \bar{e}\gamma^\mu e + b_5' A_\mu \bar{e}\gamma^\mu \gamma^5 e \\
& + c_1 \Phi^+ \bar{e}\nu + c_1' \Phi^+ \bar{e}\gamma^5 \nu + c_2 \Phi^- \bar{\nu} e + c_2' \Phi^- \bar{\nu}\gamma^5 e \\
& + c_3 \Phi_3 \bar{e}e + c_3' \Phi_3 \bar{e}\gamma^5 e + c_4 \Phi_3 \bar{\nu}\nu + c_4' \Phi_3 \bar{\nu}\gamma^5 \nu \\
& + c_0 \varphi \bar{e}e + c_0' \varphi \bar{e}\gamma^5 e + c_5 \varphi \bar{\nu}\nu + c_5' \varphi \bar{\nu}\gamma^5 \nu \Big\},
\end{aligned} \quad (4.7.1)$$

where we have used the usual definitions $\Phi^\pm = (\Phi_1 \pm i\Phi_2)/\sqrt{2}$, and similarly for $W_\mu^\pm$ and $u^\pm$. Here we have assumed the usual electric charges of the particles and charge conservation in each term. In particular there is no coupling of the photon to the neutrini. A more general situation is considered in the following section. For simplicity we only consider one family of leptons, the 'electron' and the 'neutrino', which have arbitrary masses and fulfill the Dirac equations

$$\slashed{\partial} e = -im_e e, \quad \partial_\mu(\bar{e}\gamma^\mu) = im_e \bar{e}$$
$$\slashed{\partial} \nu = -im_\nu \nu, \quad \partial_\mu(\bar{\nu}\gamma^\mu) = im_\nu \nu. \quad (4.7.2)$$

We here consider Dirac neutrini only, in order to have the same structure of the couplings for leptons and quarks. *We do not assume chiral fermions in (4.7.1), instead, we will get them out as a consequence of second order gauge invariance.* Of course, this small theory is not gauge invariant to third order due to the axial anomalies. As usual, this defect disappears if the quark degrees of freedom are added (see next section).

Since the fermions are not transformed by d_Q, first order gauge invariance immediately gives $b_5' = 0$ assuming the electron mass to be non-vanishing. This fact that the photon has no axial-vector coupling can be traced back to the absence of a scalar partner for the photon, that means to its vanishing mass. The massive gauge fields must have axial-vector couplings. The other coupling constants are restricted at first order as follows

$$c_1 = i\frac{m_e - m_\nu}{m_W} b_1, \quad c_1' = i\frac{m_e + m_\nu}{m_W} b_1'$$

$$c_2 = i\frac{m_\nu - m_e}{m_W} b_2, \quad c_2' = i\frac{m_e + m_\nu}{m_W} b_2'$$

Coupling to leptons   173

$$c_3 = 0, \qquad c_3' = 2i\frac{m_e}{m_Z}b_3'$$

$$c_4 = 0, \qquad c_4' = 2i\frac{m_\nu}{m_Z}b_4'. \qquad (4.7.3)$$

The resulting divergence form of $d_Q T_1^F$ gives the following Q-vertex:

$$T_{1/1}^{F\mu} = ig\{b_1 u^+ \bar{e}\gamma^\mu \nu + b_1' u^+ \bar{e}\gamma^\mu \gamma^5 \nu$$

$$+ b_2 u^- \bar{\nu}\gamma^\mu e + b_2' u^- \bar{\nu}\gamma^\mu \gamma^5 e + b_3 u_3 \bar{e}\gamma^\mu e + b_3' u_3 \bar{e}\gamma^\mu \gamma^5 e$$

$$+ b_4 u_3 \bar{\nu}\gamma^\mu \nu + b_4' u_3 \bar{\nu}\gamma^\mu \gamma^5 \nu + b_5 u_4 \bar{e}\gamma^\mu e\}. \qquad (4.7.4)$$

In the discussion of gauge invariance of second order tree graphs there is a slight modification. There is a new source of anomalies when we contract $T_{1/1}^{F\mu}$ with T_1^F. The resulting fermionic contractions $\sim S_m(x-y)$ or S_m^{ret} give rise to anomalies if only one derivative is applied:

$$i\partial_\mu^x \gamma^\mu S_m^{\text{ret}}(x-y) = m S_m^{\text{ret}}(x-y) - \delta(x-y).$$

Since the derivative comes from

$$d_Q T_1^F(x) = \partial_\mu^x T_{1/1}^{F\mu},$$

every Fermi field in (4.7.4) generates an anomaly if it is contracted with another Fermi field in $T_1^F(y)$ according to (4.7.1). But note that the contractions of $T_{1/1}^F$ (4.7.4) with T_1 (4.2.8) do not produce any anomaly. Things are greatly simplified by the fact that there are no normalization terms with fermionic field operators possible, neither in T_2 nor in $T_{2/1}, T_{2/2}$. This is due to the fact that a graph with two external Fermi operators $\bar{\psi}\psi$ has singular order $\omega \le -1$. In addition, the latter implies that there are no terms $\sim \partial_\mu \delta(x-y)$.

The anomalies coming from $T_{1/1}^\mu$ (4.3.17) and T_1^F can be calculated as before. Only the terms with derivative ∂^μ in (4.3.17) generate anomalies. This leads to the following list of terms

$$T_{1/1}^\mu|_{\text{an}} = ig\Big\{\sin\Theta(u_1 W_2^\nu - u_2 W_1^\nu)\partial^\mu A_\nu$$

$$+ \sin\Theta(u_2 A^\nu - u_0 W_2^\nu)\partial^\mu W_{1\nu} + \sin\Theta(u_0 W_1^\nu - u_1 A^\nu)\partial^\mu W_{2\nu}$$

$$+ \cos\Theta\Big[(u_2 Z^\nu - u_3 W_2^\nu)\partial^\mu W_{1\nu} + (u_3 W_1^\nu - u_1 Z^\nu)\partial^\mu W_{2\nu}$$

$$+ (u_1 W_2^\nu - u_2 W_1^\nu)\partial^\mu Z_\nu\Big]$$

$$+ \sin\Theta(u_4 u_1 \partial^\mu \tilde{u}_2 + u_2 u_4 \partial^\mu \tilde{u}_1 + u_1 u_2 \partial^\mu \tilde{u}_0)$$

$$+ \cos\Theta(u_2 u_3 \partial^\mu \tilde{u}_1 + u_3 u_1 \partial^\mu \tilde{u}_2 + u_1 u_2 \partial^\mu \tilde{u}_3)$$

$$+ \sin\Theta u_4(\Phi_2 \partial^\mu \Phi_1 - \Phi_1 \partial^\mu \Phi_2) + \Big(\cos\Theta - \frac{1}{2\cos\Theta}\Big)u_3(\Phi_2 \partial^\mu \Phi_1 - \Phi_1 \partial^\mu \Phi_2) +$$

$$+\frac{1}{2\cos\Theta}\Big[(u_2\Phi_1 - u_1\Phi_2)\partial^\mu\Phi_3 + u_1\Phi_3\partial^\mu\Phi_2 - u_2\Phi_3\partial^\mu\Phi_1\Big]$$
$$+\frac{1}{2}u_1(\varphi\partial^\mu\Phi_1 - \Phi_1\partial^\mu\varphi) + \frac{1}{2}u_2(\varphi\partial^\mu\Phi_2 - \Phi_2\partial^\mu\varphi)$$
$$+\frac{1}{2\cos\Theta}u_3(\varphi\partial^\mu\Phi_3 - \Phi_3\partial^\mu\varphi)\Big\}. \tag{4.7.5}$$

In the sum of all anomalies the coefficient of every Wick monomial must add up to 0 in order to have gauge invariance.

The parameters in (4.7.1) can now be determined as follows. Assuming $b_1, b_2 \neq 0$, we find from the coefficients of $u_3\Phi_3\bar{e}\gamma^5 e$ and $u_3\Phi_3\bar{\nu}\gamma^5\nu$

$$c'_0 = 0, \quad c'_5 = 0, \tag{4.7.6}$$

and from $u^+ A_\nu \bar{e}\gamma^\nu \nu$ the electric charge

$$e = g b_5 = g \sin\Theta. \tag{4.7.7}$$

Then, from $u_3\Phi_3\bar{e}e$ and $u_3\varphi_0\bar{e}\gamma^5 e$ we get the two relations

$$2b'_3 c'_3 = \frac{ic_0}{2\cos\Theta}, \quad 2b'_3 c_0 = -\frac{ic'_3}{2\cos\Theta}.$$

By (4.7.3) this leads to

$$c_0 = \frac{m_e}{2m_W} \tag{4.7.8}$$

$$b'_3 = \frac{\varepsilon_2}{4\cos\Theta}, \quad \varepsilon_2 = \pm 1. \tag{4.7.9}$$

A possible trivial solution $c_0 = 0 = b'_3$ is excluded by later conditions. Similarly we find from $u_3\Phi_3\bar{\nu}\nu$ and $u_3\varphi_0\bar{\nu}\gamma^5\nu$

$$2b'_4 c'_4 = \frac{ic_5}{2\cos\Theta}, \quad 2b'_4 c_5 = -\frac{ic'_4}{2\cos\Theta}, \tag{4.7.10}$$

which, with (4.7.3), yields

$$c_5 = \frac{m_\nu}{2m_W} \tag{4.7.11}$$

$$b'_4 = \frac{\varepsilon_3}{4\cos\Theta}, \quad \varepsilon_3 = \pm 1. \tag{4.7.12}$$

¿From $u_1 W_2^\mu \bar{e}\gamma_\mu\gamma^5 e$ and $u_1 W_2^\mu \bar{\nu}\gamma_\mu\gamma^5\nu$ we obtain

$$b_1 b'_2 + b'_1 b_2 = b'_3 \cos\Theta = -b'_4 \cos\Theta, \tag{4.7.13}$$

which determines the sign ε_3 in (4.7.12):

$$b'_3 = -b'_4 = \frac{\varepsilon_2}{4\cos\Theta}. \tag{4.7.14}$$

¿From $u^+ \Phi_3 \bar{e}\nu$ we find

$$c_1 = 2b'_1(c'_3 + c'_4).$$

If we use (4.7.3) on both sides we arrive at

$$b_1 = \varepsilon_2 b_1'. \tag{4.7.15}$$

The same reasoning with $u^- \Phi_3 \bar{\nu} e$ gives

$$b_2 = \varepsilon_2 b_2'. \tag{4.7.16}$$

Substituting these results into (4.7.13) we get

$$b_1 b_2 = \frac{1}{8} = b_1' b_2'. \tag{4.7.17}$$

¿From $u_1 W_2^\mu \bar{\nu} \gamma_\mu \nu$ we now find

$$b_4 \cos \Theta = -b_1 b_2 - b_1' b_2' = -\frac{1}{4}. \tag{4.7.18}$$

Finally, $u_1 W_2^\mu \bar{e} \gamma_\mu e$ gives the relation

$$b_3 \cos \Theta = b_1 b_2 + b_1' b_2' - \sin \Theta b_5,$$

which determines

$$b_3 = \frac{1}{4 \cos \Theta} - \sin \Theta \tan \Theta. \tag{4.7.19}$$

Now we are ready to write down the leptonic coupling, as far as it is restricted by gauge invariance alone:

$$\begin{aligned}T_1^F = ig \Big\{ & b_1 W_\mu^+ \bar{e} \gamma^\mu (1 + \varepsilon_2 \gamma^5) \nu + b_2 W_\mu^- \bar{\nu} \gamma^\mu (1 + \varepsilon_2 \gamma^5) e \\
& + \frac{1}{4 \cos \Theta} Z_\mu \bar{e} \gamma^\mu (1 + \varepsilon_2 \gamma^5) e - \sin \Theta \tan \Theta Z_\mu \bar{e} \gamma^\mu e \\
& - \frac{1}{4 \cos \Theta} Z_\mu \bar{\nu} \gamma^\mu (1 + \varepsilon_2 \gamma^5) \nu + \sin \Theta A_\mu \bar{e} \gamma^\mu e \\
& + i \frac{m_e - m_\nu}{m_W} b_1 \Phi^+ \bar{e} \nu + i \frac{m_e + m_\nu}{m_W} b_1 \varepsilon_2 \Phi^+ \bar{e} \gamma^5 \nu \\
& - i \frac{m_e - m_\nu}{m_W} b_2 \Phi^- \bar{\nu} e + i \frac{m_e + m_\nu}{m_W} b_2 \varepsilon_2 \Phi^- \bar{\nu} \gamma^5 e \\
& + \frac{i \varepsilon_2}{2 m_W} m_e \Phi_3 \bar{e} \gamma^5 e - i \varepsilon_2 \frac{m_\nu}{2 m_W} \Phi_3 \bar{\nu} \gamma^5 \nu \\
& + \frac{m_e}{2 m_W} \varphi \bar{e} e + \frac{m_\nu}{2 m_W} \varphi \bar{\nu} \nu \Big\}. \end{aligned} \tag{4.7.20}$$

We have verified that with these parameters all further conditions for other Wick monomials are satisfied. This completes the proof of gauge invariance for second order tree graphs. The importance of the result (4.7.20) lies in the chiral coupling $\sim (1 + \varepsilon_2 \gamma^5)$ of the fermions. The sign ε_2 is conventional. We see that causal gauge invariance is the origin of maximal parity violation in weak interactions and for the universality of the couplings (i.e. there is only one independent coupling constant). b_1 and b_2 can be further restricted as follows: Pseudounitarity implies $b_1^* = b_2$; absorbing the phase of b_1 by a redefinition of the field operator $\mathrm{e}(x)$ we have $b_1 = b_2$. Inserting this into (4.7.17) we get the usual values

$$b_1 = b_2 = \frac{1}{2\sqrt{2}}. \tag{4.7.21}$$

4.8 More fermionic families

We start from the following generalization of the simple leptonic electroweak coupling to more than one family

$$T_1^F = ig \sum_{j,k} \Big\{ W_\mu^+ b_{jk}^1 \bar{\mathrm{e}}_j \gamma^\mu \nu_k + W_\mu^+ b_{jk}^{\prime 1} \bar{\mathrm{e}}_j \gamma^\mu \gamma^5 \nu_k \tag{4.8.1a}$$

$$+ W_\mu^- b_{jk}^2 \bar{\nu}_j \gamma^\mu \mathrm{e}_k + W_\mu^- b_{jk}^{\prime 2} \bar{\nu}_j \gamma^\mu \gamma^5 \mathrm{e}_k \tag{4.8.1b}$$

$$+ Z_\mu b_{jk}^3 \bar{\mathrm{e}}_j \gamma^\mu \mathrm{e}_k + Z_\mu b_{jk}^{\prime 3} \bar{\mathrm{e}}_j \gamma^\mu \gamma^5 \mathrm{e}_k \tag{4.8.1c}$$

$$+ Z_\mu b_{jk}^4 \bar{\nu}_j \gamma^\mu \nu_k + Z_\mu b_{jk}^{\prime 4} \bar{\nu}_j \gamma^\mu \gamma^5 \nu_k \tag{4.8.1d}$$

$$+ A_\mu b_{jk}^5 \bar{\mathrm{e}}_j \gamma^\mu \mathrm{e}_k + A_\mu b_{jk}^{\prime 5} \bar{\mathrm{e}}_j \gamma^\mu \gamma^5 \mathrm{e}_k \tag{4.8.1e}$$

$$+ A_\mu b_{jk}^6 \bar{\nu}_j \gamma^\mu \nu_k + A_\mu b_{jk}^{\prime 6} \bar{\nu}_j \gamma^\mu \gamma^5 \nu_k \tag{4.8.1f}$$

$$+ \Phi^+ c_{jk}^1 \bar{\mathrm{e}}_j \nu_k + \Phi^+ c_{jk}^{\prime 1} \bar{\mathrm{e}}_j \gamma^5 \nu_k + \Phi^- c_{jk}^2 \bar{\nu}_j \mathrm{e}_k + \Phi^- c_{jk}^{\prime 2} \bar{\nu}_j \gamma^5 \mathrm{e}_k \tag{4.8.1g}$$

$$+ \Phi_3 c_{jk}^3 \bar{\mathrm{e}}_j \mathrm{e}_k + \Phi_3 c_{jk}^{\prime 3} \bar{\mathrm{e}}_j \gamma^5 \mathrm{e}_k + \Phi_3 c_{jk}^4 \bar{\nu}_j \nu_k + \Phi_3 c_{jk}^{\prime 4} \bar{\nu}_j \gamma^5 \nu_k \tag{4.8.1h}$$

$$+ \varphi c_{jk}^0 \bar{\mathrm{e}}_j \mathrm{e}_k + \varphi c_{jk}^{\prime 0} \bar{\mathrm{e}}_j \gamma^5 \mathrm{e}_k + \varphi c_{jk}^5 \bar{\nu}_j \nu_k + \varphi c_{jk}^{\prime 5} \bar{\nu}_j \gamma^5 \nu_k \Big\}. \tag{4.8.1i}$$

Here we have used the same notation as before. All products of field operators throughout are normally ordered (Wick monomials). W, Z, A denote the gauge fields and $\Phi^\pm = (\Phi_1 \pm i\Phi_2)/\sqrt{2}$, Φ_3 the unphysical scalars. φ is the physical scalar, but $\mathrm{e}_j(x)$ stands for the electron-, muon-, tau-fields, as well as for the quark fields d, s, b, and $\nu_k(x)$ represents the corresponding neutrini and the other quark fields u, c, t. In the last section we have only considered the coupling to leptons. There the terms (4.8.1f) are missing because the neutrini have vanishing electric charge, but here, for the quark couplings, we must include them. We also assume that the asymptotic Fermi fields fulfill the Dirac equations

$$\slashed{\partial} \mathrm{e}_j = -i m_j^e \mathrm{e}_j, \quad \partial_\mu (\bar{\mathrm{e}}_j \gamma^\mu) = i m_j^e \bar{\mathrm{e}}_j$$

$$\slashed{\partial} \nu_k = -i m_k^\nu \nu_k, \quad \partial_\mu (\bar{\nu}_k \gamma^\mu) = i m_k^\nu \bar{\nu}_k, \tag{4.8.2}$$

with arbitrary non-vanishing unequal masses $m_j^e \neq m_k^e \neq 0$, $m_j^\nu \neq m_k^\nu \neq 0$ for all $j \neq k$. We do not use further information about the multiplet structure of the fermions.

The gauge variations of the asymptotic gauge fields are given by

$$d_Q A^\mu = i\partial^\mu u_0, \quad d_Q W_{1,2}^\mu = i\partial^\mu u_{1,2}, \quad d_Q Z^\mu = i\partial^\mu u_3,$$

and for the Higgs and unphysical scalar fields by

$$d_Q \varphi = 0, \quad d_Q \Phi_{1,2} = im_W u_{1,2}, \quad d_Q \Phi_3 = im_Z u_3$$

and finally for the fermionic ghosts

$$d_Q u_a = 0, \quad a = 1, 2, 3, 4$$

$$d_Q \tilde{u}_4 = -i\partial_\mu A^\mu, \quad d_Q \tilde{u}_{1,2} = -i(\partial_\mu W_{1,2}^\mu + m_W \Phi_{1,2})$$

$$d_Q \tilde{u}_3 = -i(\partial_\mu Z^\mu + m_Z \Phi_3). \tag{4.8.3}$$

The gauge variations of the fermionic matter fields vanish.

First order gauge invariance means that the gauge variation of

$$T_1 = T_1^0 + \ldots + T_1^{11} + T_1^F$$

has divergence form

$$d_Q T_1(x) = i\partial_\mu T_{1/1}^\mu. \tag{4.8.4}$$

To verify this for the fermionic coupling T_1^F we calculate the gauge variation of (4.8.1) and take out the derivatives of the ghost fields. In the additional terms with derivatives on the matter fields we use the Dirac equations (4.8.2):

$$d_Q T_1^F = -g \sum_{j,k} \partial_\mu \Big\{ u^+ (b_{jk}^1 \bar{e}_j \gamma^\mu \nu_k + b_{jk}'^1 \bar{e}_j \gamma^\mu \gamma^5 \nu_k) +$$

$$+ u^- (b_{jk}^2 \bar{\nu}_j \gamma^\mu e_k + b_{jk}'^2 \bar{\nu}_j \gamma^\mu \gamma^5 e_k) \tag{a}$$

$$+ u_3 (b_{jk}^3 \bar{e}_j \gamma^\mu e_k + b_{jk}'^3 \bar{e}_j \gamma^\mu \gamma^5 e_k + b_{jk}^4 \bar{\nu}_j \gamma^\mu \nu_k + b_{jk}'^4 \bar{\nu}_j \gamma^\mu \gamma^5 \nu_k) \tag{b}$$

$$+ u_4 (b_{jk}^5 \bar{e}_j \gamma^\mu e_k + b_{jk}'^5 \bar{e}_j \gamma^\mu \gamma^5 e_k + b_{jk}^6 \bar{\nu}_j \gamma^\mu \nu_k + b_{jk}'^6 \bar{\nu}_j \gamma^\mu \gamma^5 \nu_k) \Big\} \tag{c}$$

$$+ g \Big\{ iu^+ [b_{jk}^1 (m_j^e - m_k^\nu) \bar{e}_j \nu_k + b_{jk}'^1 (m_j^e + m_k^\nu) \bar{e}_j \gamma^5 \nu_k] \tag{d}$$

$$+ iu^- [b_{jk}^2 (m_j^\nu - m_k^e) \bar{\nu}_j e_k + b_{jk}'^2 (m_j^\nu + m_k^e) \bar{\nu}_j \gamma^5 e_k]$$

$$+ iu_3 [b_{jk}^3 (m_j^e - m_k^e) \bar{e}_j e_k + b_{jk}'^3 (m_j^e + m_k^e) \bar{e}_j \gamma^5 e_k$$

$$+ b_{jk}^4 (m_j^\nu - m_k^\nu) \bar{\nu}_j \nu_k + b_{jk}'^4 (m_j^\nu + m_k^\nu) \bar{\nu}_j \gamma^5 \nu_k]$$

$$+ iu_4 [b_{jk}^5 (m_j^e - m_k^e) \bar{e}_j e_k + b_{jk}'^5 (m_j^e + m_k^e) \bar{e}_j \gamma^5 e_k$$

$$+ b_{jk}^6 (m_j^\nu - m_k^\nu) \bar{\nu}_j \nu_k + b_{jk}'^6 (m_j^\nu + m_k^\nu) \bar{\nu}_j \gamma^5 \nu_k] \Big\}$$

$$+ m_W [u^+ c_{jk}^1 \bar{e}_j \nu_k + u^+ c_{jk}'^1 \bar{e}_j \gamma^5 \nu_k + u^- c_{jk}^2 \bar{\nu}_j e_k + u^- c_{jk}'^2 \bar{\nu}_j \gamma^5 e_k]$$

$$+m_Z u_3 [c_{jk}^3 \bar{e}_j e_k + c_{jk}'^3 \bar{e}_j \gamma^5 e_k + c_{jk}^4 \bar{\nu}_j \nu_k + c_{jk}'^4 \bar{\nu}_j \gamma^5 \nu_k]\}. \tag{4.8.5}$$

Now, to have first order gauge invariance, the terms (d) until the end of (4.8.5) which are not of divergence form must cancel. This implies

$$b_{jk}^{\prime 5} = 0 = b_{jk}^{\prime 6}, \; \forall j, k, \quad b_{jk}^5 = 0 = b_{jk}^6 \text{ for } j \neq k \tag{4.8.6}$$

and

$$c_{jk}^1 = \frac{i}{m_W}(m_j^e - m_k^\nu) b_{jk}^1, \quad c_{jk}'^1 = \frac{i}{m_W}(m_j^e + m_k^\nu) b_{jk}'^1$$

$$c_{jk}^2 = \frac{i}{m_W}(m_j^\nu - m_k^e) b_{jk}^2, \quad c_{jk}'^2 = \frac{i}{m_W}(m_j^\nu + m_k^e) b_{jk}'^2$$

$$c_{jk}^3 = \frac{i}{m_Z}(m_j^e - m_k^e) b_{jk}^3, \quad c_{jk}'^3 = \frac{i}{m_Z}(m_j^e + m_k^e) b_{jk}'^3$$

$$c_{jk}^4 = \frac{i}{m_Z}(m_j^\nu - m_k^\nu) b_{jk}^4, \quad c_{jk}'^4 = \frac{i}{m_Z}(m_j^\nu + m_k^\nu) b_{jk}'^4. \tag{4.8.7}$$

The result (4.8.6) means that the photon has no axial-vector coupling and no mixing in the vector coupling. This is due to the fact that it has no scalar partner because it is massless.

Now we turn to second order gauge invariance. As we have seen in the last sections, the main problem here is whether the anomalies in the tree graphs cancel out. These anomalies come from two sources. First, if the terms (a)-(c) in (4.8.5) are combined with the terms in (4.8.1) by a fermionic contraction we get the causal propagator $S_m(x-y)$. Its retarded part S_m^{ret}, after distribution splitting, gives rise to a local term

$$i\partial_\mu^x \gamma^\mu S_m^{\text{ret}}(x-y) = m S_m^{\text{ret}}(x-y) - \delta(x-y). \tag{4.8.8}$$

This δ-term is the anomaly. Secondly, we can perform a bosonic contraction between the terms in (4.8.1) and the terms (4.7.5) which are the anomaly-producing part in $d_Q T_1^{A\Phi}$, coming from the Yang-Mills and scalar couplings.

To give a representative example, we calculate the anomalies with external field operators $u_3 \Phi_3 \bar{e}_j \gamma^5 e_k$. Commuting the first term in (4.8.5) (b) with the second term in (4.8.1h) we get

$$-i u_3 b_{jk'}^3 \bar{e}_j \gamma^\mu [e_{k'}(x), \bar{e}_{j'}(y)] \gamma^5 e_k \Phi_3 c_{j'k}'^3. \tag{4.8.9}$$

After splitting, by (4.8.8) there results the anomaly $\sim -i b^3 c'^3 \delta(x-y)$ involving the matrix product of b^3 with c'^3. Combining the two terms with reversed order, we obtain $c'^3 b^3$ with a different sign so that both terms together yield the commutator $i[c'^3, b^3]$. Similarly, the second term in (4.8.5) (b) together with the first term in (4.8.1h) gives the anticommutator $\{c^3, b'^3\}$, because the γ^5 is at a different place. An anomaly of the second source comes from the last term in (4.7.5) contracted by the two φ-fields with the second term in (4.8.1i):

$$-\frac{i}{2\cos\Theta} u_3 \Phi_3 [\partial^\mu \varphi(x), \varphi(y)] \bar{e}_j \gamma^5 e_k c_{jk}'^0.$$

Altogether we obtain the following matrix equation

$$-\frac{1}{2\cos\Theta}c'^0 = i[c'^3, b^3] + i\{c^3, b'^3\}. \qquad (4.8.10)$$

We now give the complete list of all second order conditions. We specify the corresponding external legs, then the origin of the terms is pretty clear. Every combination of external field operators has a corresponding one with an additional γ^5. To save space we do not write down the external legs once more for the γ^5-term.

: $u_4\varphi\bar{e}e$: $[c^0, b^5] = 0$, $\quad [c'^0, b^5] = 0$
: $u_4\Phi_3\bar{e}e$: $[c^3, b^5] = 0$, $\quad [c'^3, b^5] = 0$
: $u_4 Z\bar{e}\gamma e$: $[b^3, b^5] = 0$, $\quad [b'^3, b^5] = 0$
: $u_4\varphi\bar{\nu}\nu$: $[c^5, b^6] = 0$, $\quad [c'^5, b^6] = 0$
: $u_4\Phi_3\bar{\nu}\nu$: $[c^4, b^6] = 0$, $\quad [c'^4, b^6] = 0$
: $u_4 Z\bar{\nu}\gamma\nu$: $[b^4, b^6] = 0$, $\quad [b'^4, b^6] = 0$ $\qquad (4.8.11)$
: $u^+ A\bar{e}\gamma\nu$: $\sin\Theta b^1 = b^5 b^1 - b^1 b^6$, $\quad \sin\Theta b'^1 = b^5 b'^1 - b'^1 b^6$
: $u^- A\bar{\nu}\gamma e$: $\sin\Theta b^2 = b^2 b^5 - b^6 b^2$, $\quad \sin\Theta b'^2 = b'^2 b^5 - b^6 b'^2$ $\qquad (4.8.12)$
: $u_4\Phi^+\bar{e}\nu$: $\sin\Theta c^1 = b^5 c^1 - c^1 b^6$, $\quad \sin\Theta c'^1 = b^5 c'^1 - c'^1 b^6$
: $u_4\Phi^-\bar{\nu}e$: $\sin\Theta c^2 = c^2 b^5 - b^6 c^2$, $\quad \sin\Theta c'^2 = c'^2 b^5 - b^6 c'^2$ $\qquad (4.8.13)$
: $u_3 W^+\bar{e}\gamma\nu$: $\cos\Theta b^1 = b^3 b^1 + b'^3 b'^1 - b^1 b^4 - b'^1 b'^4$,
$\qquad \cos\Theta b'^1 = b'^3 b^1 + b^3 b'^1 - b'^1 b^4 - b^1 b'^4$
: $u_3 W^-\bar{\nu}\gamma e$: $\cos\Theta b^2 = b^2 b^3 + b'^2 b'^3 - b^4 b^2 - b'^4 b'^2$,
$\qquad \cos\Theta b'^2 = b'^2 b^3 + b^2 b'^3 - b'^4 b^2 - b^4 b'^2$ $\qquad (4.8.14)$
: $u^+ W^-\bar{e}\gamma e$: $\cos\Theta b^3 = b^1 b^2 + b'^1 b'^2 - \sin\Theta b^5$, $\quad \cos\Theta b'^3 = b'^1 b^2 + b^1 b'^2$
: $u^- W^+\bar{\nu}\gamma\nu$: $\cos\Theta b^4 = -b^2 b^1 - b'^2 b'^1 - \sin\Theta b^6$,
$\qquad \cos\Theta b'^4 = -b'^2 b^1 - b^2 b'^1$ $\qquad (4.8.15)$
: $u^+\Phi_3\bar{e}\nu$: $c^1 = 2(b'^1 c'^4 + c'^3 b'^1 + c^3 b^1 - b^1 c^4)$,
$\qquad c'^1 = 2(b'^1 c^4 + c^3 b'^1 + c'^3 b^1 - b^1 c'^4)$
: $u^-\Phi_3\bar{\nu}e$: $c^2 = 2(-b'^2 c'^3 - c'^4 b'^2 - c^4 b^2 + b^2 c^3)$,
$\qquad c'^2 = 2(-b'^2 c^3 - c^4 b'^2 - c'^4 b^2 + b^2 c'^3)$ $\qquad (4.8.16)$
: $u^+\varphi\bar{e}\nu$: $\delta_1 c^1 = i(c^0 b^1 + c'^0 b'^1 + b'^1 c'^5 - b^1 c^5)$,
$\qquad \delta_1 c'^1 = i(c'^0 b^1 + c^0 b'^1 + b'^1 c^5 - b^1 c'^5)$
: $u^-\varphi\bar{\nu}e$: $\delta_1 c^2 = i(c^5 b^2 + c'^5 b'^2 + b'^2 c'^0 - b^2 c^0)$,
$\qquad \delta_1 c'^2 = i(c'^5 b^2 + c^5 b'^2 + b'^2 c^0 - b^2 c'^0)$ $\qquad (4.8.17)$
: $u_3\Phi_3\bar{e}e$: $\delta_3 c^0 = i[b^3, c^3] - i\{b'^3, c'^3\}$, $\quad \delta_3 c'^0 = i[b^3, c'^3] - i\{b'^3, c^3\}$
: $u_3\varphi\bar{e}e$: $\delta_3 c^3 = i[c^0, b^3] + i\{c'^0, b'^3\}$,
$\qquad \delta_3 c'^3 = i[c'^0, b^3] + i\{c^0, b'^3\}$ $\qquad (4.8.18)$
: $u_3\Phi_3\bar{\nu}\nu$: $\delta_3 c^5 = i[b^4, c^4] - i\{b'^4, c'^4\}$, $\quad \delta_3 c'^5 = i[b^4, c'^4] - i\{b'^4, c^4\}$
: $u_3\varphi\bar{\nu}\nu$: $\delta_3 c^4 = i[c^5, b^4] + i\{c'^5, b'^4\}$,
$\qquad \delta_3 c'^4 = i[c'^5, b^4] + i\{c^5, b'^4\}$ $\qquad (4.8.19)$
: $u^+\Phi^-\bar{e}e$: $\delta_1 c^0 = i(b^1 c^2 - b'^1 c'^2) - ic^3/2$, $\quad \delta_1 c'^0 = i(b^1 c'^2 - b'^1 c^2) - ic'^3/2$
: $u^-\Phi^+\bar{e}e$: $\delta_1 c^0 = -i(c^1 b^2 + c'^1 b'^2) + ic^3/2$,
$\qquad \delta_1 c'^0 = -i(c'^1 b^2 + c^1 b'^2) + ic'^3/2$ $\qquad (4.8.20)$
: $u^+\Phi^-\bar{\nu}\nu$: $\delta_1 c^5 = i(b^2 c^1 - b'^2 c'^1) - ic^4/2$, $\quad \delta_1 c'^5 = i(b^2 c'^1 - b'^2 c^1) - ic'^4/2$
: $u^-\Phi^+\bar{\nu}\nu$: $\delta_1 c^5 = -i(c^2 b^1 + c'^2 b'^1) + ic^4/2$,

$$\delta_1 c'^5 = -i(c'^2 b^1 + c^2 b'^1) + ic'^4/2 \tag{4.8.21}$$
$$: u_3 \Phi^+ \bar{e}\nu : \delta_4 c^1 = b^3 c^1 - b'^3 c'^1 - c^1 b^4 - c'^1 b'^4,$$
$$\delta_4 c'^1 = -b'^3 c^1 + b^3 c'^1 - c^1 b'^4 - c'^1 b^4$$
$$: u_3 \Phi^- \bar{\nu}e : \delta_4 c^2 = -b^4 c^2 + b'^4 c'^2 + c^2 b^3 + c'^2 b'^3,$$
$$\delta_4 c'^2 = b'^4 c^2 - b^4 c'^2 + c^2 b'^3 + c'^2 b^3, \tag{4.8.22}$$

where

$$\delta_1 = \frac{1}{2}, \quad \delta_3 = \frac{1}{2\cos\Theta}, \quad \delta_4 = \cos\Theta - \frac{1}{2\cos\Theta}. \tag{4.8.23}$$

The terms with these δ's and with the electroweak mixing angle obviously come from (4.7.5). There are further combinations of external field operators which have not been written down, because they give no new condition.

In case of one family, assuming $b^6 = 0$ and taking pseudounitarity into account (4.6.21), the corresponding system of scalar equations has a unique solution, which agrees with the lepton coupling of the standard model. The solution of the above matrix equations (4.8.12-22) is not so simple. We start from the equations (4.8.12). If we write these equations with matrix elements, using the fact that b^5 and b^6 are diagonal (4.8.6), we easily conclude that b^5 and b^6 are actually multiples of the unit matrix

$$b^5 = \alpha \mathbf{1}, \quad b^6 = (\alpha - \sin\Theta)\mathbf{1}. \tag{4.8.24}$$

Here α is a free parameter (the electric charge of the upper quarks or leptons) and we have assumed that the matrices b^1, b'^1, b^2, b'^2 are nonsingular. This is not a serious limitation because we shall see that these matrices are essentially unitary. This first consequence of gauge invariance is the universality of the electromagnetic coupling: the members $e_k(x)$ and $\nu_k(x)$ of different generations all couple in the same way to the photon, with a constant charge difference $q_e - q_\nu = g\sin\Theta$, which is the electronic charge.

Next we turn to the conditions (4.8.18). It is convenient to introduce the diagonal mass matrices

$$m^e = \text{diag}(m^e_j), \quad m^\nu = \text{diag}(m^\nu_j), \quad j = 1, \ldots n_g,$$

where n_g is the number of generations. Then (4.8.7) can be written as follows

$$c^1 = \frac{i}{m_W}(m^e b^1 - b^1 m^\nu),$$

$$c'^1 = \frac{i}{m_W}(m^e b'^1 + b'^1 m^\nu), \tag{4.8.25}$$

etc. Then the first two equations in (4.8.18) read

$$c'^0 = \frac{1}{\delta_3 m_Z}\Big(2b'^3 m^e b^3 - 2b^3 m^e b'^3$$

$$+ m^e b'^3 b^3 + m^e b^3 b'^3 - b^3 b'^3 m^e - b'^3 b^3 m^e\Big)$$

$$c^0 = \frac{1}{\delta_3 m_Z}\left(2b'^3 m^e b'^3 - 2b^3 m^e b^3 + (b^3)^2 m^e + m^e(b^3)^2 + (b'^3)^2 m^e + m^e(b'^3)^2\right). \tag{4.8.26}$$

Substituting this into the last two equations of (4.8.18), we arrive at the following coupled matrix equations for b^3, b'^3:

$$\delta_3^2(m^e b'^3 + b'^3 m^e) = 3b'^3 m^e (b^3)^2 + 3(b^3)^2 m^e b'^3 + 3(b'^3)^2 m^e b'^3 + 3b'^3 m^e (b'^3)^2$$
$$- 3b^3 m^e b'^3 b^3 - 3b^3 b'^3 m^e b^3 - 3b'^3 b^3 m^e b^3 - 3b^3 m^e b^3 b'^3$$
$$+ m^e b'^3 (b^3)^2 + m^e b^3 b'^3 b^3 + m^e (b'^3)^3 + m^e (b^3)^2 b'^3$$
$$+ (b^3)^2 b'^3 m^e + b^3 b'^3 b^3 m^e + b'^3 (b^3)^2 m^e + (b'^3)^3 m^e \tag{4.8.27}$$

$$\delta_3^2(m^e b^3 - b^3 m^e) = 3(b'^3)^2 m^e b^3 + 3(b^3)^2 m^e b^3 + 3b'^3 m^e b'^3 b^3 + 3b'^3 m^e b^3 b'^3$$
$$- 3b^3 m^e (b^3)^2 - 3b^3 b'^3 m^e b'^3 - 3b'^3 b^3 m^e b'^3 - 3b^3 m^e (b'^3)^2$$
$$+ m^e (b^3)^3 - (b^3)^3 m^e + m^e (b'^3)^2 b^3 - (b'^3)^2 b^3 m^e$$
$$+ m^e b^3 (b'^3)^2 - b^3 (b'^3)^2 m^e + m^e b'^3 b^3 b'^3 - b'^3 b^3 b'^3 m^e. \tag{4.8.28}$$

The coupled cubic equations (4.8.27-28) have many solutions in general. To determine the solutions in the neighborhood of the standard model, we substitute

$$b^3 = \delta_3(\beta \mathbf{1} + x), \quad b'^3 = \delta_3(\beta' \mathbf{1} + y), \quad \beta, \beta' \in \mathbf{C}, \tag{4.8.29}$$

and assume the matrices x, y to be small so that only terms linear in x and y must be taken with in (4.8.27-28). Then the equations collapse to the simple form

$$2\beta'(1 - 4\beta'^2)m + (1 - 12\beta'^2)(my + ym) = 0 \tag{4.8.30}$$
$$(1 - 12\beta'^2)(mx - xm) = 0. \tag{4.8.31}$$

Now, (4.8.30) yields a unique solution if $\beta' = O(1)$ is assumed

$$\beta' = \frac{\varepsilon_2}{2}, \quad \varepsilon_2 = \pm 1, \quad y = 0, \tag{4.8.32}$$

where the last result follows by writing the vanishing anticommutator $\{m^e, y\}$ with matrix elements, using $m_j^e > 0, \forall j$. Then (4.8.31) implies

$$(m_i^e - m_k^e)x_{ik} = 0, \quad \text{no sum over } i, k \tag{4.8.33}$$

that means x and b^3 are diagonal, taking into account that the masses m_j^e are not degenerate. All matrices in (4.8.26) commute, thus

$$c'^0 = 0 \tag{4.8.34}$$

$$c^0 = \frac{4m^e}{\delta_3 m_Z}(b'^3)^2 = \frac{\delta_3}{m_Z} m^e, \tag{4.8.35}$$

and (4.8.18) gives

$$c_3 = 0, \quad c'_3 = i\varepsilon_2 \frac{\delta_3}{m_Z} m^e. \tag{4.8.36}$$

The same reasoning can be carried through for (4.8.19) which leads to
$$b'^4 = -\varepsilon_2 \frac{\delta_3}{2}, \quad c^4 = 0, \tag{4.8.37}$$
for the sign see below, and
$$c'^4 = -i\varepsilon_2 \frac{\delta_3}{m_Z} m^\nu, \quad c'^5 = 0 \tag{4.8.38}$$
$$c^5 = \frac{\delta_3}{m_Z} m^\nu. \tag{4.8.39}$$

With this knowledge we turn to (4.8.16). Substituting c^1 in the first equation by (4.8.25) we arrive at
$$m^e(b^1 - \varepsilon_2 b'^1) = (b^1 - \varepsilon_2 b'^1)m^\nu,$$
leading to
$$b'^1 = \varepsilon_2 b^1, \tag{4.8.40}$$
assuming non-degenerate masses again. In the same way the second equation in (4.8.16) yields
$$b'^2 = \varepsilon_2 b^2. \tag{4.8.41}$$
This is the chiral coupling of all fermion generations. The sign in (4.8.37) follows from (4.8.15).

Finally, from the four conditions (4.8.15) it is easy to conclude
$$b^2 = \frac{1}{8}(b^1)^{-1} \tag{4.8.42}$$
$$b^3 = \frac{1}{\cos\Theta}\left(\frac{1}{4} - \alpha\sin\Theta\right) \tag{4.8.43}$$
$$b^4 = \frac{1}{\cos\Theta}\left(-\frac{1}{4} - \alpha\sin\Theta + \sin^2\Theta\right). \tag{4.8.44}$$
This means that x in (4.8.29) is actually zero, so that *there is no other solution in the neighborhood of the standard model*. But solutions "far away" are not excluded. All values of the b's and c's agree with the standard model for an arbitrary number n_g of generations. It is not hard to check that with the results so determined all other second order conditions of gauge invariance are satisfied. To finish this discussion we notice that pseudo-unitarity implies
$$b^{1+} = b^2 = \frac{1}{8}(b^1)^{-1},$$
hence
$$b^1 = \frac{1}{2\sqrt{2}}V, \quad b^2 = \frac{1}{2\sqrt{2}}V^+, \tag{4.8.45}$$
where V is an arbitrary unitary matrix. This is the so-called Cabibbo - Kobayashi - Maskawa (CKM) mixing matrix for the quark coupling. The same mixing is possible in the leptonic couplings. The recently observed signals of neutrino oscillations show that this mixing probably occurs (see problems 4.9 and 4.10).

4.9 Gauge invariance at third order: axial anomalies

Adler, Bell and Jackiw discovered that there exists a possibility to violate gauge invariance at third order in the triangular graphs (see FQED, Sect.5.3). The anomalous graphs contain one axial-vector and two vector couplings (VVA) or three axial-vector couplings (AAA) of the fermions and three external gauge fields. To have a more compact notation, we collect all fermionic matter fields into a big vector $\psi = (e, \mu, \tau, \nu_e, \nu_\mu, \nu_\tau)$ or $= (d, s, b, u, c, t)$, respectively. The gauge fields A, W^+, W^-, Z are denoted by A_a^μ with $a = 0, +, -, 3$. Then the coupling between fermions and gauge fields in (4.8.1) can be written as

$$T_1^{FA} = ig\left\{\overline{\psi}\gamma_\mu M_a \psi A_a^\mu + \overline{\psi}\gamma_\mu \gamma^5 M'_a \psi A_a^\mu\right\}, \tag{4.9.1}$$

where M_a stands for the following matrices of matrices:

$$M_+ = \begin{pmatrix} 0 & b^1 \\ 0 & 0 \end{pmatrix}, \quad M_- = \begin{pmatrix} 0 & 0 \\ b^2 & 0 \end{pmatrix} \tag{4.9.2}$$

$$M_3 = \begin{pmatrix} b^3 & 0 \\ 0 & b^4 \end{pmatrix}, \quad M_0 = \begin{pmatrix} b^5 & 0 \\ 0 & b^6 \end{pmatrix}, \tag{4.9.3}$$

and similarly for the axial-vector couplings, denoted by a prime.

Each triangular graph gives rise to two diagrams which differ by a permutation of two vertices. Therefore, we have to compute the traces

$$\text{tr}(M_a M_b M_c) + \text{tr}(M_a M_c M_b) = \text{tr}(M_a \{M_b, M_c\}),$$

where one or three M's must be axial-vector couplings with a prime. It is well-known that the cancellation of the axial anomalies relies on the compensation of these traces in the sum of the leptonic and hadronic contributions. In this way third order gauge invariance gives a further restriction of the quark coupling. To work this out in detail, we consider the following cases.

A. Case $(a = 0, b = +, c = -)_{VVA}$:

$$\text{tr}(M_0\{M'_+, M_-\}) + \text{tr}(M_0\{M_+, M'_-\}) =$$
$$= \text{tr}\left[b^5(b^1 b'^2 + b'^1 b^2) + b^6(b^2 b'^1 + b'^2 b^1)\right]$$

Using the results of the last section, this is equal to

$$= n_g \frac{\varepsilon_2}{4}(2\alpha - \sin\Theta), \tag{4.9.4}$$

where n_g is the number of generations, i.e. the dimension of the matrices b^k, and α is the charge of the fermions in (4.8.24). For leptons we have $b^6 = 0$, because the neutrini have no electric charge, so that

$$\alpha_L = \sin\Theta. \tag{4.9.5}$$

Consequently, to compensate the triangular anomaly proportional to (4.9.4), one needs the compensation between leptons and quarks. We assume equal number of families in the lepton and quark sectors. Then for three colours of quarks one must have

$$2\alpha_L - \sin\Theta + 3(2\alpha_Q - \sin\Theta) = 0, \qquad (4.9.6)$$

which implies

$$\alpha_Q = \frac{1}{3}\sin\Theta \qquad (4.9.7)$$

by (4.9.5).

B. Case $(0,3,3)_{VVA}$:

In this case the trace is simply given by

$$\text{tr}\,(M_0\{M_3', M_3\}) = \text{tr}\,[b^5(b'^3 b^3 + b^3 b'^3) + b^6(b'^4 b^4 + b^4 b'^4)]$$

$$= n_g \varepsilon_2 \frac{\delta_3}{\cos\Theta}\left(\frac{1}{4} - \sin^2\Theta\right)(2\alpha - \sin\Theta). \qquad (4.9.8)$$

Due to the same factor $(2\alpha - \sin\Theta)$ as in (4.9.4) the mechanism of compensation is the same.

C. Case $(3,+,-)_{VVA}$:

Here the trace is equal to

$$\text{tr}\,(M_3'\{M_+, M_-\} + M_3\{M_+', M_-\} + M_3\{M_+, M_-'\}) =$$
$$= \text{tr}\,[b_3' b_1 b_2 + b_4' b_2 b_1 + b^3(b^1 b'^2 + b'^1 b^2) + b^4(b^2 b'^1 + b'^2 b^1)]$$
$$= n_g \frac{\varepsilon_2 \sin\Theta}{4\cos\Theta}(\sin\Theta - 2\alpha), \qquad (4.9.9)$$

with the same consequences as before.

D. Case $(3,+,-)_{AAA}$:

Now we have to compute the trace

$$\text{tr}\,(M_3'\{M_+', M_-'\}) = \text{tr}\,[b'^3 b'^1 b'^2 + b'^4 b'^2 b'^1] = 0.$$

E. Case $(0,0,3)_{VVA}$:

Here the relevant trace is equal to

$$\text{tr}\,(M_0^2 M_3') = \text{tr}\,[(b^5)^2 b'^3 + (b^6)^2 b'^4]$$

$$= n_g \varepsilon_2 \frac{\delta_3}{2}\sin\Theta(2\alpha - \sin\Theta).$$

Due to the same factor as in case A, the compensation between leptons and quarks is the same. This is also true in the next case:

F. Case $(3,3,3)_{VVA}$:

$$\text{tr}\,(M_3^2 M_3') = \text{tr}\,[(b^3)^2 b'^3 + (b^4)^2 b'^4]$$

$$= n_g \varepsilon_2 \frac{\delta_3}{\cos\Theta}\left(\sin^3\Theta - \frac{\sin\Theta}{2}\right)(2\alpha - \sin\Theta).$$

G. Case $(3,3,3)_{AAA}$:
This final case is trivial:

$$\text{tr}\,(M'_3)^3 = \text{tr}\,[(b'^3)^3 + (b'^4)^3] = 0.$$

Summing up the axial anomalies completely cancel in each generation, if and only if α_Q has the value (4.9.7).

4.10 Problems

4.1 Calculate the quartic normalization terms (4.1.28) in the Abelian Higgs model.

4.2 Derive the gauge theory with two massive and one massless gauge field by the method of Sect.4.4
   Result:
   One obtains the electroweak theory with mixing angle $\Theta = \pi/2$. The mass of the Z-boson goes to infinity..

4.3 In the old Fermi theory of weak interactions the coupling is described by an effective current-current interaction

$$L_{\text{int}} = -\frac{G_F}{\sqrt{2}} J^\mu(x) J^+_\mu(x) \tag{4.10.1}$$

where

$$J_\mu = \bar{e}\gamma_\mu(1+\gamma_5)\nu + \ldots \tag{4.10.2}$$

Relate the Fermi constant G_F to the electroweak coupling constant g.
   Result:

$$\frac{G_F}{\sqrt{2}} = \frac{g^2}{8m_W^2}. \tag{4.10.3}$$

4.4 The decay of the W^+-boson into two fermions is described by the first term in (4.7.20). Calculate the decay width Γ in the rest frame of the W^+.
   a) Take the following initial state for the W-boson

$$\Phi_i^W = \int d^3p\, w^\nu(\boldsymbol{p})\varepsilon_{j\nu}(\boldsymbol{p}) a'^+_j(\boldsymbol{p})\Omega. \tag{4.10.4}$$

Here $w^\nu(\boldsymbol{p})$ is a wave packet which is peaked at the initial momentum $\boldsymbol{p}_i$ of the W and $\varepsilon_{\nu j}$ is the initial polarization vector (see (1.5.4)), there is no sum over j. The emission operator with prime defined by

$$a'^+_j = \frac{1}{\sqrt{2}}(a^+_{1j} + ia^+_{2j})$$

is adapted to $W_\mu^+ = (W_{1\mu} + iW_{2\mu})/\sqrt{2}$. Verify that this leads to the following contribution to the S-matrix element

$$(\ldots, W_\mu^+ \Phi_i^W) = (2\pi)^{-3/2} \int d^3 p\, w^\nu(\boldsymbol{p}) \frac{\varepsilon_{\nu j}\varepsilon_{\mu j}}{\sqrt{2E(\boldsymbol{p})}} e^{-ipx}. \qquad (4.10.5)$$

b) The Fermi fields are given by (1.8.3) and (1.8.12). Represent the final fermion states by

$$\Phi_f^e = b_s^+(\boldsymbol{p}_f)\Omega \quad \text{and} \quad \Phi_f^\nu = d_\sigma^+(\boldsymbol{q}_f)\Omega.$$

Show that the first term in (4.7.20) between these states yields

$$T_{if} = \frac{ig}{2\sqrt{2}(2\pi)^{9/2}} \int \frac{d^3 p}{\sqrt{2E}}\, w^\nu(\boldsymbol{p})\bar{u}_s(\boldsymbol{p}_f)\varepsilon_{\nu j}(\boldsymbol{p})\gamma^\mu \varepsilon_{\mu j}(\boldsymbol{p}) \times$$

$$\times (1-\gamma^5) v_\sigma(\boldsymbol{q}_f) e^{i(p_f + q_f - p)x}. \qquad (4.10.6)$$

c) Integrate T_{fi} with switching function $g(x)$ over $d^4 x$ and average over the initial polarizations $j = 1, 2, 3$ by means of (1.5.7).
Result:

$$S_{fi} = \int d^3 p\, w^\nu(\boldsymbol{p}) M_{fi,\nu} \hat{g}(p_f + q_f - p), \qquad (4.10.7)$$

$$M_{fi,\nu} = \frac{ig}{6\sqrt{2}(2\pi)^{5/2}} \frac{1}{\sqrt{2E(\boldsymbol{p})}} \bar{u}_s(\boldsymbol{p}_f) \left(-\gamma_\nu + \frac{p_\nu}{m_W^2}\slashed{p}\right) \times$$

$$\times (1-\gamma^5) v_\sigma(\boldsymbol{q}_f). \qquad (4.10.8)$$

d) Calculate the absolute square of S_{fi} using a complete set of final states $f_{s\sigma k}(\boldsymbol{p}_f, \boldsymbol{q}_f)$. In the total transition probability

$$\sum_f p_{fi} = \int d^3 p_f d^3 q_f d^3 p\, d^3 p'\, M_{fi,\nu} M^*_{fi,\mu} \hat{g}^*(p_f + q_f - p') \times$$

$$\times \hat{g}(p_f + q_f - p) w^\nu(\boldsymbol{p}) w^\mu(\boldsymbol{p}')^* \qquad (4.10.9)$$

we may neglect the p, p'-dependence in M_{fi}. Study the function

$$F^\nu(P) = \int d^3 p\, \hat{g}(P-p) w^\nu(\boldsymbol{p}) \qquad (4.10.10)$$

in x-space and show that

$$\int F^\nu(P) F^\mu(P)^* d^4 P = (2\pi)^3 \int d^4 x\, |g(x)|^2 \tilde{w}^\nu(x) \tilde{w}^\mu(x)^*, \qquad (4.10.11)$$

where

$$\tilde{w}^\nu(x) = (2\pi)^{-3/2} \int d^3 p\, w^\nu(\boldsymbol{p}) e^{-ipx}.$$

Choose $g(x) = 1$ for $0 \le x^0 \le T$ and

$$\breve{w}^\nu(x) = w_1^\nu(x + x_1 + vt)$$

with the normalization

$$\int d^3x\, w_1^\nu(x) w_{1\nu}(x)^* = 1 - 3 = -2$$

which corresponds to normalization to 1 in each component. Since the integrand in (4.10.11) is sharply peaked at $P = p_i$, this leads to

$$F^\nu(P) F^\mu(P)^* = -\frac{(2\pi)^3}{2} g^{\nu\mu} \delta(P - p_i) T$$

and to the decay rate

$$\Gamma = \frac{1}{T} \sum_f p_{fi} = -\frac{(2\pi)^3}{2} \int d^3 p_f d^3 q_f\, M_{fi,\nu} M_{fi}^{\nu*} \delta(p_f + q_f - p_i). \quad (4.10.12)$$

e) Compute $|M_{fi}|^2$ herein from (4.10.8) using

$$v_\sigma(q_f) \bar{v}_\sigma(q_f) = \frac{\not{q}_f - m_\nu}{2E(q_f)}$$

$$\bar{u}_s(p) \not{q} u_s(p) = \frac{pq}{2E_p}$$

which follows from (1.8.6).
Result:

$$|M_{fi}|^2 = \frac{g^2}{144(2\pi)^5 E_i E(p_f) E(q_f)} \left[-\left(2 + \frac{p_i^4}{m_W^4}\right) p_f q_f + \right.$$

$$\left. + 2\frac{p_i q_f}{m_W^2} \left(\frac{p_i^2}{w_W^2} - 1\right) p_i p_f \right]. \quad (4.10.13)$$

f) For $p_i = (m_W, \mathbf{0})$ the remaining phase space integral in (4.10.12) is of the form

$$I = \int d^3 p\, d^3 q\, \frac{pq}{E_p E_q} \delta(E_p + E_q - m_W) \delta^3(\mathbf{p} - \mathbf{q}).$$

Calculate it *without assuming small* m_e, m_ν, so that the result can also be used for quarks.
Result:

$$I = \pi(m_W^2 - m_e^2 - m_\nu^2) \sqrt{1 - 2\frac{m_e^2 + m_\nu^2}{m_W^2} + \frac{(m_e^2 - m_\nu^2)^2}{m_W^4}}. \quad (4.10.14)$$

Using (4.10.12) this gives the following decay width

$$\Gamma = \frac{y^2}{48\pi}\left(m_W^2 - \frac{m_e^2 + m_\nu^2}{m_W}\right)\sqrt{1 - 2\frac{m_e^2 + m_\nu^2}{m_W^2} + \frac{(m_e^2 - m_\nu^2)^2}{m_W^4}}. \qquad (4.10.15)$$

g) Determine the total width of the W-boson taking the three colours of quarks into account.

4.5 Calculate the decay width of the Z-boson into an electron-positron pair.
Result:

$$\Gamma = \frac{m_Z}{12\pi\sqrt{2}}(g_v^2 + g_a^2)\sqrt{1-4y}\left(1 + 2y\frac{g_v^2 - 2g_a^2}{g_v^2 + g_a^2}\right), \qquad (4.10.16)$$

where $y = m_e^2/m_Z^2$ and g_v and g_a are the vector and axial-vector coupling constants which can be identified in (4.7.20).

4.6 Show that in the reaction $e^- e^+ \to \mu^- \mu^+$ there is a forward-backward asymmetry for the μ^- travelling forward or backward relative to the incident e^- direction.

4.7 With three fermionic generations the CKM matrix V (4.8.45) is a unitary 3×3 matrix. How many free real parameters has this matrix if one uses the fact that phases can be absorbed in the field operators ?
Result:
A convenient parametrization is given by

$$V = \begin{pmatrix} c_1 & s_1 c_3 & s_1 s_3 \\ -s_1 c_2 & c_1 c_2 c_3 - s_2 s_3 e^{i\delta} & c_1 c_2 s_3 + s_2 c_3 e^{i\delta} \\ -s_1 s_2 & c_1 s_2 c_3 + c_2 s_3 e^{i\delta} & c_1 s_2 s_3 - c_2 c_3 e^{i\delta} \end{pmatrix} \qquad (4.10.17)$$

with $c_j = \cos\vartheta_j$, $s_j = \sin\vartheta_j$. Show that the angles can be restricted to the ranges

$$0 \le \vartheta_j \le \frac{\pi}{2}, \quad -\pi \le \delta \le \pi. \qquad (4.10.18)$$

4.8 Verify that for $\delta \ne 0$ in (4.10.17) one has CP- or time reversal violation (assuming the CPT-theorem).

4.9 Neutrino oscillations: According to (4.8.1a) and (4.8.45), a charged lepton $\alpha = e, \mu, \tau$ couples to a superposition of neutrino mass eigenstates ν_j

$$\nu_\alpha = \sum_{j=1}^{3} V_{\alpha j} \nu_j, \qquad (4.10.19)$$

where V is the unitary mixing matrix in (4.8.45). Only these flavor states ν_α can be observed in experiments. On the other hand, the time evolution of the free Dirac fields $\nu_j(t)$ is diagonal in the mass eigenstates. Determine the transition amplitude from a flavor state $\nu_\alpha(0)$ to $\nu_\beta(t)$.

4.10 Calculate the probability of a transition $\nu_\alpha(0) \to \nu_\beta(t)$, using the fact that the neutrino masses are small $m_j \ll |p|$. Assume a mass hierarchy

$m_{\nu_e} \ll m_{\nu_\mu} \ll m_{\nu_\tau}$. Show that the transition probability oscillates with a period proportional to the mass difference $m_{\nu_\tau} - m_{\nu_e}$.

4.11 Consider the gauge theory with three massive gauge fields. Show with the relations in Sect.4.4 that all masses must be equal and one gets the $su(2)$-theory.

4.12 Show that a gauge theory with four massive gauge fields is impossible.

Hint: Owing to the antisymmetry of f_{abc} there are four independent coupling constants $f_{123}, f_{124}, f_{134}$ and f_{234}. Examine the relations in Sect.4.4 and show that there is no non-trivial solution with all four masses positive.

4.13 Construct the representation (4.4.69) in case of the electroweak theory.

Hint: This is a 4-dimensional *real* representation of $su(2) + u(1)$. The 3-dimensional $su(2)$ part is equivalent to the Lie algebra of the rotation group $SO(3)$.

5. Spin-2 gauge theory

In the last two chapters we have seen the strength of the gauge principle: causal gauge invariance determines gauge theories with massless and massive gauge vector fields in all details. To make a crucial test of the principle we go one step further to tensor fields. We will consider symmetric tensor fields because we have gravity in mind, but otherwise we assume to be ignorant of Einstein's general relativity. Instead we ask our standard question: which self-couplings of tensor fields are consistent with causal gauge invariance? The answer comes out very satisfactory. As in the case of Yang-Mills theory in Sect.3.2, the coupling is unique up to divergence- and co-boundary couplings. This unique self-coupling is the same as the one derived from the Einstein - Hilbert Lagrangian (Sect.5.3). This is a consequence of first order gauge invariance. In second order quartic couplings are generated in the same way as in Sect.3.4 for massless vector fields. These quartic couplings also agree with the ones derived from the Einstein - Hilbert Lagrangian. Whether this holds in arbitrary order is not known at present. But on the basis of these results it is already clear that causal quantum gauge invariance is a universal principle of nature: it governs all fundamental interactions. For this reason we prefer it as the guiding principle compared to other ideas.

The method works equally well in the case of massive tensor fields. For this reason we include the construction of massive spin-2 gauge theory. At present it is not clear whether this "massive gravity" has physical relevance. But it has a very interesting property: in the limit of vanishing graviton mass it does not go over into the massless theory. So this leads to an alternative theory of gravity which is worth while to be investigated. The reader not interested in massive gravity can skip the corresponding sections without getting difficulties for understanding the normal massless case.

There is a new aspect in quantum gravity: the couplings are necessarily of the non-normalizable type. In the framework of causal perturbation theory this means that the number of free normalization constants increases with the order n of the perturbation expansion. It remains to decide which of these many constants have observable consequences and if so, how can they be fixed by physical conditions. These problems are not analyzed here. But a problem of ultraviolet divergences does not exist in the causal approach.

5.1 Causal gauge invariance with massless tensor fields

Our fundamental free asymptotic fields in Minkowski space are a symmetric tensor field $h^{\mu\nu}(x)$ discussed in Sect.1.7 and ghost and anti-ghost fields $u^\mu(x)$, $\tilde{u}^\nu(x)$ discussed in Sect.1.6. We do not assume that $h^{\mu\nu}$ has something to do with the metric tensor $g^{\mu\nu}$ of general relativity. Of course, as discussed in Sect.1.7, the ten independent components of $h^{\mu\nu}$ contain more than the five independent components of a spin-2 field. The additional freedom can be reduced in the classical theory by a gauge condition $h^{\mu\nu}_{,\nu} = 0$ and a trace condition $h^\mu_\mu = h = 0$ (see (1.7.3-4)), but these conditions are disregarded in the construction of the quantum gauge theory. They enter at a later stage when physical states are defined, as discussed in Sect.1.7.

The h-field is quantized as follows

$$[h^{\alpha\beta}(x), h^{\mu\nu}(y)] = -ib^{\alpha\beta\mu\nu} D(x-y) \tag{5.1.1}$$

with

$$b^{\alpha\beta\mu\nu} = \tfrac{1}{2}(\eta^{\alpha\mu}\eta^{\beta\nu} + \eta^{\alpha\nu}\eta^{\beta\mu} - \eta^{\alpha\beta}\eta^{\mu\nu}). \tag{5.1.2}$$

$D(x)$ is the mass-zero Jordan-Pauli distribution and $\eta^{\mu\nu}$ the Minkowski metric. From now on we reserve the letter $g^{\mu\nu}$ for another non-flat metric. The indices are still raised and lowered with respect to the Minkowski metric. In complete analogy to spin-1 theories one introduces a gauge charge

$$Q \stackrel{\text{def}}{=} \int_{x^0=t} d^3x\, h^{\alpha\beta}_{,\beta} \overleftrightarrow{\partial}_0 u_\alpha. \tag{5.1.3}$$

In order to get a nilpotent Q

$$Q^2 = 0, \tag{5.1.4}$$

we have to quantize the ghost fields with anticommutators

$$\{u^\mu(x), \tilde{u}^\nu(y)\} = i\eta^{\mu\nu} D(x-y) \tag{5.1.5}$$

and zero otherwise. All asymptotic fields fulfill the wave equation

$$\Box h^{\mu\nu}(x) = 0 = \Box u^\alpha(x) = \Box \tilde{u}^\beta(x). \tag{5.1.6}$$

The gauge charge Q (5.1.3) defines a gauge variation according to

$$d_Q F \stackrel{\text{def}}{=} QF - (-1)^{n_F} FQ, \tag{5.1.7}$$

where n_F is the number of ghost minus anti-ghost fields in the Wick monomial F. We get the following gauge variations of the fundamental fields

$$d_Q h^{\mu\nu} = -\frac{i}{2}(u^\mu_{,\nu} + u^\nu_{,\mu} - \eta^{\mu\nu} u^\alpha_{,\alpha}) \tag{5.1.8}$$

$$= -ib^{\mu\nu\alpha\beta} u^\alpha_{,\beta}$$

$$d_Q h = iu^\mu_{,\mu}, \quad d_Q h^{\mu\nu}_{,\nu} = 0 \tag{5.1.9}$$

$$d_Q u^\mu = 0, \quad d_Q \tilde{u}^\mu = ih^{\mu\nu}_{,\nu}. \tag{5.1.10}$$

The asymptotic fields will be used to construct the time-ordered products T_n in the adiabatically switched S-matrix

$$S(g) = 1 + \sum_{n=1}^{\infty} \frac{1}{n!} \int d^4x_1 \ldots d^4x_n \, T_n(x_1, \ldots, x_n) g(x_1) \ldots g(x_n), \tag{5.1.11}$$

where $g(x)$ is a Schwartz test function. As always, the T_n's are expressed by normally ordered products of asymptotic fields. It is our essential point that gauge invariance of the S-matrix can be directly formulated with help of the T_n's. First order gauge invariance means that $d_Q T_1$ is a divergence

$$d_Q T_1(x) = i\partial_\mu T^\mu_{1/1}(x). \tag{5.1.12}$$

The $T^\mu_{1/1}$ appearing here is the Q-vertex. The definition of n-th order gauge invariance then reads as follows

$$d_Q T_n = [Q, T_n] = i \sum_{l=1}^{n} \frac{\partial}{\partial x^\mu_l} T^\mu_{n/l}(x_1, \ldots x_l \ldots x_n). \tag{5.1.13}$$

Here $T^\mu_{n/l}$ is the time-ordered product with a Q-vertex at x_l, while all other $n-1$ vertices are ordinary vertices T_1.

As a first attempt of a spin-2 theory let us consider the general trilinear coupling of the h-field and ghosts without derivatives

$$T_1 = a_1 hhh + a_2 hh^{\mu\nu} h_{\mu\nu} + a_3 h^{\mu\nu} h_{\nu\alpha} h^\alpha_\mu$$

$$+ b_1 u^\mu \tilde{u}_\mu h + b_2 u^\mu \tilde{u}^\nu h_{\mu\nu}. \tag{5.1.14}$$

As always, $h = h^\mu_\mu$ is the trace of the tensor field. *All products in (5.1.14) and throughout are normally ordered products, we always omit the double dots.* We also make a general ansatz for the Q-vertex which has ghost number one

$$T^\mu_{1/1} = c_1 u^\mu hh + c_2 u^\mu h^{\alpha\beta} h_{\alpha\beta} + c_3 u_\alpha h^{\alpha\mu} h + c_4 u^\alpha h_{\alpha\beta} h^{\beta\mu} + c_5 u^\mu \tilde{u}_\alpha u^\alpha. \tag{5.1.15}$$

This form is restricted by nilpotency

$$d_Q^2 T_1 = d_Q \partial_\mu T^\mu_{1/1} = 0. \tag{5.1.16}$$

If we insert (5.1.15) and set the coefficients of all Wick monomials equal to zero, we get a linear homogeneous system of equations for the c_j in (5.1.15). This system is overdetermined and has the trivial solution only. Consequently, the simple coupling (5.1.14) without derivatives does not lead to a gauge theory. With one derivative we cannot form Lorentz scalars, therefore, we are forced to consider the non-normalizable case of two derivatives.

The non-normalizable is a consequence of the expression for the singular order ω of a graph with n_h external gravitons and $n_u = n_{\tilde{u}}$ external ghosts which is given by

$$\omega \leq 4 - n_h - n_u - n_{\tilde{u}} - d + n. \tag{5.1.17}$$

Here d is the number of derivatives on the external fields and n is the order of perturbation theory. Apart from the latter, the expression is the same as (2.7.2) for Yang-Mills theories. In fact, the inductive step in the proof is completely identical. As in (2.7.11) it gives the result

$$\omega \leq \omega_1 + \omega_2 + 2l + a - 4 \tag{5.1.18}$$

for the combination of two subgraphs with ω_1, ω_2 by performing l contractions with totally a derivatives on the inner lines. This suggests the following ansatz for ω

$$\omega \leq 4 + 2l + a + \alpha n. \tag{5.1.19}$$

The 4 is a consequence of translation invariance. The term αn (with α still unknown) is possible because the order of perturbation theory is additive. If α comes out $= 0$, the theory is normalizable; for $\alpha < 0$ it is super-normalizable and for $\alpha > 0$ non-normalizable.

We want to express $2l$ in (5.1.19) by the number of external legs. Assuming n_1 graviton and n_2 ghost vertices in the graph, $n_1 + n_2 = n$, the number of external graviton and ghost lines is equal to

$$n_h = 3n_1 + n_2 - 2l_1, \quad n_u = n_{\tilde{u}} = n_2 - l_2, \tag{5.1.20}$$

where l_1 and l_2 are the numbers of hh- and $u\tilde{u}$-contractions, respectively, $l_1 + l_2 = l$. This implies

$$2l = 2(l_1 + l_2) = 3n_1 + n_2 - n_h + 2n_2 - n_u - n_{\tilde{u}}$$
$$= 3n - n_h - n_u - n_{\tilde{u}}.$$

Since the total number of derivatives is $2n = d + a$, we finally obtain

$$\omega \leq 4 - n_h - n_u - n_{\tilde{u}} - d + (5 + \alpha)n. \tag{5.1.21}$$

There remains to determine the constant α which is most easily obtained from $n = 1$. All first order vertices must always have $\omega = 0$, as can be checked in all previous theories. This gives

$$\omega = 4 - 3 - 2 + 5 + \alpha = 0,$$

so that $\alpha = -4$. This finishes the derivation of (5.1.17).

If we add two derivatives in (5.1.14) and (5.1.15) in all possible ways, the theory becomes very big. It is therefore important to use every mean to keep it smaller. One such tool is a result of Sect.3.3. According to theorem 3.2 the addition of a divergence to T_1 leads to a physically equivalent theory. This enables us to simplify the coupling T_1 by adding suitable divergences.

5.2 First order gauge invariance and descent equations

It is our aim to find solutions of the gauge invariance condition (5.1.12)

$$[Q, T(x)] = i\partial_\alpha T^\alpha(x), \tag{5.2.1}$$

where T and T^α are Wick polynomials in the fields introduced in the last section. Solving this problem as it stands by making a general ansatz is rather involved. Therefore we first reformulate it in a different way. Since d_Q and the space-time derivative ∂_α commute it follows from nilpotency $d_Q^2 = 0$ that

$$\partial_\alpha d_Q T^\alpha = 0.$$

If the appropriate form of the Poincaré lemma is true, this implies

$$d_Q T^\alpha = i\partial_\beta T^{\alpha\beta} \tag{5.2.2}$$

with antisymmetric $T^{\alpha\beta}$. In the same way we get

$$d_Q T^{\alpha\beta}] = i\partial_\gamma T^{\alpha\beta\gamma} \ldots \tag{5.2.3}$$

with totally antisymmetric $T^{\alpha\beta\gamma}$ and so on. These are the so-called descent equations, they define the co-homological formulation of gauge invariance.

However there are two obstacles for applying the usual Poincaré lemma: first our co-cycles are polynomials and second we are working on the mass shell; all fields obey the wave equation. If only the first obstacle would be present then we could apply the so-called algebraic Poincaré lemma (N.Dragon, Schladming lectures, hep-th/9602163), but unfortunately this nice result breaks down if we work on shell. As a consequence the above argument does not prove the descent equations. It is a highly non-trivial fact that, nevertheless, the descent equations hold in our situation. In the spin-2 case the chain of equations stops after (5.2.3). Working backwards from (5.2.3) to (5.2.1) is the descent procedure which we now carry out for the massless spin-2 case.

The descent procedure starts from $T^{\alpha\beta\gamma}$ which must contain three ghost fields u and two derivatives and is totally antisymmetric. To exclude trivial couplings (see Sect.3.3.3) we require that it does not contain a co-boundary $d_Q B$ for some $B \neq 0$. Therefore we first analyze the structure of co-boundaries depending on u and derivatives of it.

It is easy to prove the following identity

$$\partial_\mu \partial_\nu u_\varrho = i d_Q \Big[\partial_\mu h_{\nu\varrho} + \partial_\nu h_{\mu\varrho} - \partial_\varrho h_{\mu\nu} - \frac{1}{2}(\eta_{\nu\varrho}\partial_\mu h + \eta_{\mu\varrho}\partial_\nu h - \eta_{\mu\nu}\partial_\varrho h)\Big]. \tag{5.2.4}$$

This implies that all derivatives of u of order greater or equal to 2 are co-boundaries. This excludes higher derivative couplings which are sometimes considered in gravity theories. Normal products of u's with some $d_Q b$ are also co-boundaries because $d_Q u = 0$. Since $T^{\alpha\beta\gamma}$ has 3 factors u^μ there can only be zero or two first derivatives of u. As verified above (5.1.14) the first possibility gives no solution in the massless case, but it will appear in the massive one in Sect.5.4.

The differentiated ghost fields still contain co-boundaries due to the identity

$$\partial_\alpha u_\beta = (\partial_\alpha u_\beta - \partial_\beta u_\alpha) + id_Q\left(h_{\alpha\beta} - \tfrac{1}{2}\eta_{\alpha\beta}h\right). \tag{5.2.5}$$

Only the antisymmetric derivatives

$$u_{\alpha\beta} \equiv \tfrac{1}{2}(\partial_\alpha u_\beta - \partial_\beta u_\alpha) \tag{5.2.6}$$

are not co-boundaries. Then there are the following two possibilities only:

$$u_{\alpha\mu}u_{\beta\mu}u^\gamma, \quad u_{\alpha\beta}u_{\gamma\mu}u^\mu, \tag{5.2.7}$$

which have to be antisymmetrized in α, β, γ. In the second expression the u without derivative has the index μ which is summed over, in contrast to the first expression. All products are normally ordered.

Having determined the generic expression of $T^{\alpha\beta\gamma}$ in terms of antisymmetric derivatives we can return to ordinary partial derivatives in the further computations due to the identity (5.2.5). Then we start the descent procedure with the expression

$$T^{\alpha\beta\gamma} = a_1(\partial^\beta u^\alpha u^\mu \partial_\mu u^\gamma - \partial^\alpha u^\beta u^\mu \partial_\mu u^\gamma - \partial^\beta u^\gamma u^\mu \partial_\mu u^\alpha - \partial^\gamma u^\alpha u^\mu \partial_\mu u^\beta$$
$$+\partial^\alpha u^\gamma u^\mu \partial_\mu u^\beta + \partial^\gamma u^\beta u^\mu \partial_\mu u^\alpha) + a_2(\partial^\alpha u^\mu \partial_\mu u^\beta u^\gamma - \partial^\beta u^\mu \partial_\mu u^\alpha u^\gamma$$
$$-\partial^\gamma u^\mu \partial_\mu u^\beta u^\alpha - \partial^\alpha u^\mu \partial_\mu u^\gamma u^\beta + \partial^\gamma u^\mu \partial_\mu u^\alpha u^\beta + \partial^\beta u^\mu \partial_\mu u^\gamma u^\alpha). \tag{5.2.8}$$

Next we have to compute $\partial_\gamma T^{\alpha\beta\gamma}$ and this is equal to $-id_Q T^{\alpha\beta}$ by (5.2.3). To determine $T^{\alpha\beta}$ requires an "integration" d_Q^{-1}. As always in calculus this integration can be achieved by making a suitable ansatz for $T^{\alpha\beta}$ and fixing the free parameters. The following 5 parameter expression will do:

$$T^{\alpha\beta} = b_1 u^\mu \partial_\mu u_\nu \partial^\beta h^{\alpha\nu} + b_2 u^\mu \partial_\nu u^\alpha \partial_\mu h^{\beta\nu} + b_3 u^\alpha \partial_\nu u^\mu \partial_\mu h^{\beta\nu}$$
$$+\frac{b_4}{2}\partial_\mu u^\alpha \partial_\nu u^\beta h^{\mu\nu} + b_5 \partial_\mu u^\mu \partial_\nu u^\alpha h^{\beta\nu} - (\alpha \leftrightarrow \beta). \tag{5.2.9}$$

Substituting this into (5.2.3) leads to

$$b_1 = -2a_1, \ b_2 = -2a_2 = -2a_1, \ b_3 = 2a_1, \ b_4 = -4a_1, \ b_5 = -2a_1.$$

An overall factor is arbitrary, we take $a_1 = -1$ which gives

$$T^{\alpha\beta} = 2(u^\mu \partial_\mu u_\nu \partial^\beta h^{\alpha\nu} + u^\mu \partial_\nu u^\alpha \partial_\mu h^{\beta\nu} - u^\alpha \partial_\nu u^\mu \partial_\mu h^{\beta\nu}$$
$$+\partial_\mu u^\alpha \partial_\nu u^\beta h^{\mu\nu} + \partial_\nu u^\nu \partial_\mu u^\alpha h^{\beta\mu}) - (\alpha \leftrightarrow \beta). \tag{5.2.10}$$

Next in the same way we compute $\partial_\beta T^{\alpha\beta}$ and make an ansatz for T^α. The latter now has to contain ghost-antighost couplings also. The precise form can be taken from the following final result:

Spin-2 gauge theory

$$T^\alpha = 4u^\mu \partial_\mu h_{\beta\nu} \partial^\beta h^{\alpha\nu} - 2u^\mu \partial_\mu h^{\beta\nu} \partial^\alpha h_{\beta\nu} - 2u^\alpha \partial^\beta h^{\mu\nu} \partial_\mu h_{\beta\nu} - 4\partial_\nu u_\beta \partial_\mu h^{\alpha\beta} h^{\mu\nu}$$

$$+ 4\partial_\nu u^\nu \partial^\mu h^{\alpha\beta} h_{\beta\mu} + u^\alpha \partial_\beta h_{\mu\nu} \partial^\beta h^{\mu\nu} - 2\partial_\nu u^\nu h_{\mu\beta} \partial^\alpha h^{\mu\beta} -$$

$$-\frac{1}{2} u^\alpha \partial_\mu h \partial^\mu h + \partial_\nu u^\nu h h^\alpha + u^\nu \partial_\nu h \partial^\alpha h - 2\partial_\nu u^\mu h^{\mu\nu} \partial^\alpha h$$

$$+4\partial^\nu u_\mu \partial^\alpha h^{\mu\beta} h_{\beta\nu} - 4\partial^\nu u^\mu \partial_\mu h^{\alpha\beta} h_{\beta\nu} - 2u^\mu \partial_\mu u^\nu \partial^\alpha \tilde{u}_\nu + 2u^\mu \partial_\nu u^\alpha \partial_\mu \tilde{u}^\nu$$

$$-2u^\alpha \partial_\nu u^\mu \partial_\mu \tilde{u}^\nu + 2\partial_\nu u^\nu \partial_\mu u^\alpha \tilde{u}^\mu + 2u^\mu \partial_\mu \partial_\nu u^\nu \tilde{u}^\alpha - 2u^\alpha \partial_\mu \partial_\nu u^\mu \tilde{u}^\nu. \quad (5.2.11)$$

The last step calculating $\partial_\alpha T^\alpha$ and setting it equal to $-id_Q T$ gives the trilinear coupling of massless gravity

$$T = -h^{\alpha\beta} \partial_\alpha h \partial_\beta h + 2h^{\alpha\beta} \partial_\alpha h_{\mu\nu} \partial_\beta h^{\mu\nu} + 4h_{\alpha\beta} \partial_\nu h^{\beta\mu} \partial_\mu h^{\alpha\nu}$$

$$+ 2h_{\alpha\beta} \partial_\mu h^{\alpha\beta} \partial^\mu h - 4h_{\alpha\beta} \partial_\nu h^{\alpha\mu} \partial^\nu h^\beta_\mu$$

$$-4u^\mu \partial_\beta \tilde{u}_\nu \partial_\mu h^{\nu\beta} + 4\partial_\nu u^\beta \partial_\mu \tilde{u}_\beta h^{\mu\nu} - 4\partial_\nu u^\nu \partial_\mu \tilde{u}^\beta h^{\beta\mu} + 4\partial_\nu u^\mu \partial_\mu \tilde{u}_\beta h^{\nu\beta}. \quad (5.2.12)$$

The two terms in the second line which have the two derivatives contracted are divergences. To show this we derive a useful identity which we state for arbitrary massive free fields:

Lemma: Let f_1, f_2, f_3 satisfy the Klein-Gordon equations with masses m_1, m_2, m_3 respectively, then

$$2\partial_\alpha f_1 \partial^\alpha f_2 f_3 = \partial^\alpha \Big(\partial_\alpha f_1 f_2 f_3 + f_1 \partial_\alpha f_2 f_3 - f_1 f_2 \partial_\alpha f_3 \Big)$$

$$+ (m_1^2 + m_2^2 - m_3^2) f_1 f_2 f_3. \quad (5.2.13)$$

Proof: Using repeatedly the Klein-Gordon equation we get

$$\partial_\alpha f_1 \partial^\alpha f_2 f_3 = \partial_\alpha (f_1 \partial^\alpha f_2 f_3) + m_2^2 f_1 f_2 f_3 - f_1 \partial_\alpha f_2 \partial^\alpha f_3 \quad (a)$$

$$= \partial_\alpha (\partial^\alpha f_1 f_2 f_3) + m_1^2 f_1 f_2 f_3 - \partial_\alpha f_1 f_2 \partial^\alpha f_3$$

$$= \partial_\alpha (\partial^\alpha f_1 f_2 f_3) - \partial^\alpha (f_1 f_2 \partial_\alpha f_3) + m_1^2 f_1 f_2 f_3 + f_1 \partial_\alpha f_2 \partial^\alpha f_3 - m_3^2 f_1 f_2 f_3. \quad (b)$$

Adding (a) and (b) gives (5.2.13).

Since we have massless fields in (5.2.12), the coupling terms in the second line are indeed divergences. Remembering the discussion in Sect.3.3.3 (Thm. 3.3.1) these terms can be omitted because they do not alter the S-matrix. At the moment we keep these divergence term for the discussion of the result in Sect.5.5. There we show that the coupling (5.2.12) corresponds to Einstein's general relativity in the classical limit.

5.3 Massive tensor fields

Now we assume that the tensor field has mass m and satisfies the Klein-Gordon equation :
$$(\Box + m^2)h^{\mu\nu} = 0. \tag{5.3.1}$$
As in the case of the massive vector field the corresponding ghost field u^μ and the antighost field $\tilde{u}^\nu$ must have the same mass:
$$(\Box + m^2)u^\mu = 0 = (\Box + m^2)\tilde{u}^\mu. \tag{5.3.2}$$
These fields are quantized as follows
$$[h^{\alpha\beta}(x), h^{\mu\nu}(y)] = -ib^{\alpha\beta\mu\nu}D_m(x-y) \tag{5.3.3}$$
$$\{u^\mu(x), \tilde{u}^\nu(y)\} = i\eta^{\mu\nu}D_m(x-y) \tag{5.3.4}$$

Like in the spin-1 case (1.5.36) (1.5.38) the gauge variation of the massive tensor field and the ghost u^μ has the same form as in the massless case:
$$d_Q h^{\mu\nu} = -\frac{i}{2}(u^{\mu,\nu} + u^{\nu,\mu} - \eta^{\mu\nu}u^\alpha_{,\alpha}) \tag{5.3.5}$$
$$d_Q u^\mu = 0. \tag{5.3.6}$$
But $d_Q \tilde{u}^\nu$ must be changed in such a way that d_Q is nilpotent. This requires the introduction of a vector field $v^\mu(x)$ with the same mass
$$(\Box + m^2)v^\mu = 0$$
which we call vector-graviton field or v-field for short. It is quantized according to
$$[v^\mu(x), v^\nu(y)] = \frac{i}{2}\eta^{\mu\nu}D_m(x-y), \tag{5.3.7}$$
the factor $1/2$ is a convenient convention. This field appears in the gauge variation of $\tilde{u}^\mu$
$$d_Q \tilde{u}^\mu = i(\partial_\nu h^{\mu\nu} + mv^\mu) \tag{5.3.8}$$
in a similar way as the scalar field Φ in (1.5.39). Finally
$$d_Q v^\mu = -\frac{i}{2}mu^\mu \tag{5.3.9}.$$

It is not hard to verify nilpotency $d_Q^2 = 0$. Using the commutation rules above one can show as in Prop.1.4.1 that Q is expressed in x-space as follows
$$Q = \int_{x^0=t} d^3x \left[\partial_\nu h^{\mu\nu}(x) + mv^\mu(x)\right] \overleftrightarrow{\partial}_0 u_\mu(x). \tag{5.3.10}$$

This is completely analogous to (1.5.31) in the spin-1 case of massive vector fields.

Spin-2 gauge theory

To understand this gauge structure better we construct a Hilbert space representation. For this purpose we express the various fields by means of emission and absorption operators. We follow the discussion of the massless case in Sect.1.7 as close as possible. We decompose $h^{\alpha\beta}$ into its traceless part and the trace h

$$h^{\alpha\beta}(x) = H^{\alpha\beta}(x) + \frac{1}{4}g^{\alpha\beta}h(x). \tag{5.3.11}$$

From (5.3.3) we obtain the following commutation relations

$$[h(x), h(y)] = 4iD_m(x-y) \tag{5.3.12}$$

$$[H^{\alpha\beta}(x), H^{\mu\nu}(y)] = -it^{\alpha\beta\mu\nu}D_m(x-y), \tag{5.3.13}$$

with

$$t^{\alpha\beta\mu\nu} \stackrel{\text{def}}{=} \tfrac{1}{2}(\eta^{\alpha\mu}\eta^{\beta\nu} + \eta^{\alpha\nu}\eta^{\beta\mu} - \tfrac{1}{2}\eta^{\alpha\beta}\eta^{\mu\nu}) = t^{\mu\nu\alpha\beta} \tag{5.3.14}$$

and

$$[H^{\alpha\beta}(x), h(y)] = 0. \tag{5.3.15}$$

We claim that the fields in (5.3.12) (5.3.13) can be represented as follows

$$H^{\alpha\beta}(x) = (2\pi)^{-3/2} \int \frac{d^3k}{\sqrt{2\omega}} \left(a_{\alpha\beta}(\boldsymbol{k})e^{-ikx} + \eta^{\alpha\alpha}\eta^{\beta\beta}a^+_{\alpha\beta}(\boldsymbol{k})e^{ikx} \right), \tag{5.3.16}$$

where $a_{\alpha\beta} = a_{\beta\alpha}$ is symmetric and satisfies the commutation relation

$$[a_{\alpha\beta}(\boldsymbol{k}), a^+_{\mu\nu}(\boldsymbol{k}')] = \eta^{\alpha\alpha}\eta^{\beta\beta}t^{\alpha\beta\mu\nu}\delta(\boldsymbol{k}-\boldsymbol{k}'). \tag{5.3.17}$$

The trace part is given by

$$h(x) = (2\pi)^{-3/2} \int \frac{d^3k}{\sqrt{2\omega}} \left(a(\boldsymbol{k})e^{-ikx} - a^+(\boldsymbol{k})e^{ikx} \right) \tag{5.3.18}$$

with

$$[a(\boldsymbol{k}), a^+(\boldsymbol{k}')] = 4\delta(\boldsymbol{k}-\boldsymbol{k}'). \tag{5.3.19}$$

Since the right-hand side is positive, the h-sector of Fock space can be constructed in the usual way by applying products of a^+'s to the vacuum.

The situation is not so simple in the H-sector because the righthand side of (5.3.17) is not a diagonal matrix. We perform a linear transformation of the diagonal operators $a_{\alpha\alpha}$ and $a^+_{\alpha\alpha}$ in such a way that the new operators are usual annihilation and creation operators

$$[\tilde{a}_{\alpha\alpha}(\boldsymbol{k}), \tilde{a}^+_{\beta\beta}(\boldsymbol{k}')] = \delta_{\alpha\beta}\delta(\boldsymbol{k}-\boldsymbol{k}'). \tag{5.3.20}$$

The following transformation does the job:

$$a_{00} = \tfrac{1}{2}(\tilde{a}_{11} + \tilde{a}_{22} + \tilde{a}_{33})$$

$$a_{11} = \tfrac{1}{2}(-\tilde{a}_{11} + \tilde{a}_{22} + \tilde{a}_{33})$$

$$a_{22} = \tfrac{1}{2}(\tilde{a}_{11} - \tilde{a}_{22} + \tilde{a}_{33})$$

$$a_{33} = \tfrac{1}{2}(\tilde{a}_{11} + \tilde{a}_{22} - \tilde{a}_{33}). \qquad (5.3.21)$$

We note that $\tilde{a}_{00}$ does not appear because one pair of absorption and emission operators is superfluous due to the trace condition $H^\alpha{}_\alpha = 0$. In fact, from (5.3.21) we see

$$\sum_{j=1}^{3} a_{jj} = a_{00}. \qquad (5.3.22)$$

The Fock representation can now be constructed as usual by means of $\tilde{a}_{11}^+, \tilde{a}_{22}^+, \tilde{a}_{33}^+$ and $a_{\alpha\beta}^+$ with $\alpha \neq \beta$.

The other fields have the following representation in terms of emission and absorption operators:

$$u^\mu(x) = (2\pi)^{-3/2} \int \frac{d^3k}{\sqrt{2E_k}} \left(c_2^\mu(\boldsymbol{k}) e^{-ikx} - \eta^{\mu\mu} c_1^\mu(\boldsymbol{k})^+ e^{ikx} \right) \qquad (5.3.23)$$

$$\tilde{u}^\mu(x) = (2\pi)^{-3/2} \int \frac{d^3k}{\sqrt{2E_k}} \left(-c_1^\mu(\boldsymbol{k}) e^{-ikx} - \eta^{\mu\mu} c_2^\mu(\boldsymbol{k})^+ e^{ikx} \right) \qquad (5.3.24)$$

$$v^\mu(x) = (2\pi)^{-3/2} \int \frac{d^3k}{2\sqrt{E_k}} \left(b^\mu(\boldsymbol{k}) e^{-ikx} - \eta^{\mu\mu} b^\mu(\boldsymbol{k})^+ e^{ikx} \right) \qquad (5.3.25)$$

with the following (anti)commutation relations

$$\{c_j^\mu(\boldsymbol{k}), c_l^\nu(\boldsymbol{k}')^+\} = \delta_{jl} \delta_\nu^\mu \delta^3(\boldsymbol{k} - \boldsymbol{k}'), \qquad (5.3.26)$$

$$[b^\mu(\boldsymbol{k}), b^\nu(\boldsymbol{k}')^+] = \delta_\nu^\mu \delta^3(\boldsymbol{k} - \boldsymbol{k}'). \qquad (5.3.27)$$

Then the gauge charge Q (5.3.10) can be written in momentum space as follows

$$Q = \int d^3k \left(A^\alpha(\boldsymbol{k})^+ c_2^\gamma(\boldsymbol{k}) - B^\alpha(\boldsymbol{k}) c_1^\gamma(\boldsymbol{k})^+ \right) \eta_{\alpha\gamma}, \qquad (5.3.28)$$

where

$$A^\alpha = \eta^{\alpha\alpha} \eta^{\beta\beta} a^{\alpha\beta}(\boldsymbol{k}) k^\beta - \frac{k^\alpha}{4} d(\boldsymbol{k}) - im_1 \eta^{\alpha\alpha} b^\alpha \qquad (5.3.29)$$

$$B^\alpha = (a^{\alpha\beta}(\boldsymbol{k}) k_\beta + \frac{k^\alpha}{4} d(\boldsymbol{k}) + im_1 b^\alpha) \eta^{\alpha\alpha}, \qquad (5.3.30)$$

$$m_1 = \frac{m}{\sqrt{2}}. \qquad (5.3.31)$$

The adjoint is given by

$$Q^+ = \int d^3k \left(c_2^\beta(\boldsymbol{k})^+ A^\alpha(\boldsymbol{k}) - c_1^\beta(\boldsymbol{k}) B^\alpha(\boldsymbol{k})^+ \right) \eta_{\alpha\beta}. \qquad (5.3.32)$$

According to (1.4.21) (1.4.26) the physical Hilbert space is expressed by means of the gauge charge Q in the following two equivalent forms:

$$\mathcal{H}_{\text{phys}} = \text{Ker}Q/\text{Ran}Q \qquad (5.3.33)$$

Spin-2 gauge theory

$$\mathcal{H}_\text{phys} - \text{Ker}(QQ^+ + Q^+Q). \tag{5.3.34}$$

The first form leaves open the choice of the representative in the equivalence classes modulo Ran Q. Since we work in a concrete representation, the second form (5.3.34) is better. We must study the selfadjoint operator

$$\{Q, Q^+\} = \int d^3k\, d^3k' \left(A^\alpha(\boldsymbol{k})^+ A^\beta(\boldsymbol{k}')\{c_2^\gamma(\boldsymbol{k}), c_2^\delta(\boldsymbol{k}')^+\} \right.$$

$$+ B^\beta(\boldsymbol{k}')^+ B^\alpha(\boldsymbol{k})\{c_1^\delta(\boldsymbol{k}'), c_1^\gamma(\boldsymbol{k})^+\} + c_2^\delta(\boldsymbol{k}')^+ c_2^\gamma(\boldsymbol{k})[A^\beta(\boldsymbol{k}'), A^\alpha(\boldsymbol{k})^+]$$

$$\left. + c_1^\gamma(\boldsymbol{k})^+ c_1^\delta(\boldsymbol{k}')[B^\alpha(\boldsymbol{k}), B^\beta(\boldsymbol{k}')^+] \right) \eta_{\alpha\gamma}\eta_{\beta\delta}. \tag{5.3.35}$$

We restrict to the graviton sector because the ghost sector is unphysical:

$$\{Q, Q^+\}|_\text{graviton} = \int d^3k \sum_{\alpha=0}^{3} \left(A^{\alpha+}A^\alpha + B^{\alpha+}B^\alpha \right). \tag{5.3.36}$$

It is convenient to introduce time-like and space-like components:

$$A^0 = k_0 (a^{00} - a^0_\| - \frac{d}{4} - \frac{im_1}{k_0} b^0),$$

$$A^j = k_0 (-a^{0j} + a^j_\| - \frac{k^j}{k_0}\frac{d}{4} + \frac{im_1}{k_0} b^j),$$

$$B^0 = k_0 (a^{00} + a^0_\| + \frac{d}{4} + \frac{im_1}{k_0} b^0),$$

$$B^j = k_0 (-a^{0j} - a^j_\| - \frac{k^j}{k_0}\frac{d}{4} - \frac{im_1}{k_0} b^j), \tag{5.3.37}$$

where

$$a^\mu_\| = \frac{k_j}{k_0} a^{\mu j}. \tag{5.3.38}$$

We choose a Lorentz frame where $k^\mu = (\omega, 0, 0, k^3)$. Then we get for the integrand in (5.3.36)

$$\sum_{\alpha=0}^{3} \left(A^{\alpha+}A^\alpha + B^{\alpha+}B^\alpha \right) = 2\omega^2 \Big\{ a^{00+}a^{00} + \frac{k_3^2}{\omega^2} a^{00+} a^{03} +$$

$$+ a^{01+}a^{01} + a^{02+}a^{02} + a^{03+}a^{03} +$$

$$+ \frac{k_3^2}{\omega^2}\left[a^{13+}a^{13} + a^{23+}a^{23} + a^{33+}a^{33} \right] +$$

$$+ \frac{d^+ d}{8} - \frac{m_1^2}{16\omega^2} d^+ d + \frac{im_1}{4\omega} d^+ b^0 - \frac{im_1}{4\omega} b^{0+} d +$$

$$+ \frac{im_1 k_3}{\omega^2} a^{03+} b^0 - \frac{im_1 k_3}{\omega^2} b^{0+} a^{03} +$$

$$+\frac{im_1k_3}{\omega^2}[a^{13+}b^1 + a^{23+}b^2 + a^{33+}b^3 - b^{1+}a^{13} - b^{2+}a^{23} - b^{3+}a^{33}]+$$

$$+\frac{m_1^2}{\omega^2}\left[b^{0+}b^0 + \sum_{j=1}^{3}b^{j+}b^j\right]\biggr\}. \tag{5.3.39}$$

Since a^{12+} does not appear inhere, the states $a^{12+}\Omega$ where Ω is the Fock vacuum certainly belong to the kernel of (5.3.36) and, hence, are in the physical subspace. We have still to substitute the diagonal operators $a^{\mu\mu}$ by means of (5.3.21) by the operators $\tilde{a}^{jj}$ which generate the Fock states. Then the quadratic form (5.3.39) can be represented in matrix notation A^+XA where A^+ stands for the emission operators

$$A^+ = (\tilde{a}_{11}^+, \tilde{a}_{22}^+, \tilde{a}_{33}^+, \sqrt{2}a_{01}^+, \sqrt{2}a_{02}^+, \sqrt{2}a_{03}^+, \sqrt{2}a_{13}^+, \sqrt{2}a_{23}^+, \tfrac{1}{2}d^+, b_0^+, b_1^+, b_2^+, b_3^+), \tag{5.3.40}$$

the numerical factors are necessary in order to get the states correctly normalized due to (5.3.17) (5.3.19) (5.3.20) and (5.3.27).

According to (5.3.39) X is the hermitian 13×13 matrix

$$X = \begin{pmatrix}
\tfrac{1}{2}(\omega^2+k_3^2) & \tfrac{1}{2}(\omega^2+k_3^2) & \tfrac{1}{2}m_1^2 & 0 & 0 & 0 & 0 & 0 & 0 & 0 & 0 & 0 & im_1k_3 \\
\tfrac{1}{2}(\omega^2+k_3^2) & \tfrac{1}{2}(\omega^2+k_3^2) & \tfrac{1}{2}m_1^2 & 0 & 0 & 0 & 0 & 0 & 0 & 0 & 0 & 0 & im_1k_3 \\
\tfrac{1}{2}m_1^2 & \tfrac{1}{2}m_1^2 & \tfrac{1}{2}(\omega^2+k_3^2) & 0 & 0 & 0 & 0 & 0 & 0 & 0 & 0 & 0 & -im_1k_3 \\
0 & 0 & 0 & 2\omega^2 & 0 & 0 & 0 & 0 & 0 & 0 & 0 & 0 & 0 \\
0 & 0 & 0 & 0 & 2\omega^2 & 0 & 0 & 0 & 0 & 0 & 0 & 0 & 0 \\
0 & 0 & 0 & 0 & 0 & 2(\omega^2+k_3^2) & 0 & 0 & 2im_1k_3 & 0 & 0 & 0 & 0 \\
0 & 0 & 0 & 0 & 0 & 0 & 2k_3^2 & 0 & 0 & 0 & 2im_1k_3 & 0 & 0 \\
0 & 0 & 0 & 0 & 0 & 0 & 0 & 2k_3^2 & 0 & 0 & 0 & 2im_1k_3 & 0 \\
0 & 0 & 0 & 0 & 0 & -2im_1k_3 & 0 & 0 & z & \tfrac{i}{2}m_1\omega & 0 & 0 & 0 \\
0 & 0 & 0 & 0 & 0 & 0 & 0 & 0 & -\tfrac{i}{2}m_1\omega & 2m_1^2 & 0 & 0 & 0 \\
-im_1k_3 & 0 & 0 & 0 & 0 & 0 & -2im_1k_3 & 0 & 0 & 0 & 2m_1^2 & 0 & 0 \\
0 & -im_1k_3 & 0 & 0 & 0 & 0 & 0 & -2im_1k_3 & 0 & 0 & 0 & 2m_1^2 & 0 \\
-im_1k_3 & -im_1k_3 & im_1k_3 & 0 & 0 & 0 & 0 & 0 & 0 & 0 & 0 & 0 & 2m_1^2
\end{pmatrix},$$

$$\tag{5.3.41}$$

Spin-2 gauge theory

where
$$z = \frac{\omega^2}{4} - \frac{m_1^2}{8}. \tag{5.3.42}$$

The eigenvalues of this matrix are:

$$\lambda_1 = \lambda_2 = \lambda_3 = \lambda_4 = \lambda_5 = 0$$
$$\lambda_6 = \lambda_7 = \lambda_8 = \lambda_9 = \lambda_{10} = 2(m_1^2 + k_3^2)$$
$$\lambda_{11} = \frac{3m_1^2}{2} + k_3^2 \tag{5.3.43}$$
$$\lambda_{12} = \frac{1}{16}\left(33m_1^2 + 34k_3^2 - \sqrt{m_1^4 + 900k_3^2(m_1^2 + k_3^2)}\right)$$
$$\lambda_{13} = \frac{1}{16}\left(33m_1^2 + 34k_3^2 + \sqrt{m_1^4 + 900k_3^2(m_1^2 + k_3^2)}\right).$$

The 5 eigenvectors with eigenvalue 0 determine the kernel. They can be easily calculated from the matrix (5.3.41). Together with the previously found state $a_{12}^+ \Omega$ we have 6 physical massive graviton modes:

$$\psi_1 = a_{12}^+ \Omega \tag{5.3.44}$$

$$\psi_2 = c_2 \left(\frac{im_1}{2k_3}(\tilde{a}_{11} + \tilde{a}_{22} - 2\tilde{a}_{33}) + b_3\right)^+ \Omega \tag{5.3.45}$$

$$c_2 = \left(1 + \frac{3m_1^2}{2k_3^2}\right)^{-1/2}$$

$$\psi_3 = c_3 \left(\frac{im_1}{k_3} a_{23} + b_2\right)^+ \Omega \tag{5.3.46}$$

$$c_3 = \left(1 + \frac{m_1^2}{k_3^2}\right)^{-1/2}$$

$$\psi_3 = c_3 \left(\frac{im_1}{k_3} a_{13} + b_1\right)^+ \Omega \tag{5.3.47}$$

$$c_3 = \left(1 + \frac{m_1^2}{k_3^2}\right)^{-1/2}$$

$$\psi_5 = c_5 \left(\frac{im_1}{m_1^2 + 2k_3^2}(k_3 a_{03} + 4\omega d) + b_0\right)^+ \Omega \tag{5.3.48}$$

$$c_5 = \left(1 + m_1^2 \frac{k_3^2 + 16\omega^2}{(m_1^2 + 2k_3^2)^2}\right)^{-1/2}.$$

$$\psi_6 = \frac{1}{\sqrt{2}}(\tilde{a}_{11} - \tilde{a}_{22})^+ \Omega. \tag{5.3.49}$$

All physical states are normalized to 1.

The states ψ_1 and ψ_6 agree exactly with the two transverse physical modes of massless gravitons (compare (1.7.44)). The other four massive physical graviton modes converge in the massless limit to the four vector graviton states $b_\mu^+ \Omega$, $\mu = 0, 1, 2, 3$. That means the vector graviton field describes physical degrees of freedom in our Hilbert space representation. One can construct another representation where the physical states are all in the tensor sector generated by $h^{\mu\nu}$. But this representation does not have a smooth limit for graviton mass $m \to 0$, so we do not consider it.

5.4 Massive gravity

Now we derive the gauge invariant coupling of massive spin-2 fields by the descent method in complete analogy to the massless case in Sect.5.2. In fact we will heavily use the previous calculations. The starting expression (5.2.8) must only be modified by a simple mass term

$$T_m^{\alpha\beta\gamma} = T^{\alpha\beta\gamma} + am^2 u^\alpha u^\beta u^\gamma. \tag{5.4.1}$$

This mass term is needed to compensate mass terms from the Klein-Gordon equation when we compute

$$\partial_\gamma T_m^{\alpha\beta\gamma} = (\partial_\gamma T^{\alpha\beta\gamma})_0 - m^2 u^\alpha u^\mu \partial_\mu u^\beta + m^2 u^\beta u^\mu \partial_\mu u^\alpha -$$
$$- m^2 u^\mu \partial_\mu u^\beta u^\alpha + m^2 u^\mu \partial_\mu u^\alpha u^\beta$$
$$+ am^2 (\partial_\gamma u^\alpha u^\beta u^\gamma + u^\alpha \partial_\gamma u^\beta u^\gamma + u^\alpha u^\beta \partial_\gamma u^\gamma). \tag{5.4.2}$$

Here the subscript 0 means always the exact zero mass expression from Sect.5.2, but with all fields massive, of course. The additional mass terms come from those terms in (5.2.8) which have a derivative ∂^γ such that with the ∂_γ in (5.4.2) a wave operator is obtained. Now most of the mass terms in (5.4.2) cancel if we choose $a = -2$.

The last term $u^\alpha u^\beta \partial_\gamma u^\gamma$ in (5.4.2) survives, it gives rise to a deformation of $T^{\alpha\beta}$ (5.2.10):

$$T_m^{\alpha\beta} = T^{\alpha\beta} + 2m(u^\beta \partial_\mu u^\mu v^\alpha - u^\alpha \partial_\mu u^\mu v^\beta). \tag{5.4.3}$$

Indeed, $-id_Q$ of the additional mass terms cancels the last mass terms in (5.4.2). In the same manner we compute

$$\partial_\beta T_m^{\alpha\beta} = (\partial_\beta T^{\alpha\beta})_0 - 2m^2 u^\mu \partial_\mu u_\nu h^{\nu\alpha}$$
$$+ 2m(\partial_\beta u^\beta \partial_\mu u^\mu v^\alpha + u^\beta \partial_\mu \partial_\beta u^\mu v^\alpha + u^\beta \partial_\mu u^\mu \partial_\beta v^\alpha -$$
$$- \partial_\beta u^\alpha \partial_\mu u^\mu v^\beta - u^\alpha \partial_\mu \partial_\beta u^\mu v^\beta - u^\alpha \partial_\mu u^\mu \partial_\beta v^\beta), \tag{5.4.4}$$

where the first term in the bracket vanishes by antisymmetry. Calculating $-id_Q T^\alpha$ from the massless expression (5.2.11) we now get additional terms mv^ν from $-id_Q \tilde{u}^\nu$. Consequently, (5.4.4) is equal to

$$-id_Q T^\alpha + 2m(u^\mu \partial_\mu u_\nu \partial^\alpha v^\nu - u^\mu \partial_\nu u^\alpha \partial_\mu v^\nu + u^\alpha \partial_\nu u^\mu \partial_\mu v^\nu -$$
$$- \partial_\nu u^\nu \partial_\mu u^\alpha v^\mu - u^\mu \partial_\mu \partial_\nu u^\nu v^\alpha + u^\alpha \partial_\mu \partial_\nu u^\mu v^\nu)$$
$$+ 2m(u^\beta \partial_\mu \partial_\beta u^\mu v^\alpha + u^\beta \partial_\mu u^\mu \partial_\beta v^\alpha - \partial_\beta u^\alpha \partial_\mu u^\mu v^\beta - u^\alpha \partial_\mu \partial_\beta u^\mu v^\beta - u^\alpha \partial_\mu u^\mu \partial_\beta v^\beta)$$
$$- 2m^2 u^\mu \partial_\mu u^\nu h^{\nu\alpha} =$$
$$= -id_Q T^\alpha + 2m u^\mu (\partial_\mu u^\nu \partial^\alpha v^\nu - \partial_\nu u^\alpha \partial_\mu v^\nu + \partial_\beta u^\beta v^\alpha)$$
$$+ 2m u^\alpha (\partial_\nu u^\mu \partial_\mu v^\nu - \partial_\mu u^\mu \partial_\beta v^\beta) - 2m^2 u^\mu \partial_\mu u^\nu h^{\nu\alpha}. \tag{5.4.5}$$

Spin-2 gauge theory

This implies

$$T_m^\alpha = T^\alpha + 4u^\mu \partial_\mu v_\nu \partial^\alpha v^\nu - 2u^\alpha \partial_\mu v_\nu \partial^\mu v^\nu$$
$$+ 4m(u^\alpha \partial_\mu v_\nu h^{\mu\nu} - u^\mu \partial_\mu v_\nu h^{\alpha\nu}) - m^2 u^\alpha (h_{\mu\nu} h^{\mu\nu} - \frac{1}{2} h^2). \quad (5.4.6)$$

In the last step we compute

$$\partial_\alpha T_m^\alpha = (\partial_\alpha T^\alpha)_0 + m^2 \partial_\nu u^\nu h^{\mu\beta} h_{\mu\beta} - \frac{m^2}{2} \partial_\nu u^\nu h^2 + 2m^2 \partial_\nu u_\mu h^{\mu\nu} h$$
$$- 4m^2 \partial_\nu u^\mu h_{\mu\beta} h^{\nu\beta} + 2m^2 u^\mu \partial_\mu u^\nu \tilde{u}_\nu$$
$$+ 4\partial_\alpha u^\mu \partial^\alpha v_\nu \partial_\mu v^\nu - 4m^2 u^\mu \partial_\mu v_\nu v^\nu - 2\partial_\alpha u^\alpha \partial_\mu v_\nu \partial^\mu v^\nu$$
$$+ 4m(\partial_\alpha u^\alpha \partial_\mu v_\nu h^{\mu\nu} + u^\alpha \partial_\mu v_\nu \partial_\alpha h^{\mu\nu} - \partial_\alpha u^\mu \partial_\mu v_\nu h^{\alpha\nu} - u^\mu \partial_\mu v_\nu \partial_\alpha h^{\alpha\nu}). \quad (5.4.7)$$

where the first m^2-terms come from the Klein-Gordon equation. This is equal to

$$= -i(d_Q T)_0 + 4m(u^\mu \partial_\beta v_\nu \partial_\mu h^{\nu\beta} - \partial_\nu u^\beta \partial_\mu v_\beta h^{\mu\nu} + \partial_\nu u^\nu \partial_\mu v_\beta h^{\mu\beta} -$$
$$- \partial_\nu u^\mu \partial_\mu v_\beta h^{\nu\beta}) - im^2 d_Q \left(\frac{4}{3} h_{\mu\nu} h^{\mu\beta} h^\nu_\beta - h^{\mu\beta} h_{\mu\beta} h + \frac{1}{6} h^3\right)$$
$$- 4im d_Q (\partial_\mu v^\nu u^\mu \tilde{u}_\nu) + 4i d_Q (h^{\mu\alpha} \partial_\alpha v^\nu \partial_\mu v_\nu), \quad (5.4.8)$$

as can be easily verified. When we compute d_Q of the four ghost coupling terms in the last line of (5.2.12) with massive fields, we obtain just the additional $4m(.)$-bracket in the first line of (5.4.8). Consequently, (5.4.8) is a co-boundary $-id_Q T_m$ where

$$T_m = T + m^2 \left(\frac{4}{3} h^{\mu\nu} h_{\mu\beta} h^{\nu\beta} - h^{\mu\beta} h_{\mu\beta} h + \frac{1}{6} h^3\right)$$
$$+ 4m u^\mu \tilde{u}^\nu \partial_\mu v_\nu - 4h^{\mu\alpha} \partial_\alpha v^\nu \partial_\mu v_\nu. \quad (5.4.9)$$

is the trilinear coupling of massive gravity:

The mass depending terms in T_m are expected, but the last coupling term independent of m

$$T_v = -4h^{\mu\nu} \partial_\nu v^\lambda \partial_\mu v_\lambda. \quad (5.4.10)$$

is a big surprise. In the limit $m \to 0$ this term survives. That means, *the massless limit of massive gravity is different from massless gravity.* There remains the (now massless) vector graviton field v^μ coupled to the tensor field $h^{\mu\nu}$. Even without any experimental sign of a non-zero mass of the graviton, there is now an alternative gauge theory for gravity with more degrees of freedom given by the v-field. For this reason we will continue to study the massive theory parallel to the massless one in the following sections.

5.5 Expansion of the Einstein-Hilbert Lagrangian

In this section we make contact to classical general relativity. For this purpose we leave quantum field theory aside and discuss classical field theory only, for details see any text on general relativity. We take the metric tensor $g_{\mu\nu}(x)$ as the fundamental field, but the indices are raised and lowered with $g^{\nu\varrho}$ itself, defined as the inverse

$$g_{\mu\nu} g^{\nu\varrho} = \delta^\varrho_\mu. \tag{5.5.1}$$

One also introduces the determinant

$$g = \det g_{\mu\nu}. \tag{5.5.2}$$

Our starting point is the Einstein-Hilbert action given by

$$S_{EH} = -\frac{2}{\kappa^2} \int d^4x \, \sqrt{-g} R, \quad \kappa^2 = 32\pi G, \tag{5.5.3}$$

where G is Newton's constant according to (2.5.45). R is the scalar curvature

$$R = g^{\mu\nu} R_{\mu\nu} \tag{5.5.4}$$

which follows from the Ricci tensor

$$R_{\mu\nu} = \partial_\alpha \Gamma^\alpha_{\mu\nu} - \partial_\nu \Gamma^\alpha_{\mu\alpha} + \Gamma^\alpha_{\alpha\beta} \Gamma^\beta_{\mu\nu} - \Gamma^\alpha_{\nu\beta} \Gamma^\beta_{\alpha\mu}, \tag{5.5.5}$$

where

$$\Gamma^\alpha_{\beta\gamma} = \tfrac{1}{2} g^{\alpha\mu} (g_{\beta\mu,\gamma} + g_{\mu\gamma,\beta} - g_{\beta\gamma,\mu}) \tag{5.5.6}$$

are the Christoffel symbols.

The variation of (5.5.3) is given by

$$S_{EH}[g+\varepsilon f] - S_{EH}[g] = \varepsilon \int d^4x \left(\frac{\partial}{\partial g^{\mu\nu}} \sqrt{-g} g^{\alpha\beta} \right) R_{\alpha\beta} f^{\mu\nu}(x) +$$

$$+ \int d^4x \, \sqrt{-g} g^{\alpha\beta} \Big(R_{\alpha\beta}[g+\varepsilon f] - R_{\alpha\beta}[g] \Big) + O(\varepsilon^2). \tag{5.5.7}$$

By calculating in geodesic coordinates one finds that the last term vanishes. Since

$$\frac{\partial}{\partial g^{\mu\nu}} \sqrt{-g} g^{\alpha\beta} = \frac{1}{2\sqrt{-g}} g g_{\mu\nu} g^{\alpha\beta} + \sqrt{-g} \delta^\alpha_\mu \delta^\beta_\nu$$

$$= \sqrt{-g} \left(-\tfrac{1}{2} g_{\mu\nu} g^{\alpha\beta} + \delta^\alpha_\mu \delta^\beta_\nu \right),$$

we finally obtain

$$S_{EH}[g+\varepsilon f] - S_{EH}[g] = \varepsilon \int d^4x \, \sqrt{-g} \left(-\tfrac{1}{2} g_{\alpha\beta} R + R_{\alpha\beta} \right) f^{\alpha\beta}(x) + O(\varepsilon^2). \tag{5.5.8}$$

This implies Einstein's field equations in vacuum

$$R_{\alpha\beta} - \tfrac{1}{2} g_{\alpha\beta} R = 0. \tag{5.5.9}$$

For this reason the Lagrangian

$$L_{EH} = -\frac{2}{\kappa^2} \sqrt{-g} R \tag{5.5.10}$$

can be taken as starting point of the classical theory.

A glance to (5.5.5-6) shows that the first two terms in (5.5.5) contain second derivatives of the fundamental tensor field $g_{\mu\nu}$. This defect can be removed by splitting off a divergence. We rewrite the first term in (5.5.5) as follows

$$\sqrt{-g} g^{\mu\nu} \Gamma^{\alpha}_{\mu\nu,\alpha} = (\sqrt{-g} g^{\mu\nu} \Gamma^{\alpha}_{\mu\nu})_{,\alpha} - \Gamma^{\alpha}_{\mu\nu} (\sqrt{-g} g^{\mu\nu})_{,\alpha} \tag{5.5.11}$$

and calculate the last derivative with the help of

$$g^{\mu\nu}{}_{,\alpha} = -\Gamma^{\mu}_{\beta\alpha} g^{\beta\nu} - \Gamma^{\nu}_{\alpha\beta} g^{\beta\mu}.$$

Proceeding with the second term in the same way we find

$$\sqrt{-g} R = \sqrt{-g} G - \left(\sqrt{-g} g^{\mu\nu} \Gamma^{\alpha}_{\mu\nu} - \sqrt{-g} g^{\mu\alpha} \Gamma^{\nu}_{\mu\nu} \right)_{,\alpha} \tag{5.5.12}$$

where

$$G = g^{\mu\nu} \left(\Gamma^{\alpha}_{\nu\beta} \Gamma^{\beta}_{\mu\alpha} - \Gamma^{\alpha}_{\mu\nu} \Gamma^{\beta}_{\alpha\beta} \right). \tag{5.5.13}$$

Since the divergence in (5.5.12) does not matter in the variational principle, we can go on with the Lagrangian

$$L = -\frac{2}{\kappa^2} \sqrt{-g} g^{\mu\nu} \left(\Gamma^{\alpha}_{\nu\beta} \Gamma^{\beta}_{\mu\alpha} - \Gamma^{\alpha}_{\mu\nu} \Gamma^{\beta}_{\alpha\beta} \right), \tag{5.5.14}$$

which contains first derivatives of g only.

For the following it is convenient to remove the square root $\sqrt{-g}$ by introducing the so-called Goldberg variables

$$\tilde{g}^{\mu\nu} = \sqrt{-g} g^{\mu\nu},$$
$$\tilde{g}_{\mu\nu} = (-g)^{-1/2} g_{\mu\nu}. \tag{5.5.15}$$

Using

$$\partial_\varrho g = g \tilde{g}_{\alpha\beta} \tilde{g}^{\alpha\beta}{}_{,\varrho}$$
$$\partial_\varrho g^{\mu\nu} = (-g)^{-1/2} \left(\tilde{g}^{\mu\nu}{}_{,\varrho} - \tfrac{1}{2} \tilde{g}^{\mu\nu} \tilde{g}_{\alpha\beta} \tilde{g}^{\alpha\beta}{}_{,\varrho} \right),$$
$$\partial_\varrho g_{\mu\nu} = \sqrt{-g} \left(\tfrac{1}{2} \tilde{g}_{\mu\nu} \tilde{g}_{\alpha\beta} \tilde{g}^{\alpha\beta}{}_{,\varrho} - \tilde{g}_{\mu\alpha} \tilde{g}_{\nu\beta} \tilde{g}^{\alpha\beta}{}_{,\varrho} \right) \tag{5.5.16}$$

in (5.5.6) we obtain

$$\Gamma^{\alpha}_{\beta\gamma} = \tfrac{1}{2} \left(\tfrac{1}{2} \delta^{\alpha}_{\beta} \tilde{g}_{\mu\nu} \tilde{g}^{\mu\nu}{}_{,\gamma} + \tfrac{1}{2} \delta^{\alpha}_{\gamma} \tilde{g}_{\mu\nu} \tilde{g}^{\mu\nu}{}_{,\beta} - \tilde{g}_{\beta\mu} \tilde{g}^{\alpha\mu}{}_{,\gamma} - $$

$$-\tilde{g}_{\gamma\mu}\tilde{g}^{\alpha\mu}{}_{,\beta}+\tilde{g}^{\alpha\varrho}\tilde{g}_{\gamma\mu}\tilde{g}_{\beta\nu}\tilde{g}^{\mu\nu}{}_{,\varrho}-\tfrac{1}{2}\tilde{g}^{\alpha\varrho}\tilde{g}_{\beta\gamma}\tilde{g}_{\mu\nu}\tilde{g}^{\mu\nu}{}_{,\varrho}\Big). \tag{5.5.17}$$

This enables us to express the Lagrangian L (5.5.14) by $\tilde{g}_{\mu\nu}$. It is simple to compute the second term

$$\tilde{g}^{\mu\nu}\Gamma^{\alpha}_{\mu\nu}\Gamma^{\beta}_{\alpha\beta} = -\tfrac{1}{2}\tilde{g}^{\mu\alpha}{}_{,\mu}\,\tilde{g}_{\nu\beta}\tilde{g}^{\nu\beta}{}_{,\alpha}. \tag{5.5.18}$$

But the first term in (5.5.14) requires the collection of many terms, until one arrives at the following simple result

$$\tilde{g}^{\mu\nu}\Gamma^{\alpha}_{\nu\beta}\Gamma^{\beta}_{\mu\alpha} = \frac{1}{4}\Big(-2\tilde{g}_{\mu\nu}\tilde{g}^{\mu\nu}{}_{,\alpha}\,\tilde{g}^{\alpha\beta}{}_{,\beta}+2\tilde{g}_{\alpha\beta}\tilde{g}^{\alpha\mu}{}_{,\nu}\,\tilde{g}^{\beta\nu}{}_{,\mu}-$$

$$-\tilde{g}_{\alpha\varrho}\tilde{g}_{\beta\sigma}\tilde{g}^{\mu\nu}\tilde{g}^{\varrho\beta}{}_{,\mu}\,\tilde{g}^{\sigma\alpha}{}_{,\nu}+\tfrac{1}{2}\tilde{g}^{\alpha\beta}\tilde{g}_{\mu\nu}\tilde{g}^{\mu\nu}{}_{,\alpha}\,\tilde{g}_{\varrho\sigma}\tilde{g}^{\varrho\sigma}{}_{,\beta}\Big). \tag{5.5.19}$$

Then the total Lagrangian is given by

$$L = \frac{1}{\kappa^2}\Big(-\tilde{g}_{\alpha\beta}\tilde{g}^{\alpha\mu}{}_{,\nu}\,\tilde{g}^{\beta\nu}{}_{,\mu}+\tfrac{1}{2}\tilde{g}_{\alpha\varrho}\tilde{g}_{\beta\sigma}\tilde{g}^{\varrho\beta}{}_{,\mu}\,\tilde{g}^{\alpha\sigma}{}_{,\nu}\,\tilde{g}^{\mu\nu}-$$

$$-\frac{1}{4}\tilde{g}_{\mu\nu}\tilde{g}^{\mu\nu}{}_{,\alpha}\,\tilde{g}_{\varrho\sigma}\tilde{g}^{\varrho\sigma}{}_{,\beta}\,\tilde{g}^{\alpha\beta}\Big). \tag{5.5.20}$$

To make contact with quantum field theory on Minkowski space we consider the situation in scattering theory where at large distances the metric is given by the Minkowski metric $\eta^{\mu\nu}$. Then we write the metric tensor as a sum

$$\tilde{g}^{\mu\nu}(x) = \eta^{\mu\nu} + \kappa h^{\mu\nu}(x). \tag{5.5.21}$$

We do not assume that the new dynamical field $h^{\mu\nu}(x)$ is small in some sense, it only goes to zero at large distances because of the asymptotically flat situation. The indices of $h^{\mu\nu}$ are ordinary Lorentz indices which can be raised and lowered with the Minkowski metric. Then the inverse of (5.5.21) is given by

$$\tilde{g}_{\mu\nu}(x) = \eta_{\mu\nu} - \kappa h_{\mu\nu}(x) + \kappa^2 h_{\mu\alpha}h^{\alpha}{}_{\nu} - \ldots. \tag{5.5.22}$$

Substituting these expressions into (5.5.20), the Lagrangian L becomes an infinite sum

$$L = \sum_{n=0}^{\infty}\kappa^n L^{(n)}. \tag{5.5.23}$$

Here is the proliferation of couplings which can be traced back to the infinite series (5.5.22). The three terms in (5.5.20) give the following contributions

$$L_1 = \Big(-\eta_{\alpha\beta}+\kappa h_{\alpha\beta}-\kappa^2 h_{\alpha\varrho}h^{\varrho}_{\beta}+\ldots\Big)h^{\alpha\mu}{}_{,\nu}\,h^{\beta\nu}{}_{,\mu} \tag{5.5.24}$$

$$L_2 = \tfrac{1}{2}\Big(\eta_{\alpha\varrho}-\kappa h_{\alpha\varrho}+\kappa^2 h_{\alpha\alpha'}h^{\alpha'}_{\varrho}-\ldots\Big)\Big(\eta_{\beta\sigma}-\kappa h_{\beta\sigma}+\kappa^2 h_{\beta\beta'}h^{\beta'}_{\sigma}-\ldots\Big)$$

$$\times (\eta^{\mu\nu}+\kappa h^{\mu\nu})h^{\varrho\beta}{}_{,\mu}\,h^{\alpha\sigma}{}_{,\nu} \tag{5.5.25}$$

$$L_3 = \frac{1}{4}\left(-\eta_{\mu\nu} + \kappa h_{\mu\nu} - \kappa^2 h_{\mu\alpha'} h_\nu^{\alpha'} - \ldots\right)\left(\eta_{\varrho\sigma} - \kappa h_{\varrho\sigma} + \kappa^2 h_{\varrho\beta'} h_\sigma^{\beta'} - \ldots\right)$$
$$\times (\eta^{\alpha\beta} + \kappa h^{\alpha\beta}) h^{\mu\nu}{}_{,\alpha}\, h^{\varrho\sigma}{}_{,\beta}. \tag{5.5.26}$$

The lowest order
$$L^{(0)} = \frac{1}{2} h^{\alpha\beta}{}_{,\mu}\, h^{\mu}_{\alpha\beta} - h^{\alpha\beta}{}_{,\mu}\, h^{\mu}_{\alpha,\beta} - \frac{1}{4} h_{,\alpha}\, h^{,\alpha}, \tag{5.5.27}$$

where $h = h^\mu{}_\mu$, defines the free theory. Indeed, the corresponding Euler-Lagrange equation reads
$$\Box h^{\alpha\beta} - \tfrac{1}{2}\eta^{\alpha\beta}\Box h - h^{\alpha\mu,\beta}{}_{,\mu} - h^{\beta\mu,\alpha}{}_{,\mu} = 0. \tag{5.5.28}$$

Both (5.5.27) and (5.5.28) are invariant under the classical gauge transformation
$$h^{\alpha\beta} \to h'^{\alpha\beta} = h^{\alpha\beta} + f^{\alpha,\beta} + f^{\beta,\alpha} - \eta^{\alpha\beta} f^\mu{}_{,\mu}. \tag{5.5.29}$$

The gauge can be specified by the Hilbert condition
$$h'^{\alpha\beta}{}_{,\beta} = 0. \tag{5.5.30}$$

This can be achieved by choosing the solution of the inhomogeneous wave equation
$$\Box f^\alpha = -h^{\alpha\beta}{}_{,\beta} \tag{5.5.31}$$

as gauge function in (5.5.29). In the Hilbert gauge the equation of motion (5.5.28) gets simplified
$$\Box h^{\alpha\beta} - \tfrac{1}{2}\eta^{\alpha\beta}\Box h = 0.$$

Taking the trace we conclude
$$\Box h = 0, \quad \Box h^{\alpha\beta} = 0, \tag{5.5.32}$$

so that we precisely arrive at the free tensor field as it was introduced in Sect.1.7.

The first order coupling $O(\kappa)$ in (5.5.24-26) can easily be computed
$$L^{(1)} = -\frac{1}{4} h^{\alpha\beta} h_{,\alpha}\, h_{,\beta} + \frac{1}{2} h^{\mu\nu} h^{\alpha\beta}{}_{,\mu}\, h_{\alpha\beta,\nu} + h_{\alpha\beta} h^{\alpha\mu}{}_{,\nu}\, h^{\beta\nu}{}_{,\mu} +$$
$$+ \frac{1}{2} h_{\mu\nu} h^{\mu\nu}{}_{,\alpha}\, h^{,\alpha} - h_{\mu\nu} h^{\alpha\mu}{}_{,\varrho}\, h^{\nu,\varrho}_\alpha. \tag{5.5.33}$$

The first three terms herein agree precisely with the coupling T_1^h (5.2.12) which was derived in Sect. 5.2 from quantum gauge invariance. Note the overall factor 1/4 which will be important later in Sect. 5.13. The last two terms are divergences as noticed before (5.2.13) and, therefore, have no importance for the physics of the theory.

For later use we also compute the second order $O(\kappa^2)$ in (5.5.24-26)
$$L^{(2)} = -h_{\alpha\varrho} h^\varrho_\beta h^{\alpha\mu}{}_{,\nu}\, h^{\beta\nu}{}_{,\mu} - \frac{1}{2} h_{\varrho\beta} h^\beta_\sigma h^{\varrho\sigma}{}_{,\alpha}\, h^{,\alpha} -$$

$$-\frac{1}{4}h_{\mu\nu}h^{\mu\nu}{}_{,\alpha}\,h_{\varrho\sigma}h^{\varrho\sigma,\alpha}+\frac{1}{2}h_{\mu\nu}h^{\mu\nu}{}_{,\alpha}\,h^{\alpha\beta}h_{,\beta}-h_{u\varrho}h^{\varrho}_{\sigma,\mu}\,h^{\alpha\sigma}{}_{,\nu}\,h^{\mu\nu}\ |$$

$$+h_{\varrho\beta}h^{\beta}_{\sigma}h^{\alpha\varrho}{}_{,\mu}\,h^{\sigma,\mu}_{\alpha}+\frac{1}{2}h_{\alpha\varrho}h_{\beta\sigma}h^{\alpha\sigma}{}_{,\mu}\,h^{\beta\varrho,\mu}. \tag{5.5.34}$$

This four-graviton coupling which goes beyond the linearized Einstein's theory will be recovered at the end of Sect.5.7 from second order gauge invariance. It is known that the *classical* theory is uniquely determined to all orders by $L^{(2)}$ (see the bibliographical notes). For this reason it is important to deduce (5.5.34) from *quantum* gauge theory.

Finally we discuss the relation between classical covariance under general coordinate transformations and quantum gauge invariance. Let us consider an infinitesimal coordinate transformation

$$x'^{\mu}=x^{\mu}+\varepsilon^{\mu}. \tag{5.5.35}$$

Then the metric tensor transforms according to

$$g'^{\mu\nu}(x')=g^{\varrho\sigma}(x)\frac{\partial x'^{\mu}}{\partial x^{\varrho}}\frac{\partial x'^{\nu}}{\partial x^{\sigma}}$$

$$=g^{\varrho\sigma}(x)\left(\delta^{\mu}_{\varrho}+\frac{\partial\varepsilon^{\mu}}{\partial x^{\varrho}}\right)\left(\delta^{\nu}_{\sigma}+\frac{\partial\varepsilon^{\nu}}{\partial x^{\sigma}}\right)$$

$$=g^{\mu\nu}(x)+g^{\mu\sigma}\frac{\partial\varepsilon^{\mu}}{\partial x^{\sigma}}+g^{\varrho\nu}\frac{\partial\varepsilon^{\mu}}{\partial x^{\varrho}}+O(\varepsilon^{2}). \tag{5.5.36}$$

Writing the metric with Minkowski background

$$g^{\mu\mu}=\eta^{\mu\nu}+\kappa\tilde{h}^{\mu\nu} \tag{5.5.37}$$

as in (5.5.21), we obtain the following infinitesimal coordinate transformation

$$\tilde{h}'^{\mu\nu}=\tilde{h}^{\mu\nu}+\partial^{\mu}\varepsilon^{\nu}+\partial^{\nu}\varepsilon^{\mu}. \tag{5.5.38}$$

Here the trace term $\sim\eta^{\mu\nu}$ in (5.5.29) is absent because we do not use the Goldberg variables (5.5.21).Correspondingly, in the quantum gauge transformation the trace term is lacking as well, so that (5.5.38) has the same form as the quantum gauge transformation. However the meaning of the symbols is very different: ε^{μ} is a real function, whereas u^{μ} is a quantized Fermi field. We consider quantum theory as more fundamental than classical field theory. A quantized Fermi field u^{μ} cannot have a real function ε^{μ} as its classical limit. From this we get doubts about that arbitrary coordinate transformations have physical significance. This will be an important point in the last chapter about non-geometric general relativity.

5.6 Expansion in the massive case

Now we asks ourselves which modification of the Einstein-Hilbert Lagrangian (5.5.10) gives after expansion around flat background the mass terms of the coupling (5.4.9). There are not many candidates, the right one is Einstein's Lagrangian with cosmological term

$$L_E = -\frac{2}{\kappa^2}\sqrt{-g}(R+2\Lambda), \quad \kappa^2 = 32\pi G, \tag{5.6.1}$$

G is Newton's constant and $g = \det(g_{\mu\nu})$. For the sign of the cosmological constant Λ we have adopted the convention used in astrophysics. Again we work with the Goldberg variables

$$\tilde{g}^{\mu\nu} = \sqrt{-g}g^{\mu\nu}, \quad \tilde{g}_{\mu\nu} = (-g)^{-1/2}g_{\mu\nu}. \tag{5.6.2}$$

and write the metric tensor as

$$\tilde{g}^{\mu\nu} = \eta^{\mu\nu} + \kappa h^{\mu\nu} \tag{5.6.3}$$

$$\tilde{g}_{\mu\nu} = \eta_{\mu\nu} - \kappa h_{\mu\nu} + \kappa^2 h_{\mu\alpha}h^{\alpha\nu} - \kappa^3 h_{\mu\alpha}h^{\alpha\beta}h_{\beta\nu} + \ldots$$
$$= \eta_{\mu\lambda}(\delta^\lambda_\nu - \kappa h^\lambda_\nu + \kappa^2 h^\lambda_\alpha h^\alpha_\nu - \kappa^3 h^\lambda_\alpha h^{\alpha\beta}h_{\beta\nu} + \ldots). \tag{5.6.4}$$

In the massive case this expansion around Minkowski background is purely formal because Minkowski space is not a solution of Einstein's equations with cosmological term. This solution is the de-Sitter or anti-de-Sitter metric. But locally all those metrics are Minkowskian. The quantum field theory we are investigating describes the quantum fluctuations around this local Minkowskian background.

We take the determinant of (5.6.4)

$$(-g)^{-4/2}g = \det \eta_{\mu\lambda} \exp \mathrm{Tr} \log(\delta^\lambda_\nu - \kappa h^\lambda_\nu + \kappa^2 h^\lambda_\alpha h^\alpha_\nu - \ldots).$$

Expanding the logarithm we find

$$(-g)^{-1} = \exp \mathrm{Tr}\left(-\kappa h^\lambda_\nu + \frac{\kappa^2}{2}h^\lambda_\alpha h^\alpha_\nu - \frac{\kappa^3}{3}h^\lambda_\alpha h^\alpha_\beta h^\beta_\nu + \kappa^4 h^\lambda_\alpha h^{\alpha\beta}h_{\beta\varrho}h^\varrho_\nu \ldots\right) \tag{5.6.5}$$

Taking the trace in the exponent and the power $-\frac{1}{2}$ of both sides we get

$$\sqrt{-g} = \exp\left(\frac{\kappa}{2}h - \frac{\kappa^2}{4}h^\nu_\alpha h^\alpha_\nu + \frac{\kappa^3}{6}h^\nu_\alpha h^\alpha_\beta h^\beta_\nu - \frac{\kappa^4}{8}h^\nu_\alpha h^{\alpha\beta}h_{\beta\varrho}h^\varrho_\nu\right).$$

After expanding the exponential function also we arrive at

$$\sqrt{-g} = 1 + \frac{\kappa}{2}h - \frac{\kappa^2}{4}h^\nu_\alpha h^\alpha_\nu + \frac{\kappa^2}{8}h^2 + \frac{\kappa^3}{6}h^\nu_\alpha h^\alpha_\beta h^\beta_\nu - \frac{\kappa^3}{8}h^\nu_\alpha h^\alpha_\nu h + \frac{\kappa^3}{48}h^3+$$

$$+\frac{\kappa^4}{32}(h^\nu_\alpha h^\alpha_\nu)^2 + \frac{\kappa^4}{12}h h^\nu_\alpha h^{\alpha\beta}h_{\beta\nu} - \frac{\kappa^4}{32}h^2 h^\nu_\alpha h^\alpha_\nu + \frac{\kappa^4}{4!}h^4 - \frac{\kappa^4}{8}h^\nu_\alpha h^{\alpha\beta}h_{\beta\mu}h^\mu_\nu. \tag{5.6.6}$$

Now we substitute into the Einstein Lagrangian (5.6.1); for the gravitational part we can use the result (5.5.23):

$$L_E = -\frac{4\Lambda}{\kappa^2} - \frac{2\Lambda}{\kappa}h+$$

$$+\tfrac{1}{2}\partial_\mu h^{\alpha\beta}\partial^\mu h_{\alpha\beta} - \partial_\mu h^{\alpha\beta}\partial_\beta h^\mu_\alpha - \tfrac{1}{4}\partial_\alpha h \partial^\alpha h-$$

$$-\Lambda(\tfrac{1}{2}h^2 - h^{\alpha\beta}h_{\alpha\beta})+$$

$$+\kappa\Big[L^{(1)} - 4\Lambda\Big(\tfrac{1}{6}h^{\alpha\beta}h_{\beta\gamma}h^\gamma_\alpha - \tfrac{1}{8}h^{\alpha\beta}h_{\alpha\beta}h + \tfrac{1}{48}h^3\Big)\Big]+$$

$$+\kappa^2\Big[L^{(2)} - 4\Lambda\Big(\tfrac{1}{32}(h^{\alpha\beta}h_{\alpha\beta})^2 + \tfrac{1}{12}hh^{\alpha\beta}h_{\beta\gamma}h^\gamma_\alpha -$$

$$-\tfrac{1}{32}h^2 h^{\alpha\beta}h_{\alpha\beta} + \tfrac{1}{4!}\tfrac{h^4}{16} - \tfrac{1}{8}h^{\alpha\beta}h_{\beta\gamma}h^{\gamma\nu}h_{\nu\alpha}\Big)\Big], \tag{5.6.7}$$

where $L^{(1)}$ is given by (5.5.33) and $L^{(2)}$ by (5.5.34).

The terms $O(\kappa^{-2})$ and $O(\kappa^{-1})$ in the Lagrangian (5.6.7) are the price for the formal expansion around Minkowski background. As they stand these terms are meaningless as a classical Langrangian: the constant $-4\Lambda/\kappa^2$ cannot be integrated over d^4x and the term $O(\kappa^{-1})$ linear in h alone gives no consistent equation of motion. But our aim here is the comparison with the quantum theory and for this purpose only the orders $\sim \kappa^0, \kappa^1, \kappa^2$ are relevant. The term $O(\kappa^0)$ and quadratic in h

$$L^{(0)} = \tfrac{1}{2}\partial_\mu h^{\alpha\beta}\partial^\mu h_{\alpha\beta} - \partial_\mu h^{\alpha\beta}\partial_\beta h^\mu_\alpha - \tfrac{1}{4}\partial_\alpha h \partial^\alpha h-$$

$$-\Lambda\Big(\frac{h^2}{2} - h^{\alpha\beta}h_{\alpha\beta}\Big). \tag{5.6.8}$$

defines the free asymptotic theory. It gives the following Euler-Lagrange equation

$$\Box h^{\alpha\beta} - \partial_\mu(\partial^\beta h^{\alpha\mu} + \partial^\alpha h^{\beta\mu}) - \tfrac{1}{2}\eta^{\alpha\beta}\Box h+$$

$$+\Lambda(\eta^{\alpha\beta}h - 2h^{\alpha\beta}) = 0. \tag{5.6.9}$$

Taking the trace we find

$$\partial_\mu\partial_\alpha h^{\alpha\mu} = -\tfrac{1}{2}\Box h + \Lambda h. \tag{5.6.10}$$

Differentiating (5.6.9) by ∂_α and substituting (5.6.10) we derive the Hilbert gauge condition

$$\partial_\alpha h^{\alpha\beta} = 0. \tag{5.6.11}$$

Then (5.6.10) reduces to the Klein-Gordon equation

$$\Box h - 2\Lambda h = 0,$$

and from (5.6.9) we obtain the Klein-Gordon equation for the tensor field

Spin-2 gauge theory

$$\Box h^{\alpha\beta} - 2\Lambda h^{\alpha\beta} = 0. \tag{5.6.12}$$

This means the graviton becomes massive with mass

$$m^2 = -2\Lambda. \tag{5.6.13}$$

The constant Λ must be negative. Therefore, it has nothing to do with the positive cosmological constant used to describe dark energy.

Now comes the main point. Using (5.6.13) the cubic part $O(\kappa^1)$ in (5.6.7) gives the coupling

$$T_1^h = -\frac{1}{4}h^{\alpha\beta}\partial_\alpha h \partial_\beta h + \frac{1}{2}h^{\mu\nu}\partial_\mu h^{\alpha\beta}\partial_\nu h^{\alpha\beta} + h^{\alpha\beta}\partial_\nu h^{\alpha\mu}\partial_\mu h^{\beta\nu}$$

$$+\frac{1}{2}h^{\mu\nu}\partial_\alpha h^{\mu\nu}\partial_\alpha h - h^{\mu\nu}\partial_\varrho h^{\alpha\mu}\partial^\varrho h^{\alpha\nu}+$$

$$+2m^2\left(\frac{1}{6}h^{\alpha\beta}h_{\beta\gamma}h^{\gamma\alpha} - \frac{1}{8}h^{\alpha\beta}h^{\alpha\beta}h + \frac{1}{48}h^3\right). \tag{5.6.14}$$

This agrees exactly with the pure graviton coupling terms in (5.4.9), if we multiply with the overall factor 4 which was previously found after (5.5.33). The quartic part $O(\kappa^2)$ is equal to

$$T_2^h = -h_{\alpha\varrho}h^\varrho_\beta h^{\alpha\mu}_{,\nu}h^{\beta\nu}_{,\mu} - \tfrac{1}{2}h_{\varrho\beta}h^\beta_\gamma h^{\varrho\gamma}_{,\alpha}h^{,\alpha}$$

$$-\frac{1}{4}h_{\mu\nu}h^{\mu\nu}_{,\alpha}h_{\varrho\gamma}h^{\varrho\gamma,\alpha} + \tfrac{1}{2}h_{\mu\nu}h^{\mu\nu}_{,\alpha}h^{\alpha\beta}h_{,\beta} - h_{\alpha\varrho}h^\varrho_{\gamma,\mu}h^{\alpha\gamma}_{,\nu}h^{\mu\nu}$$

$$+h_{\varrho\beta}h^\beta_\gamma h^{\alpha\varrho}_{,\mu}h^{\gamma,\mu}_\alpha + \tfrac{1}{2}h_{\alpha\varrho}h_{\beta\gamma}h^{\alpha\gamma}_{,\mu}h^{\beta\varrho,\mu}$$

$$+2m^2\Big(\frac{1}{32}(h^{\alpha\beta}h_{\alpha\beta})^2 + \frac{1}{12}h h^{\alpha\beta}h_{\beta\gamma}h^\gamma_\alpha-$$

$$-\frac{1}{32}h^2 h^{\alpha\beta}h_{\alpha\beta} + \frac{1}{4!}\frac{h^4}{16} - \frac{1}{8}h^{\alpha\beta}h_{\beta\gamma}h^{\gamma\nu}h_{\nu\alpha}\Big). \tag{5.6.15}$$

These coupling terms we shall find later from second order gauge invariance.

5.7 Second order gauge invariance: graviton sector

In second order we must construct chronological products $T(x, y)$ and $T_\mu(x, y)$ such that

$$d_Q T(x, y) = i \frac{\partial}{\partial x^\mu} T^\mu(x, y) + x \leftrightarrow y \qquad (5.7.1)$$

is verified. The construction procedure is well-known from spin-1 gauge theory, but we recall the main steps. We first compute the causal commutators $[T(x), T(y)]$ and $[T_\mu(x), T(y)]$ and substitute the causal Pauli-Jordan distributions in the tree graph contributions by Feynman propagators $D^F(x - y)$. If on the right-hand side of (5.7.1) a wave operator ∂^2 operates on D^F, we obtain a local term $\sim \delta(x - y)$. These anomalies must be compensated by finite renormalizations.

The generic form of the anomaly is

$$A(x, y) = \delta(x - y) a(x) + [\partial_\mu^x \delta(x - y)] a^\mu(x, y) \qquad (5.7.2)$$

The total anomaly is obtained by adding the contribution $A(y, x)$ with x, y interchanged. Then the terms with $\partial \delta$ can be combined by means of the identity (3.4.21)

$$[\partial_\mu^x \delta(x - y)] f(x, y) + x \leftrightarrow y = [\partial_\mu^y f - \partial_\mu^x f] \delta(y - x)], \qquad (5.7.3)$$

which follows by smearing with symmetric test functions; this is the right test function space here, due to the symmetry of the chronological products. Then the total anomaly is equal to

$$A_{\text{tot}}(x, y) = [2a(x) + \partial_\mu^y a^\mu - \partial_\mu^x a^\mu] \delta(x - y) \equiv A(x) \delta(x - y). \qquad (5.7.4)$$

The cancellation of the anomalies is equivalent to

$$A_{\text{tot}}(x, y) = d_Q R(x, y) - i \partial_\mu R^\mu(x, y) + x \leftrightarrow y; \qquad (5.7.5)$$

here the expressions $R(x, y)$ and $R^\mu(x, y)$ are *finite renormalizations*: these are quasilocal operators:

$$R(x, y) = \delta(x - y) B(x) + \cdots \qquad (5.7.6)$$

and

$$R^\mu(x, y) = \delta(x - y) B^\mu(x) + \cdots \qquad (5.7.7)$$

where B and B^μ are some Wick polynomials and $\cdots$ are similar terms with derivatives on the delta distribution. Indeed, in this case one can eliminate the anomaly by redefinition of the chronological products

$$T(x, y) \to T(x, y) + R(x, y) \qquad (5.7.8)$$

and

$$T^\mu(x, y) \to T^\mu(x, y) + R^\mu(x, y). \qquad (5.7.9)$$

Spin-2 gauge theory

One can prove that the cancellation (5.7.5) of the anomalies is achieved if we can write the operator part $A(x)$ in (5.7.4) in the form

$$A = d_Q B - i\partial_\mu B^\mu. \tag{5.7.10}$$

In fact, the derivative terms in (5.7.2) can be combined with help of the identity (3.4.22)

$$\partial_\mu^x[B^\mu(x,y)\delta(x-y)] + x \leftrightarrow y = [\partial_\mu^x B^\mu + \partial_\mu^y B^\mu]\delta(x-y). \tag{5.7.11}$$

To make the long expressions in the following calculations more transparent we shall use the following convention: *we write the Lorentz indices of the partial derivatives downstairs (without comma) and all other Lorentz indices upstairs*. Of course, if we consider a Lorentz scalar, all indices are contracted in pairs with the Minkowski tensor $\eta_{\mu\nu}$, but we omit the η's.

We multiply the first order coupling (5.2.12) by $i/4$ to be in agreement with the expansion of the Einstein-Hilbert Lagrangian:

$$T_1 = i\Big(\underbrace{-\frac{1}{4}h^{\alpha\beta}h_\alpha h_\beta}_{h1} + \underbrace{\frac{1}{2}h^{\alpha\beta}h_\alpha^{\varrho\sigma}h_\beta^{\varrho\sigma}}_{h2} + \underbrace{h^{\alpha\beta}h_\sigma^{\beta\varrho}h_\varrho^{\alpha\sigma}}_{h3} +$$

$$+ \underbrace{\frac{1}{2}h^{\alpha\beta}h_\varrho^{\alpha\beta}h_\varrho}_{h4} \underbrace{-h^{\alpha\beta}h_\sigma^{\alpha\varrho}h_\sigma^{\beta\varrho}}_{h5} +$$

$$+ \underbrace{u_\beta^\alpha\tilde{u}_\alpha^\varrho h^{\varrho\beta}}_{u1} - \underbrace{u^\alpha\tilde{u}_\beta^\varrho h_\alpha^{\varrho\beta}}_{u2} - \underbrace{u_\alpha^\alpha\tilde{u}_\beta^\varrho h^{\varrho\beta}}_{u3} + \underbrace{u_\varrho^\alpha\tilde{u}_\beta^\alpha h^{\varrho\beta}}_{u4}\Big). \tag{5.7.12}$$

Similarly we multiply the Q-vertex T^μ (5.2.11) by $i/4$

$$iT_{1/1}^\mu = \underbrace{-\frac{1}{4}u^\mu h_\alpha^{\varrho\sigma}h_\alpha^{\varrho\sigma}}_{Q1} + \underbrace{\frac{1}{2}u_\alpha^\alpha h^{\varrho\sigma}h_\mu^{\varrho\sigma}}_{Q2} + \underbrace{\frac{1}{2}u^\alpha h_\alpha^{\varrho\sigma}h_\mu^{\varrho\sigma}}_{Q3} +$$

$$+ \underbrace{\frac{1}{8}u^\mu h_\alpha h_\alpha}_{Q4} - \underbrace{\frac{1}{4}u_\alpha^\alpha h h_\mu}_{Q5} - \underbrace{\frac{1}{4}u^\alpha h_\alpha h_\mu}_{Q6} + \underbrace{\frac{1}{2}u_\beta^\alpha h^{\alpha\beta}h_\mu}_{Q7} -$$

$$\underbrace{-u_\varrho^\alpha h_\mu^{\alpha\sigma}h^{\sigma\varrho}}_{Q8} + \underbrace{u_\varrho^\alpha h_\beta^{\alpha\mu}h^{\varrho\beta}}_{Q9} + \underbrace{u_\varrho^\alpha h_\alpha^{\mu\beta}h^{\varrho\beta}}_{Q10} -$$

$$\underbrace{-u^\alpha h_\varrho^{\mu\beta}h_\alpha^{\varrho\beta}}_{Q11} \underbrace{-u_\alpha^\alpha h_\varrho^{\mu\beta}h^{\varrho\beta}}_{Q12} + \underbrace{\frac{1}{2}u^\mu h_\varrho^{\alpha\beta}h_\alpha^{\varrho\beta}}_{Q13} -$$

$$\underbrace{-\frac{1}{2}u_{\alpha\beta}^\alpha\tilde{u}^\mu u^\beta}_{Q14} \underbrace{-\frac{1}{2}u^\beta\tilde{u}_\mu^\alpha u_\beta^\alpha}_{Q15} + \frac{1}{2}\Big(\underbrace{u^\beta\tilde{u}_\beta^\alpha u_\alpha^\mu}_{Q16} + \underbrace{u^\beta\tilde{u}_\beta^\alpha u^\mu}_{Q17} + \underbrace{u^\beta\tilde{u}^\alpha u_\alpha^\mu}_{Q18} + \underbrace{u_{\beta\alpha}^\beta\tilde{u}^\alpha u^\mu}_{Q19}\Big). \tag{5.7.13}$$

Here we have introduced short abbreviations for the terms.

We have first to calculate the causal commutator

$$iD^\mu_{2/1}(x,y) = i[T^\mu_{1/1}(x), T_1(y)]. \qquad (5.7.14)$$

The anomalies which may violate gauge invariance come from those terms in $T^\mu_{1/1}(x)$ (5.7.13) which have a derivative index μ downstairs, namely the terms $Q2, Q3, Q5, Q6, Q7, Q8$ and the term $Q9$. The latter contributes to the ghost sector which is considered in the next section. Those terms must be commuted with all terms in (5.7.12).

For illustration we consider the first commutator

$$[Q2, h1] = -\frac{1}{8} u^\alpha h^{\varrho\sigma}_\alpha [h^{\varrho\sigma}_\mu(x), h^{\alpha'\beta'}(y) h_{\alpha'} h_{\beta'}]$$

$$= -\frac{1}{8} u^\alpha h^{\varrho\sigma}_\alpha \left\{ -\frac{i}{2}(2 h_\varrho h_\sigma D_\mu - \eta^{\varrho\sigma} h_\beta h_\beta D_\mu) + 2i\eta^{\varrho\sigma} h^{\alpha'\beta'} h_{\beta'} D_{\mu\alpha'} \right\}. \qquad (5.7.15)$$

Here D_μ is the derivative of the mass-zero Jordan-Pauli distribution, for example

$$D_{\mu\alpha'} = \partial^x_\mu \partial^y_{\alpha'} D(x-y). \qquad (5.7.16)$$

In (5.7.15) we have only calculated the terms which produce anomalies. In the same way we continue with all commutators until $[Q8, h5]$ and collect the terms with the same field operators. After causal splitting and application of the derivative ∂_μ we obtain the anomalies according to the short rules

$$D_\mu \Longrightarrow \delta(x-y), \quad D_{\mu\alpha'} \Longrightarrow \partial^y_{\alpha'} \delta(x-y). \qquad (5.7.17)$$

We do not yet include the factor 2 for the term with $x \longleftrightarrow y$ interchanged, nor do we insert free normalizations terms in the splitting of $D_{\mu\alpha'}$, these operations are carried out a little later.

The resulting anomalies are grouped according to their Lorentz structure:

$$A_1 = i\Big\{ \frac{1}{8}(u^\alpha h^{\varrho\sigma}_\alpha + u_\alpha h^{\varrho\sigma})h_\varrho h_\sigma + \qquad (5.7.18)$$

$$+\frac{1}{4}(u^\alpha h_\alpha + u_\alpha h) h^{\alpha'\beta'} h_{\beta'} \partial^y_{\alpha'} - \frac{1}{4} u^\alpha_\varrho h^{\varrho\sigma} h_\alpha h_\sigma - \qquad (G1)$$

$$-\frac{1}{4}(u^\alpha h^{\varrho\sigma}_\alpha + u_\alpha h^{\varrho\sigma}) h^{\varrho'\sigma'}_\varrho h^{\varrho'\sigma'}_\sigma - \frac{1}{2}(u^\alpha h^{\varrho\sigma}_\alpha + u_\alpha h^{\varrho\sigma}) h^{\varrho\sigma}_{\beta'} h^{\alpha'\beta'} \partial_{\alpha'} +$$

$$+\frac{1}{2} u^\alpha_\varrho h^{\varrho\sigma} h^{\varrho'\sigma'}_\alpha h^{\varrho'\sigma'}_\sigma - \qquad (G2)$$

$$-\frac{1}{2}(u^\alpha h^{\varrho\sigma}_\alpha + u_\alpha h^{\varrho\sigma}) h^{\varrho\sigma}_{\varrho'} h^{\varrho'\sigma}_\sigma - (u^\alpha h^{\varrho\sigma}_\alpha + u_\alpha h^{\varrho\sigma}) h^{\alpha'\varrho} h^{\alpha'\sigma'}_\sigma \partial_{\sigma'} +$$

$$+u^\alpha_\varrho h^{\varrho\sigma} h^{\sigma\alpha'} h^{\alpha'\sigma'}_\alpha \partial_{\sigma'} + \qquad (G3)$$

$$+u^\alpha_\varrho h^{\varrho\sigma} h^{\sigma\varrho'}_{\sigma'} h^{\alpha\sigma'}_{\varrho'} - \qquad (G4)$$

216   Spin-2 gauge theory

$$-\frac{1}{4}(u^\alpha h_\alpha^{\varrho\sigma} + u_\alpha^\alpha h^{\varrho\sigma})h_{\varrho'}^{\varrho\sigma}h_{\varrho'} - \frac{1}{4}(u^\alpha h_\alpha^{\varrho\sigma} + u_\alpha^\alpha h^{\varrho\sigma})h^{\varrho\sigma}h_{\varrho'}\partial_{\varrho'} -$$

$$-\frac{1}{4}(u^\alpha h_\alpha + u_\alpha^\alpha h)h^{\alpha'\beta'}h_{\varrho'}^{\alpha'\beta'}\partial_{\varrho'} + \qquad (G5)$$

$$+\frac{1}{2}u_\varrho^\alpha h^{\varrho\sigma}h_{\varrho'}^{\alpha\sigma}h_{\varrho'} + \frac{1}{2}u_\varrho^\alpha h^{\varrho\sigma}h^{\alpha\sigma}h_{\varrho'}\partial_{\varrho'} + \qquad (G6)$$

$$+\frac{1}{2}(u^\alpha h_\alpha^{\varrho\sigma} + u_\alpha^\alpha h^{\varrho\sigma})h_{\sigma'}^{\varrho\varrho'}h_{\sigma'}^{\sigma\varrho'} + (u^\alpha h_\alpha^{\varrho\sigma} + u_\alpha^\alpha h^{\varrho\sigma})h^{\varrho\varrho'}h_{\sigma'}^{\sigma\varrho'}\partial_{\sigma'} - \qquad (G7)$$

$$-u_\varrho^\alpha h^{\varrho\sigma}h_{\sigma'}^{\sigma\varrho'}h_{\sigma'}^{\varrho'\alpha} - u_\varrho^\alpha h^{\varrho\sigma}h_{\sigma'}^{\sigma\varrho'}h^{\varrho'\alpha}\partial_{\sigma'} - u_\varrho^\alpha h^{\varrho\sigma}h^{\sigma\varrho'}h_{\sigma'}^{\varrho'\alpha}\partial_{\sigma'} + \qquad (G8)$$

$$+u_\varrho^\alpha h^{\varrho\sigma}h^{\alpha\varrho'}h_{\sigma}^{\varrho'\sigma'}\partial_{\sigma'} + u_\varrho^\alpha h^{\varrho\sigma}h_{\sigma'}^{\alpha\sigma}h^{\varrho'\sigma'}\partial_{\varrho'} + \qquad (G9)$$

$$+\frac{1}{2}u_\varrho^\alpha h^{\alpha\varrho}h^{\alpha'\sigma'}h_{\varrho'}^{\alpha'\sigma'}\partial_{\varrho'} \qquad (G10)$$

$$-\frac{1}{2}u_\varrho^\alpha h^{\alpha\varrho}h^{\alpha'\sigma'}h_{\sigma'}\partial_{\alpha'}\Big\}\delta(x-y). \qquad (G11)$$

The fields with primed indices have arguments y while the first two fields in each term have argument x.

Now we add the corresponding contributions with $x \longleftrightarrow y$ interchanged, which come from $D_{2/2}^\mu$. In the terms with $\delta(x-y)$ this gives simply a factor 2. But in the terms with derivative $\partial\delta(x-y)$ we have to use the identity (5.7.3). Let us denote the total sum of anomalies by $A_1 + A_2$. To achieve second order gauge invariance we must try to compensate the anomalies $A_1 + A_2$ by normalization terms

$$A_1 + A_2 = -d_Q N_2 + \partial_\alpha^x N_{2/1}^\alpha + \partial_\alpha^y N_{2/2}^\alpha. \qquad (5.7.19)$$

The terms N_2 are normalization terms for ordinary second order tree graphs in T_2 and $N_{2/1}^\alpha$ come from the T-product with a Q-vertex at x and similarly for $N_{2/2}^\alpha$. The latter terms were previously included in the identity (5.7.3). The key point is whether it is possible to write the anomalies (5.7.19) as a sum of co-boundaries plus divergences. It is our goal to achieve this for all Lorentz types (G1)-(G11) in (5.7.18).

Type (G4):

This is the simplest type because it contains one term only

$$A_{(4)} = 2iu_\varrho^\alpha h^{\varrho\sigma} h_{\sigma'}^{\sigma\varrho'} h_{\varrho'}^{\alpha\sigma'} \delta(x-y). \tag{5.7.20}$$

Here divergences cannot appear because they would produce more than one term. Hence, the only possibility to compensate this anomaly is by means of a co-boundary term $d_Q N_2$. In this case the ghost field u_ϱ^α must be the result of d_Q. Therefore the normalization term has the following form

$$N_1 = if_1 h^{\alpha\varrho} h^{\varrho\sigma} h_{\sigma'}^{\sigma\varrho'} h_{\varrho'}^{\alpha\sigma'} \delta(x-y). \tag{5.7.21}$$

We now compute the gauge variation

$$d_Q N_1 = if_1 \Bigg\{ -\frac{i}{2} \underbrace{\Big(u_\varrho^\alpha + u_\alpha^\varrho\Big)}_{G4 \quad G9} h^{\varrho\sigma} h_{\sigma'}^{\sigma\varrho'} h_{\varrho'}^{\alpha\sigma'} + \frac{i}{2} \underbrace{u_\mu^\mu h^{\alpha\sigma} h_{\sigma'}^{\sigma\varrho'} h_{\varrho'}^{\alpha\sigma'}}_{G3} -$$

$$-\frac{i}{2} \underbrace{\Big(u_\sigma^\varrho + u_\varrho^\sigma\Big)}_{G9 \quad G4} h^{\alpha\varrho} h_{\sigma'}^{\sigma\varrho'} h_{\varrho'}^{\alpha\sigma'} + \frac{i}{2} \underbrace{u_\mu^\mu h^{\alpha\varrho} h_{\sigma'}^{\varrho\varrho'} h_{\varrho'}^{\alpha\sigma'}}_{G3} -$$

$$-\frac{i}{2} \underbrace{\Big(u_{\varrho'\sigma'}^\sigma + u_{\sigma\sigma'}^{\varrho'}\Big)}_{G8 \quad G3} h^{\alpha\varrho} h^{\varrho\sigma} h_{\varrho'}^{\alpha\sigma'} + \frac{i}{2} \underbrace{u_{\mu\sigma'}^\mu h^{\alpha\varrho} h^{\varrho\sigma} h_\sigma^{\alpha\sigma'}}_{G3} -$$

$$-\frac{i}{2} \underbrace{\Big(u_{\varrho'\sigma'}^\alpha + u_{\alpha\varrho'}^{\sigma'}\Big)}_{G8 \quad G3} h^{\alpha\varrho} h^{\varrho\sigma} h_{\sigma'}^{\sigma\varrho'} + \frac{i}{2} \underbrace{u_{\mu\varrho'}^\mu h^{\alpha\varrho} h^{\varrho\sigma} h_\alpha^{\sigma\varrho'}}_{G3} \Bigg\} \delta. \tag{5.7.22}$$

The two terms with subscript G4 compensate the anomaly (5.7.20) if we choose

$$f_1 = 2i. \tag{5.7.23}$$

But the price is high because many terms of other Lorentz types, indicated by subscripts, are produced. Consequently, the entire compensation of local terms is a highly non-trivial task. We therefore give some details. Since there are many terms of type G3 in (5.7.22), we next turn to this type.

218   Spin-2 gauge theory

Typo (G3):

Transforming the $G3$-contribution in (5.7.18) into normal form by means of (5.7.3) we get the following anomaly

$$A_{(3)} = i\Big\{-u^\alpha h_\alpha^{\varrho\sigma} h_{\varrho'}^{\varrho\sigma'} h_{\sigma'}^{\sigma\varrho'} - u_\alpha^\alpha h^{\varrho\sigma} h_{\varrho'}^{\varrho\sigma'} h_{\sigma'}^{\sigma\varrho'} -$$
$$-(u_{\sigma'}^\alpha h_\alpha^{\varrho\sigma} + u^\alpha h_{\alpha\sigma'}^{\varrho\sigma}) h^{\alpha'\varrho} h_\sigma^{\alpha'\sigma'} + u^\alpha h_\alpha^{\varrho\sigma} (h_{\sigma'}^{\alpha'\varrho} h_\sigma^{\alpha'\sigma'} + h^{\alpha'\varrho} h_{\sigma\sigma'}^{\alpha'\sigma'}) -$$
$$-(u_{\alpha\sigma'}^\alpha h^{\varrho\sigma} + u_\alpha^\alpha h_{\sigma'}^{\varrho\sigma}) h^{\alpha'\varrho} h_\sigma^{\alpha'\sigma'} + u_\alpha^\alpha h^{\varrho\sigma} (h_{\sigma'}^{\alpha'\varrho} h_\sigma^{\alpha'\sigma'} + h^{\alpha'\varrho} h_{\sigma\sigma'}^{\alpha'\sigma'}) +$$
$$+(u_{\varrho\sigma'}^\alpha h^{\varrho\sigma} + u_\varrho^\alpha h_{\sigma'}^{\varrho\sigma}) h^{\sigma\alpha'} h_\alpha^{\alpha'\sigma'} - u_\varrho^\alpha h^{\varrho\sigma} (h_{\sigma'}^{\alpha'\sigma} h_\alpha^{\alpha'\sigma'} + h^{\alpha'\sigma} h_{\alpha\sigma'}^{\alpha'\sigma'})\Big\}\delta. \quad (5.7.24)$$

Now we need normalization terms of divergence form. They are obtained by choosing a term in (5.7.24) and taking out one of the three derivatives. For example, from the second term we read off the following divergence term

$$\partial_\nu^x N_1^\nu = g_1 \partial_\nu^x (u_\alpha^\alpha h^{\varrho\sigma} h^{\varrho\varrho'} h_{\varrho'}^{\sigma\nu} \delta). \quad (5.7.25)$$

This term is transformed into normal form by means of the other identity (3.4.22)

$$\partial_x^\sigma [F(x) G(y) \delta(x-y)] + \partial_y^\sigma [F(y) G(x) \delta(x-y)] =$$
$$= (F_\sigma G + F G_\sigma) \delta(x-y). \quad (5.7.26)$$

Thus

$$\partial_\nu^x N_1^\nu \to g_1 \Big[(u_{\alpha\nu}^\alpha h^{\varrho\sigma} + u_\alpha^\alpha h_\nu^{\varrho\sigma}) h^{\varrho\varrho'} h_{\varrho'}^{\sigma\nu} +$$
$$+ u_\alpha^\alpha h^{\varrho\sigma} (h_\nu^{\varrho\varrho'} h_{\varrho'}^{\sigma\nu} + h^{\varrho\varrho'} h_{\varrho'\nu}^{\sigma\nu})\Big]\delta. \quad (5.7.27)$$

We observe with satisfaction that the divergences do not couple different Lorentz types. Then it is not difficult to find the right terms which are necessary to cancel all anomalies. In addition to (5.7.25) we need four other terms

$$\partial_\nu^x N_2^\nu = g_2 \partial_\nu^x (u^\alpha h^{\varrho\sigma} h_\alpha^{\varrho\varrho'} h_{\varrho'}^{\sigma\nu} \delta), \quad (5.7.28)$$

$$\partial_\alpha^x N_3^\alpha = g_3 \partial_\alpha^x (u^\alpha h^{\varrho\sigma} h_\nu^{\varrho\varrho'} h_{\varrho'}^{\sigma\nu} \delta), \quad (5.7.29)$$

$$\partial_\nu^x N_4^\nu = g_4 \partial_\nu^x (u_\varrho^\alpha h^{\varrho\sigma} h^{\varrho\varrho'} h_\alpha^{\sigma\nu} \delta). \quad (5.7.30)$$

We transform all terms into the normal form as in (5.7.27) and then compare the coefficients of all Wick monomials with (5.7.24). From $u_\alpha^\alpha h^{\varrho\sigma} h^{\varrho\varrho'} h_{\varrho'\sigma'}^{\sigma\sigma'}$ we find $g_1 = i$, from $u^\alpha h_\alpha^{\varrho\sigma} h_{\varrho'}^{\varrho\varrho'} h_{\sigma'}^{\sigma\varrho'}$ we get $g_3 = -i$, and then from $u_\alpha^\alpha h^{\varrho\sigma} h_{\varrho'}^{\varrho\varrho'} h_{\sigma'}^{\sigma\varrho'}$

$$-f_1 + g_1 + g_3 = -2i. \quad (5.7.31)$$

This gives $f_1 = 2i$ in agreement with (5.7.23). The other coefficients are equal to

$$g_2 = i, \quad g_4 = -i. \quad (5.7.32)$$

The whole system of linear equations is consistently solved. The coefficients g_j correspond to normalization terms with Q-vertex and, therefore, are not physical, in contrast to the f_j which are physical.

Type (G5):

This type is quite complicated because we need two new normalization terms

$$N_2 = i f_2 h^{\varrho\sigma} h_\beta^{\sigma\alpha} h^{\varrho\alpha} h_\beta \delta \tag{5.7.33}$$

$$N_3 = i f_3 h^{\varrho\sigma} h_\beta^{\varrho\sigma} h^{\alpha\nu} h_\beta^{\alpha\nu} \delta. \tag{5.7.34}$$

The gauge variations

$$d_Q N_2 = f_2 \Big\{ (u_\sigma^\varrho + u_\varrho^\sigma) \underbrace{h_\beta^{\sigma\alpha} h^{\varrho\alpha} h_\beta}_{G6} - \underbrace{u_\mu^\mu h_\beta^{\varrho\alpha} h^{\varrho\alpha} h_\beta}_{G5} +$$

$$+ \underbrace{u_{\alpha\beta}^\sigma h^{\varrho\sigma} h^{\varrho\alpha} h_\beta}_{G6} - \tfrac{1}{2} \underbrace{u_{\mu\beta}^\mu h^{\varrho\alpha} h^{\varrho\alpha} h_\beta}_{G5} - \underbrace{u_{\beta\mu}^\mu h^{\varrho\sigma} h_\beta^{\sigma\alpha} h^{\varrho\alpha}}_{G7} \Big\} \delta \tag{5.7.35}$$

$$d_Q N_3 = f_3 \Big\{ 2 \underbrace{u_\sigma^\varrho h_\beta^{\varrho\sigma} h^{\alpha\nu} h_\beta^{\alpha\nu}}_{G10} - \underbrace{u_\mu^\mu h_\beta^{\alpha\nu} h^{\alpha\nu} h_\beta}_{G5} +$$

$$+ 2 \underbrace{u_{\sigma\beta}^\varrho h^{\varrho\sigma} h_\beta^{\alpha\nu} h^{\alpha\nu}}_{G10} - \underbrace{u_{\beta\mu}^\mu h_\beta^{\alpha\nu} h^{\alpha\nu} h}_{G5} \Big\} \delta \tag{5.7.36}$$

show to which types they contribute. For a consistent solution we need the following five divergence terms

$$\partial_\nu^x N_1^\nu = g_1 \partial_\nu^x (u_\alpha^\alpha h^{\varrho\sigma} h^{\varrho\sigma} h_\nu \delta) \tag{5.7.37}$$

$$\partial_\nu^x N_2^\nu = g_2 \partial_\nu^x (u^\alpha h^{\varrho\sigma} h_\alpha^{\varrho\sigma} h_\nu \delta) \tag{5.7.38}$$

$$\partial_\nu^x N_3^\nu = g_3 \partial_\nu^x (u_\alpha^\alpha h_\nu^{\varrho\sigma} h^{\varrho\sigma} h \delta) \tag{5.7.39}$$

$$\partial_\nu^x N_4^\nu = g_4 \partial_\nu^x (u^\alpha h_\nu^{\varrho\sigma} h^{\varrho\sigma} h_\alpha \delta) \tag{5.7.40}$$

$$\partial_\nu^x N_5^\nu = g_5 \partial_\alpha^x (u^\alpha h^{\varrho\sigma} h_\nu^{\varrho\sigma} h_\nu \delta). \tag{5.7.41}$$

To compensate the anomalies of this type we must choose the following values for the normalization constants:

$$f_2 = i, \quad f_3 = \frac{i}{2}, \tag{5.7.42}$$

$$g_1 = \frac{i}{4} = g_2 = g_3 = g_4, \quad g_5 = -\frac{i}{2}.$$

Remaining types:

For the remaining types we list all normalization terms with and without Q-vertex with their right factors. Then it is not hard for the reader to check that all anomalies (5.7.18) cancel out. For type G1 we find

$$N_4 = h^{\alpha\varrho} h^{\beta\sigma} h_\varrho^{\beta\sigma} h_{\alpha} \delta, \tag{5.7.43}$$

$$\partial_\alpha N_1^\alpha = \frac{i}{4}\partial_\alpha^x(u^\alpha h^{\varrho\sigma} h_\varrho h_\sigma \delta),$$

$$\partial_\varrho N_2^\varrho = -\frac{i}{4}\partial_\varrho^x(u_\alpha^\alpha h^{\varrho\sigma} h h_\sigma \delta),$$

$$\partial_\varrho N_3^\varrho = -\frac{i}{4}\partial_\varrho^x(u^\alpha h^{\varrho\sigma} h_\alpha h_\sigma \delta).$$

For type G2 we have to use N_4 and

$$N_5 = -2 h_\alpha^{\varrho\sigma} h^{\nu\sigma} h^{\alpha\beta} h_\beta^{\varrho\nu} \delta, \tag{5.7.44}$$

$$\partial_\nu^x N_1^\nu = \frac{i}{2}\partial_\nu^x(u_\alpha^\alpha h^{\varrho\sigma} h^{\beta\nu} h_\beta^{\varrho\sigma} \delta),$$

$$\partial_\nu^x N_2^\nu = \frac{i}{2}\partial_\nu^x(u^\alpha h_\alpha^{\varrho\sigma} h^{\beta\nu} h_\beta^{\varrho\sigma} \delta),$$

$$\partial_\nu^x N_3^\nu = -\frac{i}{2}\partial_\nu^x(u^\nu h^{\varrho\sigma} h_\varrho^{\alpha\beta} h_\sigma^{\alpha\beta} \delta).$$

In type G6 we need N_2 (5.7.33) and

$$\partial_\nu^x N_1^\nu = -\frac{i}{2}\partial_\nu^x(u_\varrho^\alpha h^{\alpha\sigma} h^{\varrho\sigma} h_\nu \delta).$$

For type G7 we have to use N_2, N_5 and

$$N_6 = 2 h^{\varrho\sigma} h^{\nu\varrho} h_\beta^{\alpha\sigma} h_\beta^{\alpha\nu} \delta, \tag{5.7.45}$$

$$N_7 = h^{\varrho\sigma} h^{\nu\alpha} h_\beta^{\varrho\alpha} h_\beta^{\nu\sigma} \delta, \tag{5.7.46}$$

$$\partial_\nu^x N_1^\nu = -i\partial_\nu^x(u^\alpha h_\alpha^{\varrho\sigma} h_\nu^{\varrho\beta} h^{\sigma\beta} \delta),$$

$$\partial_\nu^x N_2^\nu = i\partial_\nu^x(u^\nu h^{\varrho\sigma} h_\alpha^{\varrho\beta} h^{\sigma\beta} \delta),$$

$$\partial_\nu^x N_3^\nu = -i\partial_\nu^x(u_\varrho^\varrho h^{\alpha\beta} h_\nu^{\alpha\sigma} h^{\beta\sigma} \delta).$$

In type G8 we use

$$\partial_\nu^x N_1^\nu = i\partial_\nu^x(u_\varrho^\alpha h^{\alpha\beta} h_\nu^{\beta\sigma} h^{\sigma\varrho} \delta),$$

$$\partial_\nu^x N_2^\nu = i\partial_\nu^x(u_\varrho^\alpha h_\nu^{\alpha\beta} h^{\beta\sigma} h^{\sigma\varrho} \delta),$$

in addition to N_1, N_6, N_7.

Finally in type G9 we need N_1, N_5 and

$$\partial_\sigma^x N_1^\sigma = -i\partial_\sigma^x(u_\beta^\alpha h_\nu^{\varrho\sigma} h^{\beta\nu} h^{\alpha\varrho} \delta),$$

$$\partial_\sigma^x N_2^\sigma = -i\partial_\sigma^x(u_\beta^\alpha h^{\varrho\beta} h^{\nu\sigma} h_{\nu'}^{\alpha\varrho}\delta).$$

In type G10 we use N_3 and

$$\partial_\nu^x N_1^\nu = -\frac{i}{2}\partial_\nu^x(u_\varrho^\alpha h^{\alpha\varrho} h_\nu^{\beta\sigma} h^{\beta\sigma}\delta).$$

In the last type G11 we need N_4 together with

$$\partial_\nu^x N_1^\nu = \frac{i}{2}\partial_\nu^x(u_\varrho^\alpha h^{\alpha\varrho} h^{\nu\beta} h_\beta\delta).$$

For the divergence terms we have used the same symbols in the various types because these terms only contribute to one single type, so that there is no danger of confusion.

After the long computational work there comes the reward. Collecting all normalization terms $N_1, \ldots N_7$ we obtain

$$N = \Big(-2h^{\alpha\varrho}h^{\varrho\sigma}h_\nu^{\sigma\beta}h_\beta^{\alpha\nu} - h^{\varrho\sigma}h_\beta^{\sigma\alpha}h^{\varrho\alpha}h_\beta -$$

$$-\frac{1}{2}h^{\varrho\sigma}h_\beta^{\varrho\sigma}h^{\alpha\nu}h_\beta^{\alpha\nu} + h^{\alpha\varrho}h^{\beta\sigma}h_\varrho^{\sigma\beta}h_\alpha - 2h_\alpha^{\varrho\sigma}h^{\sigma\nu}h^{\alpha\beta}h_\beta^{\varrho\nu} +$$

$$+2h^{\varrho\sigma}h^{\nu\varrho}h_\beta^{\alpha\sigma}h_\beta^{\alpha\nu} + h^{\varrho\sigma}h^{\nu\alpha}h_\beta^{\varrho\alpha}h_\beta^{\nu\sigma}\Big)\delta(x-y). \qquad (5.7.47)$$

Multiplying this with the factor $1/2!$ for the second order this agrees precisely with the quartic term $L^{(2)}$ in the expansion of the Einstein-Hilbert Lagrangian in Sect.5.5 (5.5.34). In the same way as the four-gluon coupling in Yang-Mills theory was generated by gauge invariance (4.4.14), so is the four-graviton self-coupling. This clearly shows that the gauge principle also works in quantum gravity, it is a universal principle of nature. In contrast to Yang-Mills theory, this process does not terminate after second order. In order n there arise self-couplings of $n+2$ h-fields. This is the proliferation of couplings which we already noticed in the classical theory (5.5.23). In old-fashioned quantum field theory this is a serious problem, but in the framework of causal gauge invariance it is not, because the many couplings are all normalization terms determined by gauge invariance.

5.8 Second order gauge invariance: ghost sector

In this section we consider the anomalies with field operators $uu\tilde{u}h$. They arise by combining the four ghost vertices $u1$ - $u4$ in (5.7.12) with the terms in the Q-vertex (5.7.13). According to the usual mechanism we have to calculate the commutators with those terms in (5.7.13) which have a derivative ∂_μ. These are the terms Q2, Q3, Q5, Q6, Q7, Q8 where the commutator between two graviton fields must be computed. For the term Q9 we have to perform an anticommutator between ghost fields, for example

Spin-2 gauge theory

$$[Q0, u1] = -\frac{i}{2} iu^\beta u^\alpha_\beta \{\tilde{u}^\alpha_\mu(x), u^{\alpha'}_{\beta'}(y)\} \tilde{u}^{\varrho'}_{\alpha'} h^{\varrho'\beta'}$$

$$= -\frac{i}{2} u^\beta u^\alpha_\beta(x) \tilde{u}^{\varrho'}_\alpha(y) h^{\varrho'\beta'} D_{\mu\beta'}(x-y). \qquad (5.8.1)$$

According to (5.7.17) this gives the anomaly

$$-\frac{i}{2} u^\beta u^\alpha_\beta(x) \tilde{u}^{\varrho'}_\alpha(y) h^{\varrho'\beta'} \partial^y_{\beta'} \delta(x-y).$$

The resulting anomalies are again grouped according to their Lorentz structure:

$$A_1 = \frac{i}{2} \Big\{ u^\beta u^\alpha_\beta \tilde{u}^{\varrho'}_{\beta'} h^{\varrho'\beta'}_\alpha - u^\beta u^\alpha_\beta \tilde{u}^{\varrho'}_\alpha h^{\varrho'\beta'} \partial_{\beta'} + \qquad (5.8.2)$$

$$+ u^\beta u^\alpha_\beta \tilde{u}^{\varrho'}_{\beta'} h^{\varrho'\beta'} \partial_\alpha - u^\alpha h^{\varrho\sigma}_\alpha u^{\alpha'}_\varrho \tilde{u}^\sigma_{\alpha'} + u^\alpha h^{\varrho\sigma}_\alpha u^{\alpha'} \tilde{u}^\varrho_\sigma \partial_{\alpha'} + u^\alpha h^{\varrho\sigma}_\alpha u^{\alpha'}_{\alpha'} \tilde{u}^\varrho_\sigma +$$

$$+ u^\alpha_\alpha h^{\varrho\sigma} u^{\alpha'} \tilde{u}^\varrho_\sigma \partial_{\alpha'} + u^\alpha_\alpha h^{\varrho\sigma} u^{\alpha'}_\varrho \tilde{u}^\sigma_{\alpha'} - u^\alpha_\varrho h^{\varrho\sigma} u^{\alpha'} \tilde{u}^\sigma_\alpha \partial_{\alpha'} \qquad (G1)$$

$$- u^\beta u^\alpha_\beta \tilde{u}^\alpha_{\beta'} h^{\varrho'\beta'} \partial_{\varrho'} - u^\alpha h^{\varrho\sigma}_\alpha u^{\alpha'}_\varrho \tilde{u}^{\alpha'}_\sigma - u^\alpha_\varrho h^{\varrho\sigma} u^{\alpha'} \tilde{u}^\alpha_\sigma \partial_{\alpha'} +$$

$$+ u^\alpha_\varrho h^{\varrho\sigma} u^{\alpha'}_\alpha \tilde{u}^{\alpha'}_\sigma \Big\} \delta(x-y). \qquad (G2)$$

The first two fields in each term have arguments x and the last two ones have argument y. The derivatives on the δ-distribution are always with respect to y.

We first turn to the anomalies of type (G1). Adding the contributions with $x \longleftrightarrow y$ interchanged and transforming the result to normal form by means of the identity (5.7.3) we obtain

$$G1 = \frac{i}{2} \Big\{ u^\alpha_\sigma u^\beta_\alpha \tilde{u}^\varrho_\beta h^{\varrho\sigma} - u^\alpha_\beta u^\beta \tilde{u}^\varrho_\alpha h^{\varrho\sigma}_\sigma - u^\alpha u^\beta \tilde{u}^\varrho_{\sigma\beta} h^{\varrho\sigma}_\alpha +$$

$$+ u^\alpha u^\beta_\beta \tilde{u}^\varrho_{\alpha\sigma} h^{\varrho\sigma} - u^\alpha u^\beta_\sigma \tilde{u}^\varrho_\beta h^{\varrho\sigma}_\alpha + u^\alpha_\sigma u^\beta_\beta \tilde{u}^\varrho_\alpha h^{\varrho\sigma} + u^\alpha_\sigma u^\beta \tilde{u}^\varrho_{\alpha\beta} h^{\varrho\sigma} \Big\} \delta(x-y). \qquad (5.8.3)$$

It is not hard to write this result as a sum of normalization terms of divergence form:

$$G1 = \frac{i}{2} \partial_\nu \Big\{ \Big(u^\alpha u^\beta_\beta \tilde{u}^\varrho_\alpha h^{\varrho\nu} + u^\alpha u^\beta \tilde{u}^\varrho_\alpha h^{\varrho\nu}_\beta +$$

$$+ u^\nu u^\alpha \tilde{u}^\varrho_\alpha h^{\varrho\sigma}_\sigma + u^\nu_\sigma u^\alpha \tilde{u}^\varrho_\alpha h^{\varrho\sigma} \Big) \delta(x-y) \Big\} + x \longleftrightarrow y. \qquad (5.8.4)$$

Here the second identity for local terms (5.7.26) has been used.

We proceed in the same way with the anomalies of type (G2):

$$G2 = \frac{i}{2} \Big\{ u^\beta u^\alpha_\beta \tilde{u}^\alpha_{\varrho\sigma} h^{\varrho\sigma} + u^\beta u^\alpha_\beta \tilde{u}^\alpha_\sigma h^{\varrho\sigma}_\varrho - u^\beta_\beta u^\alpha_\varrho \tilde{u}^\alpha_\sigma h^{\varrho\sigma} -$$

$$- u^\beta u^\alpha_\varrho \tilde{u}^\alpha_{\beta\sigma} h^{\varrho\sigma} - u^\beta u^\alpha_\varrho \tilde{u}^\alpha_\sigma h^{\varrho\sigma}_\beta + u^\beta_\sigma u^\alpha_\beta \tilde{u}^\alpha_\sigma h^{\varrho\sigma} \Big\} \delta(x-y). \qquad (5.8.5)$$

Again the result is a divergence:

Second order gauge invariance: ghost sector   223

$$G2 = \frac{i}{2}\partial_\nu \left\{ \left(u^\beta u^\alpha_\beta \tilde{u}^\alpha_\sigma h^{\nu\sigma} - u^\nu u^\alpha_\varrho \tilde{u}^\alpha_\sigma h^{\varrho\sigma} \right) \delta(x-y) \right\} + x \longleftrightarrow y. \quad (5.8.6)$$

So far so good, but this is not yet the whole story. There exists another mechanism for producing anomalies which we now have to examine. The term $u3$ in the ghost coupling (5.7.12) contains a factor u^α_α with the same index α as upper Lorentz index and lower derivative index. By anticommuting this factor with $\tilde{u}^\mu$ the index μ can be shifted down to become a derivative ∂_μ so that an anomaly can be produced. A term with $\tilde{u}^\mu$ is Q14 in (5.7.13). Let us calculate the commutator

$$[Q14, u3] = -\frac{i}{2} u^\alpha_{\alpha\beta} u^\beta \{\tilde{u}^\mu, u^{\alpha'}_{\alpha'}\} \tilde{u}^{\varrho'}_{\beta'} h^{\varrho'\beta'}$$

$$= -\frac{1}{2} u^\alpha_{\alpha\beta} u^\beta \tilde{u}^{\varrho'}_{\beta'} h^{\varrho'\beta'} \partial^y_\mu D(x-y). \quad (5.8.7)$$

After distribution splitting and applying the external derivative ∂^μ_x we get the anomaly from $\partial^\mu_x \partial^y_\mu D^{\text{ret}} = -\delta$:

$$A_1 = \frac{1}{2} u^\alpha_{\alpha\beta} u^\beta \tilde{u}^\varrho_\sigma h^{\varrho\sigma} \delta(x-y). \quad (5.8.8)$$

To see whether this anomaly can be compensated by normalization terms, we consider the following tree graph with two ghost vertices

$$[u2, u3] = -u^\alpha h^{\varrho\beta}_\alpha \{\tilde{u}^\varrho_\beta, u^{\alpha'}_{\alpha'}\} \tilde{u}^{\varrho'}_{\beta'} h^{\varrho'\beta'}$$

$$= -iu^\alpha h^{\varrho\beta}_\alpha \tilde{u}^{\varrho'}_{\beta'} h^{\varrho'\beta'} \partial^x_\beta \partial^y_\varrho D(x-y). \quad (5.8.9)$$

We note that again the factor $u^{\alpha'}_{\alpha'}$ in $u3$ is essential here. After distribution splitting this gives the following retarded distribution

$$R_2(x,y) = -iu^\alpha h^{\varrho\beta}_\alpha \tilde{u}^{\varrho'}_{\beta'} h^{\varrho'\beta'} (\partial^x_\beta \partial^y_\varrho D^{\text{ret}}(x-y) + \eta_{\beta\varrho} C\delta), \quad (5.8.10)$$

where C is a normalization constant. If we now calculate the gauge variation, then from

$$d_Q h^{\varrho\beta}_\alpha = \ldots - \frac{i}{2} \eta^{\varrho\beta} u^\mu_{\mu\alpha}$$

we indeed obtain the same field operators as in (5.8.8):

$$d_Q R_2 = -\frac{1}{2} u^\alpha u^\mu_{\mu\alpha} \tilde{u}^{\varrho'}_{\beta'} h^{\varrho'\beta'} \eta^{\varrho\beta} (\partial^x_\beta \partial^y_\varrho D^{\text{ret}}(x-y) + \eta_{\beta\varrho} C\delta) + \ldots \quad (5.8.11)$$

However, the gauge variation of the other factors $\tilde{u}^{\varrho'}_{\beta'} h^{\varrho'\beta'}$ in (5.8.10) would produce new anomalies abbreviated by the dots in (5.8.11). Therefore, it is not practicable to choose the normalization constant $C \neq 0$, so that we take $C = 0$. It is a new phenomenon that we get a local term from another source, namely from D^{ret}, because the two derivatives collapse to a wave operator

$$d_Q R_2 = \frac{1}{2} u^\alpha u^\mu_{\mu\alpha} \tilde{u}^{\varrho'}_{\beta'} h^{\varrho'\beta'} \eta^{\varrho\beta} \delta(x-y)$$

+non-local terms. $\quad (5.8.12)$

This local term cancels the anomaly A_1 (5.8.8). This finishes the proof of second order gauge invariance.

5.9 Coupling to matter

We have already discussed matter-graviton coupling in Sect.2.5. We now want to derive the coupling (2.5.3) which was considered there from gauge invariance. We start from an arbitrary coupling between the graviton field $h^{\alpha\beta}$ and a symmetric tensor field $T^{\mu\nu}$

$$T_1^m = i\frac{\tilde{\kappa}}{2}h^{\alpha\beta}a_{\alpha\beta\mu\nu}T^{\mu\nu}. \tag{5.9.1}$$

The gauge variation of the graviton field is given by (5.1.8)

$$d_Q h^{\mu\nu} = -ib^{\mu\nu\alpha\beta}u_{\alpha,\beta} \tag{5.9.2}$$

but the matter tensor has zero gauge variation

$$d_Q T^{\mu\nu} = 0. \tag{5.9.3}$$

Then the gauge variation of (5.9.1) becomes

$$d_Q T_1^m = -\frac{\tilde{\kappa}}{2}b^{\varrho\sigma\alpha\beta}\partial_\varrho u_\sigma a_{\alpha\beta\mu\nu}T^{\mu\nu}$$

$$= -\frac{\tilde{\kappa}}{2}b^{\varrho\sigma\alpha\beta}a_{\alpha\beta\mu\nu}\Big[\partial_\varrho(u_\sigma T^{\mu\nu}) - u_\sigma\partial_\varrho T^{\mu\nu}\Big]. \tag{5.9.4}$$

For first order gauge invariance this must be a divergence, thus the last term herein must vanish.

The Lorentz tensor $a_{\alpha\beta\mu\nu}$ of rank four can be expressed by tensor products of the Minkowski metric tensor $\eta_{\alpha\beta}$. Since it is symmetric both in α, β and μ, ν, it must be of the following form

$$a_{\alpha\beta\mu\nu} = a_1(\eta_{\alpha\mu}\eta_{\beta\nu} + \eta_{\alpha\nu}\eta_{\beta\mu}) + a_2\eta_{\alpha\beta}\eta_{\mu\nu}. \tag{5.9.5}$$

Using

$$b^{\varrho\sigma\alpha\beta} = \tfrac{1}{2}(\eta^{\varrho\alpha}\eta^{\sigma\beta} + \eta^{\varrho\beta}\eta^{\sigma\alpha} - \eta^{\varrho\sigma}\eta^{\alpha\beta}),$$

gauge invariance in (5.9.4) requires

$$a_1\Big(u_\nu\partial_\mu T^{\mu\nu} + u_\mu\partial_\nu T^{\mu\nu}\Big) - (a_1 + a_2)u^\varrho\partial_\varrho T^\mu_\mu = 0$$

which gives the condition

$$\partial_\mu T^{\mu\nu} = \frac{1}{2}\Big(1 + \frac{a_2}{a_1}\Big)\partial^\nu T^\mu_\mu. \tag{5.9.6}$$

In classical field theory the energy-momentum tensor is conserved

$$\partial_\mu T^{\mu\nu} = \partial_\nu T^{\mu\nu} = 0. \tag{5.9.7}$$

Assuming this we must have $a_2 = -a_1$ because the trace T^μ_μ does not vanish in general. That means, up to a factor the a-tensor (5.9.5) agrees with the b-tensor (5.1.2)

$$T_1^m = i\frac{\tilde\kappa}{2}h^{\alpha\beta}b_{\alpha\beta\mu\nu}T^{\mu\nu}. \tag{5.9.8}$$

So the graviton field $h^{\alpha\beta}$ couples to a conserved symmetric tensor by means of the b-tensor. This is a consequence of the gauge variation (5.9.2), i.e. of our use of Goldberg variables. Until now the coupling constant $\tilde\kappa$ is still arbitrary; it follows from second order gauge invariance below that it must be equal to κ.

It is interesting to note that condition (5.9.6) can also be satisfied by a tensor $t_{\mu\nu}$ which is neither conserved nor traceless. An important example is the coupling term (5.4.10) of the vector graviton field $t_{\mu\nu} = \partial_\mu v_\lambda \partial_\nu v^\lambda$ in the massless case. Using $\partial^2 v_\lambda = 0$, (5.9.6) holds with $a_1 = 2, a_2 = -1$.

The result (5.9.8) is usually derived from Lagrangian field theory as follows. The relevant Lagrangian density in general relativity has the form $\sqrt{-g}L$ where L is the matter Lagrangian. We expand the metric tensor $g_{\alpha\beta}(x)$ in a Minkowski background in the simple form

$$g_{\alpha\beta} = \eta_{\alpha\beta} + \kappa f_{\alpha\beta}. \tag{5.9.9}$$

Then $g^{\alpha\beta}$ which is the inverse is given by

$$g^{\alpha\beta} = \eta^{\alpha\beta} - \kappa f^{\alpha\beta} + O(\kappa^2). \tag{5.9.10}$$

For simplicity we assume that L only depends on $g_{\alpha\beta}$ not on its derivatives, so that

$$\sqrt{-g}L = L_0 - \kappa \frac{\partial(\sqrt{-g}L)}{\partial g_{\alpha\beta}}\bigg|_{\eta_{\alpha\beta}} f_{\alpha\beta} + O(\kappa^2). \tag{5.9.11}$$

We claim that the partial derivative herein is again the energy-momentum tensor, up to a factor. To see this we start from the action

$$S = \int d^4x \sqrt{-g}L \tag{5.9.12}$$

and consider the variation

$$\delta S = \int \frac{\partial(\sqrt{-g}L)}{\partial g^{\alpha\beta}} \delta g^{\alpha\beta} d^4x. \tag{5.9.13}$$

We assume that the variation of the metric is due to an infinitesimal coordinate transformation $x'^\alpha = x^\alpha + \varepsilon \xi^\alpha$. This implies the following change of the metric tensor

$$g'^{\alpha\beta}(x') = g^{\mu\nu}(x)\frac{\partial x'^\alpha}{\partial x^\mu}\frac{\partial x'^\beta}{\partial x^\nu} = g^{\mu\nu}\left(\delta^\alpha_\mu + \varepsilon\frac{\partial \xi^\alpha}{\partial x^\mu}\right)\left(\delta^\beta_\nu + \varepsilon\frac{\partial \xi^\beta}{\partial x^\nu}\right)$$

$$= g^{\alpha\beta}(x) + \varepsilon g^{\alpha\nu}\frac{\partial \xi^\beta}{\partial x^\nu} + \varepsilon g^{\mu\beta}\frac{\partial \xi^\alpha}{\partial x^\mu} + O(\varepsilon^2). \tag{5.9.14}$$

We expand the l.h.s. around the original position x

$$g'^{\alpha\beta}(x+\varepsilon\xi) = g'^{\alpha\beta}(x) + \varepsilon \frac{\partial g'^{\alpha\beta}}{\partial x^\mu}\xi^\mu + O(\varepsilon^2). \tag{5.9.15}$$

In the last term $O(\varepsilon)$ we can change g' into g because this requires a correction of order $O(\varepsilon^2)$, then (5.9.14) yields

$$g'^{\alpha\beta}(x) = g^{\alpha\beta}(x) - \varepsilon \frac{\partial g^{\alpha\beta}}{\partial x^\mu}\xi^\mu + \varepsilon g^{\alpha\nu}\frac{\partial \xi^\beta}{\partial x^\nu} + \varepsilon g^{\mu\beta}\frac{\partial \xi^\alpha}{\partial x^\mu} + O(\varepsilon^2). \tag{5.9.16}$$

The terms $O(\varepsilon)$ in (5.9.16) can be written as contravariant derivatives of ξ^α which are defined by

$$\nabla^\beta \xi^\alpha = g^{\beta\varrho}\nabla_\varrho \xi^\alpha = g^{\beta\varrho}\left(\frac{\partial \xi^\alpha}{\partial x^\varrho} + \Gamma^\alpha_{\sigma\varrho}\xi^\sigma\right)$$

$$= g^{\beta\mu}\frac{\partial \xi^\alpha}{\partial x^\mu} + \tfrac{1}{2}g^{\beta\varrho}g^{\alpha\mu}\Big(g_{\sigma\mu,\varrho} + g_{\mu\varrho,\sigma} - g_{\sigma\varrho,\mu}\Big)\xi^\sigma, \tag{5.9.17}$$

where (5.5.6) has been substituted. We interchange α and β and the dummy indices ϱ, μ and obtain

$$\nabla^\alpha \xi^\beta = g^{\alpha\nu}\frac{\partial \xi^\beta}{\partial x^\nu} + \tfrac{1}{2}g^{\alpha\mu}g^{\beta\varrho}\Big(g_{\sigma\varrho,\mu} + g_{\mu\varrho,\sigma} - g_{\sigma\mu,\varrho}\Big)\xi^\sigma. \tag{5.9.18}$$

Adding the two equations yields the symmetric combination

$$\nabla^\beta \xi^\alpha + \nabla^\alpha \xi^\beta = g^{\alpha\nu}\frac{\partial \xi^\beta}{\partial x^\nu} + g^{\beta\mu}\frac{\partial \xi^\alpha}{\partial x^\mu} + g^{\alpha\mu}g^{\beta\varrho}g_{\mu\varrho,\sigma}\xi^\sigma. \tag{5.9.19}$$

In the last term we use the relation

$$g^{\alpha\mu}g^{\beta\varrho}g_{\mu\varrho,\sigma} = -g^{\alpha\mu}g^{\beta\varrho}{}_{,\sigma}g_{\mu\varrho} = -g^{\beta\alpha}{}_{,\sigma},$$

then the result agrees with (5.9.16)

$$g'^{\alpha\beta}(x) = g^{\alpha\beta}(x) + \varepsilon(\nabla^\beta \xi^\alpha + \nabla^\alpha \xi^\beta) + O(\varepsilon^2). \tag{5.9.20}$$

The bracket $O(\varepsilon)$ gives the variation $\delta g^{\alpha\beta}$ in (5.9.13).

The variation of the action δS now assumes the following form

$$\delta S = \varepsilon \int \frac{\partial(\sqrt{-g}L)}{\partial g^{\alpha\beta}}\left(\nabla^\beta \xi^\alpha + \nabla^\alpha \xi^\beta\right)d^4x. \tag{5.9.21}$$

Using the symmetry in α, β we can write

$$\delta S = \varepsilon \int T_{\alpha\beta}\nabla^\beta \xi^\alpha \sqrt{-g}d^4x, \tag{5.9.22}$$

where we have introduced the symmetric tensor $T_{\alpha\beta}$

$$\tfrac{1}{2}\sqrt{-g}T_{\alpha\beta} = \frac{\partial(\sqrt{-g}L)}{\partial g^{\alpha\beta}}. \tag{5.9.23}$$

By partial integration the vanishing of δS now implies

$$0 = -\varepsilon \int \nabla^\beta T_{\alpha\beta} \xi^\alpha \sqrt{-g} d^4x$$

the conservation laws

$$\nabla^\beta T_{\alpha\beta} = 0. \tag{5.9.24}$$

That means, up to a factor, $T_{\alpha\beta}$ is identical with the energy-momentum tensor (5.9.7). Hence, the coupling $O(\kappa)$ in (5.9.11) is of the following form

$$T_1^m \sim \frac{\kappa}{2} T^{\alpha\beta} f_{\alpha\beta}. \tag{5.9.25}$$

For comparison with (5.9.8) we must find the relation between the simple expansion (5.9.9) and the Goldberg variables

$$\sqrt{-g} g^{\mu\nu} = \eta^{\mu\nu} + \kappa h^{\mu\nu}. \tag{5.9.26}$$

For this purpose we expand the determinant g as follows

$$\sqrt{-g} = \exp \tfrac{1}{2} \operatorname{Tr} \log(\delta_\beta^\alpha + \kappa f_\beta^\alpha) =$$

$$= \exp \tfrac{1}{2} \operatorname{Tr} (\kappa f_\beta^\alpha - \tfrac{1}{2}\kappa^2 f_\varrho^\alpha f_\beta^\varrho + \ldots)$$

$$= \exp \tfrac{1}{2}(\kappa f - \tfrac{1}{2}\kappa^2 f_\varrho^\alpha f_\alpha^\varrho + \ldots)$$

$$= 1 + \frac{\kappa}{2} f - \frac{\kappa^2}{4} f_\varrho^\alpha f_\alpha^\varrho + \frac{\kappa^2}{8} f^2 + \ldots, \tag{5.9.27}$$

where $f = f_\alpha^\alpha$ denotes the trace. Now from (5.9.26)

$$\eta^{\mu\nu} + \kappa h^{\mu\nu} = \left(1 + \frac{\kappa}{2}f - \frac{\kappa^2}{4} f^{\alpha\beta} f_{\alpha\beta} + \frac{\kappa^2}{8} f^2 \right)$$

$$\times \left(\eta^{\mu\nu} - \kappa f^{\mu\nu} + \kappa^2 f^{\mu\varrho} f_\varrho^\nu \right)$$

we conclude

$$\kappa h^{\mu\nu} = -\kappa f^{\mu\nu} + \tfrac{1}{2}\kappa f \eta^{\mu\nu} + O(\kappa^2) = -\kappa f_{\alpha\beta} b^{\alpha\beta\mu\nu} + O(\kappa^2). \tag{5.9.28}$$

By means of the relation

$$b_{\alpha\beta\mu\nu} b^{\mu\nu\varrho\sigma} = \tfrac{1}{2}(\eta_\alpha^\varrho \eta_\beta^\sigma + \eta_\alpha^\sigma \eta_\beta^\varrho) \tag{5.9.29}$$

we invert (5.9.28) and find

$$f_{\varrho\sigma} = -h^{\mu\nu} b_{\mu\nu\varrho\sigma}. \tag{5.9.30}$$

This shows that (5.9.25) agrees with (5.9.8) (taking the equality of the coupling constants in advance)..

Let us now assume that the matter is described by a real massive scalar field $\varphi(x)$ with Lagrangian density

$$L = \tfrac{1}{2}(g^{\alpha\beta}\varphi_{,\alpha}\,\varphi_{,\beta} - M^2\varphi^2). \tag{5.9.31}$$

Since

$$\frac{\partial\sqrt{-g}}{\partial g^{\alpha\beta}} = -\frac{1}{2\sqrt{-g}}\frac{\partial g}{\partial g^{\alpha\beta}} = -\tfrac{1}{2}\sqrt{-g}\,g_{\alpha\beta}, \tag{5.9.32}$$

we then get

$$\frac{\partial(\sqrt{-g}L)}{\partial g^{\alpha\beta}} = \sqrt{-g}\left(-\tfrac{1}{2}g_{\alpha\beta}L + \tfrac{1}{2}\varphi_{,\alpha}\,\varphi_{,\beta}\right).$$

Going over to flat space, we find the energy-momentum tensor

$$T_{\alpha\beta} = \varphi_{,\alpha}\,\varphi_{,\beta} - \tfrac{1}{2}\eta_{\alpha\beta}\varphi_{,\mu}\,\varphi^{,\mu} + \tfrac{1}{2}M^2\eta_{\alpha\beta}\varphi^2, \tag{5.9.33}$$

which gives the first order coupling. For later use we write down the expansion of the matter Lagrangian (5.9.11) up to second order

$$\sqrt{-g}L = \tfrac{1}{2}(\eta^{\alpha\beta}\varphi_{,\alpha}\,\varphi_{,\beta} - M^2\varphi^2) + \tfrac{1}{2}\kappa h^{\alpha\beta}(\varphi_{,\alpha}\,\varphi_{,\beta} - \tfrac{1}{2}M^2\eta_{\alpha\beta}\varphi^2 +$$

$$+\frac{\kappa^2 M^2}{8}(h^{\alpha\beta}h_{\alpha\beta}\varphi^2 - \tfrac{1}{2}h^2\varphi^2) + O(\kappa^3), \tag{5.9.34}$$

where we have inserted the Goldberg variable $h_{\alpha\beta}$ which we always use in quantum theory.

The full coupling of gravitons and scalar matter reads

$$T = \kappa T_1 + i\frac{\tilde{\kappa}}{2}h^{\alpha\beta}b_{\alpha\beta\mu\nu}T^{\mu\nu} =$$

$$= \kappa T_1 + i\frac{\tilde{\kappa}}{2}\left(h^{\alpha\beta}\varphi_{,\alpha}\,\varphi_{,\beta} - \tfrac{1}{2}M^2h\varphi^2\right) \tag{5.9.35}$$

according to (5.7.12), (5.9.8) and (5.9.33). The corresponding Q-vertex follows from (5.7.13) and (5.9.4)

$$iT^\mu = i\kappa T^\mu_{1/1} - \tfrac{1}{2}\tilde{\kappa}u_\nu T^{\nu\mu} =$$

$$= i\kappa T^\mu_{1/1} - \tfrac{1}{2}\tilde{\kappa}\left(u_\nu\varphi^{,\nu}\varphi^{,\mu} - \tfrac{1}{2}u^\mu\varphi_{,\nu}\,\varphi^{,\nu} + \tfrac{1}{2}M^2 u^\mu\varphi^2\right). \tag{5.9.36}$$

As discussed after (5.7.2), anomalies in second order tree graphs are generated by those terms in $T^\mu_{1/1}$ which have a derivative index μ downstairs, namely the terms Q2, Q3, Q5, Q6, Q7, Q8 in (5.7.13) and the first term $\sim \varphi_{,\mu}$ in the matter coupling:

$$iT^\mu = \kappa\left(\tfrac{1}{2}u^\nu h^{\varrho\sigma}{}_{,\nu}h^{\varrho\sigma}{}_{,\mu} + \tfrac{1}{2}u^\nu{}_{,\nu}h^{\varrho\sigma}h^{\varrho\sigma}{}_{,\mu} - \tfrac{1}{4}u^\nu{}_{,\nu}hh_{,\mu} -\right.$$

$$\left.- \tfrac{1}{4}u^\nu h_{,\nu}h_{,\mu} + \tfrac{1}{2}u^\nu{}_{,\varrho}h^{\nu\varrho}h_{,\mu} - u^\nu{}_{,\varrho}h^{\varrho\sigma}h^{\sigma\nu}{}_{,\mu}\right) - \tfrac{1}{2}\tilde{\kappa}u^\nu\varphi_{,\nu}\,\varphi_{,\mu}. \tag{5.9.37}$$

By the simple mechanism of Sect. 5.7 we then obtain the following anomalies with scalar fields

$$A_1 = \frac{i}{4}\Big\{\kappa\tilde{\kappa}\Big(u^\nu h^{\varrho\sigma}{}_{,\nu}\,\varphi_{,\varrho}\,\varphi_{,\sigma} + u^\nu{}_{,\nu}\,h^{\varrho\sigma}\varphi_{,\varrho}\,\varphi_{,\sigma} -$$
$$-\tfrac{1}{2}M^2 u^\nu{}_{,\nu}\,h\varphi^2 - \tfrac{1}{2}M^2 u^\nu h_{,\nu}\,\varphi^2 - 2u^\nu{}_{,\varrho}\,h^{\varrho\sigma}\varphi_{,\sigma}\,\varphi_{,\nu} + M^2 u^\nu{}_{,\varrho}\,h^{\nu\varrho}\varphi^2\Big)$$
$$+\tilde{\kappa}^2\Big(2u^\nu\varphi_{,\nu}\,h^{\varrho'\sigma'}\varphi_{,\varrho'}\,\partial^y_{\sigma'} - M^2 u^\nu h\varphi_{,\nu}\,\varphi\Big)\Big\}\delta(x-y). \qquad (5.9.38)$$

By means of the identity (5.7.3) the anomalies can be written in normal form

$$A_1 + A_2 = \frac{i}{2}\Big\{\kappa\tilde{\kappa}\Big(u^\nu h^{\varrho\sigma}{}_{,\nu}\,\varphi_{,\varrho}\,\varphi_{,\sigma} + u^\nu{}_{,\nu}\,h^{\varrho\sigma}\varphi_{,\varrho}\,\varphi_{,\sigma} -$$
$$-\tfrac{1}{2}M^2 u^\nu{}_{,\nu}\,h\varphi^2 - \tfrac{1}{2}M^2 u^\nu h_{,\nu}\,\varphi^2 - 2u^\nu{}_{,\varrho}\,h^{\varrho\sigma}\varphi_{,\sigma}\,\varphi_{,\nu} + M^2 u^\nu{}_{,\varrho}\,h^{\nu\varrho}\varphi^2\Big)$$
$$+\tilde{\kappa}^2\Big(-M^2 u^\nu h\varphi\varphi_{,\nu} + u^\nu{}_{,\sigma}\,h^{\varrho\sigma}\varphi_{,\varrho}\,\varphi_\nu + u^\nu h^{\varrho\sigma}\varphi_{,\varrho}\,\varphi_{,\nu\sigma} -$$
$$-u^\nu h^{\varrho\sigma}{}_{,\sigma}\,\varphi_{,\varrho}\,\varphi_{,\nu} - u^\nu h^{\varrho\sigma}\varphi_{,\varrho\sigma}\,\varphi_{,\nu}\Big)\Big\}\delta(x-y). \qquad (5.9.39)$$

This can be written as divergence plus co-boundary, if and only if $\tilde{\kappa} = \kappa$. This follows from the combination of the terms according to their Lorentz structure.

Looking at the Lorentz indices upstairs we decompose $A_1 + A_2$ into three different types $G1, G2$ and $G3$. Without the pre-factor we have

$$G1 = \Big(u^\nu{}_{,\nu}\,h^{\varrho\sigma}\varphi_{,\varrho}\,\varphi_{,\sigma} + u^\nu h^{\varrho\sigma}{}_{,\nu}\,\varphi_{,\varrho}\,\varphi_{,\sigma} + u^\nu h^{\varrho\sigma}\varphi_{,\varrho}\,\varphi_{,\nu\sigma} -$$
$$-u^\nu{}_{,\varrho}\,h^{\varrho\sigma}\varphi_{,\sigma}\,\varphi_{,\nu} - u^\nu h^{\varrho\sigma}{}_{,\sigma}\,\varphi_{,\varrho}\,\varphi_{,\nu} - u^\nu h^{\varrho\sigma}\varphi_{,\varrho\sigma}\,\varphi_{,\nu}\Big)\delta(x-y). \qquad (5.9.40)$$

To achieve gauge invariance we must try to write this as a divergence plus a co-boundary. Using the second identity (5.7.11), we see that this is indeed a divergence

$$G1 = \partial^x_\nu(u^\nu h^{\varrho\sigma}\varphi_{,\varrho}\,\varphi_{,\sigma}\,\delta(x-y)) - \partial^x_\varrho(u^\nu h^{\varrho\sigma}\varphi_{,\nu}\,\varphi_{,\sigma}\,\delta(x-y))+$$
$$+x \longleftrightarrow y. \qquad (5.9.41)$$

The next type is

$$G2 = \Big(\frac{M^2}{2}u^\nu{}_{,\nu}\,h\varphi^2 + \frac{M^2}{2}u^\nu h_{,\nu}\,\varphi^2 + M^2 u^\nu h\varphi_{,\nu}\,\varphi\Big)\delta(x-y), \qquad (5.9.42)$$

which is also a divergence

$$G2 = \frac{M^2}{2}\partial^x_\nu(u^\nu h\varphi^2\delta(x-y)) + x \longleftrightarrow y. \qquad (5.9.43)$$

Finally, the last type

$$G3 = M^2 u^\nu{}_{,\varrho}\,h^{\nu\varrho}\varphi^2\delta(x-y) \qquad (5.9.44)$$

consists of one term only. This cannot be a divergence, but it is a co-boundary $d_Q N_2$ where

$$N_2 = \frac{i}{4}\kappa^2 M^2 \left(h^{\alpha\beta}h^{\alpha\beta}\varphi^2 - \tfrac{1}{2}h^2\varphi^2\right)\delta(x-y). \tag{5.9.45}$$

The additional quartic coupling (5.9.45) which is required by gauge invariance agrees precisely with the second order term in the expansion of the classical Lagrangian (5.9.34) if we take the factor 1/2! for the second order S-matrix into account. As before we realize that quantum gauge invariance is strong enough to determine the coupling, the classical theory is not needed. One point remains to be discussed: The local term (5.9.45) cannot be regarded as a normalization term for a second order tree graph with two graviton-matter vertices (5.9.35). But it is a normalization term of the fourth order box graph with two graviton-graviton and two graviton-matter vertices. Since the normalization constants are completely free, this is still a proper normalization.

Finally we also consider the coupling to a complex (non-hermitian) scalar field φ. This is equivalent to two commuting real scalar fields

$$\varphi = \varphi_1 + i\varphi_2. \tag{5.9.46}$$

The energy-momentum tensor (5.9.33) is then additive

$$T_{\alpha\beta} = \sum_{j=1,2}\left[\varphi_{j,\alpha}\varphi_{j,\beta} - \frac{1}{2}\eta_{\alpha\beta}(\varphi_{j,\mu}\varphi_j^{,\mu} - M^2\varphi_j^2)\right]. \tag{5.9.47}$$

Substituting

$$\varphi_1 = \tfrac{1}{2}(\varphi + \varphi^+), \quad \varphi_2 = -\tfrac{1}{2}i(\varphi - \varphi^+) \tag{5.9.48}$$

we obtain the energy-momentum tensor of the complex scalar field

$$T_{\alpha\beta} = \tfrac{1}{2}(\varphi^+_{,\alpha}\varphi_{,\beta} + \varphi_{,\alpha}\varphi^+_{,\beta}) - \tfrac{1}{2}\eta_{\alpha\beta}(\varphi^+_{,\mu}\varphi^{,\mu} - M^2\varphi^+\varphi). \tag{5.9.49}$$

Instead of (5.9.35) this leads to the following first order coupling to gravity

$$T = \kappa T_1 + i\frac{\kappa}{4}\left[h^{\alpha\beta}(\varphi^+_{,\alpha}\varphi_{,\beta} + \varphi^+_{,\beta}\varphi_{,\alpha}) - hM^2\varphi^+\varphi\right]. \tag{5.9.50}$$

5.10 Radiative corrections

In this section we are interested in radiative corrections to Newton's law of gravitation. This law was derived from quantum field theory in Sect.2.5. It is a consequence of graviton exchange between two particles described by a second order tree graph. The simplest radiative corrections to this process are self-energy insertions in the graviton line. To compute these we must calculate the graviton self-energy where we concentrate on the graviton and ghost loops, matter loops can be treated in a similar way.

For comparison with results in the literature we calculate again with the non-minimal coupling (5.7.1). The computation of second order loop graphs with massless fields has been described in detail in Sect.2.6 in the case of Yang-Mills theory. Therefore, we must only indicate the small changes which are necessary in quantum gravity. The contractions in the normal ordering are given by the following commutation relations

$$[h^{\alpha\beta}(x)^{(-)}, h^{\mu\nu}(y)^{(+)}] = -ib^{\alpha\beta\mu\nu}D_0^{(+)}(x-y) \quad (5.10.1)$$

$$\{u^\mu(x)^{(-)}, \tilde{u}^\nu(y)(+)\} = i\eta^{\mu\nu}D_0^{(+)}(x-y) = -\{\tilde{u}^\mu(x)^{(-)}, u^\nu(y)(+)\}. \quad (5.10.2)$$

In addition to the pure graviton loop there are contributions from ghost-antighost loops which arise from the ghost-graviton coupling and are essential for gauge invariance. The self-energy distribution is of the following form

$$T_2(x,y) =: h^{\alpha\beta}(x)h^{\mu\nu}(y) : t_2^{(1)}(x-y)_{\alpha\beta\mu\nu}+$$

$$+ : h^{\alpha\beta}{}_{,\varrho}(x)h^{\mu\nu}(y) : t_2^{(2)}(x-y)_{\alpha\beta\mu\nu}^{\varrho}+ : h^{\alpha\beta}(x)h^{\mu\nu}{}_{,\sigma}(y) : t_2^{(3)}(x-y)_{\alpha\beta\mu\nu}^{\sigma}+$$

$$+ : h^{\alpha\beta}{}_{,\varrho}(x)h^{\mu\nu}{}_{,\sigma}(y) : t_2^{(4)}(x-y)_{\alpha\beta\mu\nu}^{\varrho\sigma}. \quad (5.10.3)$$

The C-number distributions $t_2^{(j)}$ are in momentum space given by

$$\hat{t}_2^{(j)}(p)_{\alpha\beta\mu\nu} = P^{(j)}(p)_{\alpha\beta\mu\nu}\hat{t}(p), \quad (5.10.4)$$

where the scalar part

$$\hat{t}(p) = \frac{i}{2\pi}\log\left(\frac{-p^2-i0}{M^2}\right) \quad (5.10.5)$$

is the same as in (2.6.55). The $P^{(j)}$ are the following covariant polynomials:

$$P^{(j)}(p) = \frac{\pi\kappa^2}{960(2\pi)^4}\left[P_{\text{grav}}^{(j)}(p) + P_{\text{ghost}}^{(j)}(p)\right]. \quad (5.10.6)$$

Here

$$\left(P_{\text{grav}}^{(1)} + P_{\text{ghost}}^{(1)}\right)(p)_{\alpha\beta\mu\nu} = -80p^\alpha p^\beta p^\mu p^\nu + 60p^2(p^\alpha p^\beta \eta^{\mu\nu} + p^\mu p^\nu \eta^{\alpha\beta})+$$

$$+10p^2(p^\alpha p^\mu \eta^{\beta\nu} + p^\alpha p^\nu \eta^{\beta\mu} + p^\beta p^\mu \eta^{\alpha\nu} + p^\beta p^\nu \eta^{\alpha\mu}) - 70p^4(\eta^{\alpha\mu}\eta^{\beta\nu} + \eta^{\alpha\nu}\eta^{\beta\mu}) \tag{5.10.7}$$

$$P^{(2)}_{\text{grav}}(p)^\gamma_{\alpha\beta\mu\nu} = i\Big[-120 p_\mu p_\nu [p_\alpha \eta^\gamma_\beta + (\alpha \longleftrightarrow \beta)] +$$
$$+ 20[p_\mu p_\nu p^\gamma \eta_{\alpha\beta} + p_\alpha p_\nu p_\beta \eta^\gamma_\mu + p_\alpha p_\mu p_\beta \eta^\gamma_\nu]\Big], \tag{5.10.8}$$

$$P^{(2)}_{\text{ghost}}(p)^\gamma_{\alpha\beta\mu\nu} = ip^2\Big[[p_\alpha(-20\eta_{\mu\nu}\eta^\gamma_\beta + 130\eta^\gamma_\mu \eta_{\beta\nu} + 130\eta_{\mu\beta}\eta^\gamma_\nu] + (\alpha \longleftrightarrow \beta)] +$$
$$+ [p_\mu(-20\eta_{\alpha\beta}\eta^\gamma_\nu + 20\eta_{\alpha\nu}\eta^\gamma_\beta + 20\eta^\gamma_\alpha \eta_{\nu\beta}) + (\mu \longleftrightarrow \nu] +$$
$$+ p^\gamma[110\eta_{\mu\nu}\eta_{\alpha\beta} - 210\eta_{\nu\beta}\eta_{\mu\alpha} - 210\eta_{\mu\beta}\eta_{\alpha\nu}]\Big]. \tag{5.10.9}$$

The polynomial $P^{(3)}$ is obtained from $P^{(2)}$ by interchanging μ with α, ν with β and multiplying by -1.

The last polynomial $P^{(4)}$ is given by

$$P^{(4)}_{\text{grav}}(p)^{\gamma\varrho}_{\alpha\beta\mu\nu} = -10\Big[[p_\alpha p_\beta(8\eta^\varrho_\mu \eta^\gamma_\nu + 8\eta^\varrho_\nu \eta^\gamma_\mu) + (\alpha \longleftrightarrow \beta, \mu \longleftrightarrow \nu)] +$$
$$+ [p_\alpha p_\mu(16\eta_{\nu\beta}\eta^{\varrho\gamma} + 12\eta^\gamma_\beta \eta^\varrho_\nu + 10\eta^\varrho_\beta \eta^\gamma_\nu) + (\alpha \leftrightarrow \beta) + (\mu \leftrightarrow \nu) + (\alpha \leftrightarrow \beta, \mu \leftrightarrow \nu)] +$$
$$+ p^\varrho p^\gamma(-34\eta_{\mu\nu}\eta_{\alpha\beta} + 60\eta_{\alpha\mu}\eta_{\beta\nu} + 60\eta_{\alpha\nu}\eta_{\beta\mu}] +$$
$$+ [4p_\alpha p^\gamma(-\eta_{\mu\beta}\eta^\varrho_\nu - \eta_{\beta\nu}\eta^\varrho_\mu + \eta^\varrho_\beta \eta_{\mu\nu}) + (\alpha \longleftrightarrow \beta) +$$
$$+ (\alpha \longleftrightarrow \mu, \gamma \longleftrightarrow \varrho, \beta \longleftrightarrow \nu) + (\alpha \longleftrightarrow \nu, \gamma \longleftrightarrow \varrho, \beta \longleftrightarrow \mu)] +$$
$$+ [p_\alpha p^\varrho(-30\eta_{\beta\mu}\eta^\gamma_\nu - 30\eta_{\beta\nu}\eta^\gamma_\mu + 18\eta^\gamma_\beta \eta_{\mu\nu}) + (\alpha \longleftrightarrow \beta) +$$
$$+ (\alpha \longleftrightarrow \mu, \gamma \longleftrightarrow \varrho, \beta \longleftrightarrow \nu) + (\alpha \longleftrightarrow \nu, \gamma \longleftrightarrow \varrho, \beta \longleftrightarrow \mu)]\Big], \tag{5.10.10}$$

$$P^{(4)}_{\text{ghost}}(p)^{\gamma\varrho}_{\alpha\beta\mu\nu} = -10p^2\Big[\eta_{\alpha\beta}\eta_{\mu\nu}\eta^{\gamma\varrho} - 8\eta^{\gamma\varrho}[\eta_{\alpha\mu}\eta_{\beta\nu} + (\mu \longleftrightarrow \nu)] -$$
$$- 3[\eta_{\alpha\beta}\eta^\varrho_\nu \eta^\gamma_\mu + (\mu \longleftrightarrow \nu)] - 3[\eta_{\mu\nu}\eta^\varrho_\beta \eta^\gamma_\alpha + (\alpha \longleftrightarrow \beta)] +$$
$$+ [2\eta^\gamma_\alpha \eta_{\beta\mu}\eta^\varrho_\nu + 5\eta_{\alpha\mu}\eta^\gamma_\nu \eta^\varrho_\beta + (\mu \longleftrightarrow \nu) + (\alpha \longleftrightarrow \beta) + (\alpha \longleftrightarrow \beta, \mu \longleftrightarrow \nu)]\Big]. \tag{5.10.11}$$

All polynomials have been written in symmetric form, the symmetry properties follow from those of the field operators in (5.10.3).

We now insert the self-energy loop into an inner graviton line. Since the integration over the inner variables in x-space are convolutions, it is very convenient to go over to momentum space, then we get products of the distributions in p-space. For the total graviton propagator we thus obtain

$$D(p)^{\alpha\beta\mu\nu} = b^{\alpha\beta\mu\nu} D_0^F(p) + (2\pi)^4 D_0^F(p) b^{\alpha\beta\gamma\delta} \Pi(p)_{\gamma\delta\varrho\sigma} b^{\varrho\sigma\mu\nu} D_0^F(p) + \ldots \tag{5.10.12}$$

Here the self-energy tensor $\Pi(p)$ is given by

Radiative corrections 233

$$\Pi(p)_{\alpha\beta\mu\nu} = \frac{\kappa^2 \pi}{960(2\pi)^5}\left[P(p)^{\text{grav}}_{\alpha\beta\mu\nu} + P(p)^{\text{ghost}}_{\alpha\beta\mu\nu}\right]\log\left(\frac{-p^2 - i0}{M^2}\right), \quad (5.10.13)$$

where the quartic polynomials

$$P(p)^{\text{grav}}_{\alpha\beta\mu\nu} = -880 p^\alpha p^\beta p^\mu p^\nu - 260 p^2 (p^\alpha p^\beta \eta^{\mu\nu} + p^\mu p^\nu \eta^{\alpha\beta}) +$$

$$+160 p^2 (p^\alpha p^\mu \eta^{\beta\nu} + p^\alpha p^\nu \eta^{\beta\mu} + p^\beta p^\mu \eta^{\alpha\nu} + p^\beta p^\nu \eta^{\alpha\mu}) - 170 p^4 (\eta^{\alpha\mu}\eta^{\beta\nu} + \eta^{\alpha\nu}\eta^{\beta\mu}) +$$

$$+110 p^4 \eta^{\alpha\beta}\eta^{\mu\nu} \quad (5.10.14)$$

$$P(p)^{\text{ghost}}_{\alpha\beta\mu\nu} = 224 p^\alpha p^\beta p^\mu p^\nu + 52 p^2 (p^\alpha p^\beta \eta^{\mu\nu} + p^\mu p^\nu \eta^{\alpha\beta}) +$$

$$+2 p^2 (p^\alpha p^\mu \eta^{\beta\nu} + p^\alpha p^\nu \eta^{\beta\mu} + p^\beta p^\mu \eta^{\alpha\nu} + p^\beta p^\nu \eta^{\alpha\mu}) + 8 p^4 (\eta^{\alpha\mu}\eta^{\beta\nu} + \eta^{\alpha\nu}\eta^{\beta\mu}) +$$

$$+8 p^4 \eta^{\alpha\beta}\eta^{\mu\nu} \quad (5.10.15)$$

come from the graviton- or ghost-loops, respectively.

Now we are ready to determine the first quantum correction to Newton's law. If (5.10.12) appears between two conserved energy-momentum tensors of matter, only the terms $\sim p^4$ in (5.10.14-15) contribute, all other contributions vanish by energy-momentum conservation of the matter tensor. The factor p^4 is cancelled by the two Feynman propagators in (5.10.12). The correction is then proportional to $\log(p^2/M^2)$. In the non-relativistic limit the distributional Fourier transform of $\log(\boldsymbol{p}^2/M^2)$ yields $(-2\pi r^3)^{-1}$ up to a local potential $\log M^2 \delta^3(\boldsymbol{x})$. Substituting back the physical constants $\hbar$ and c, this gives the following gravitational potential

$$V(r) = -G\frac{mM}{r}\left(1 + \frac{G\hbar}{c^3 \pi}\frac{206}{30}\frac{1}{r^2}\right). \quad (5.10.16)$$

The length scale of the correction term is the Planck length $l_P = \sqrt{G\hbar/c^3} \approx 4 \times 10^{-33}$ cm which, unfortunately, is out of any possibility of observation. Nevertheless, it is interesting that the result (5.10.16) is unique, independent of normalization. In fact, the normalization constant M only affects the singular potential $\sim \delta^3(\boldsymbol{x})$. The same is true for all other normalization constants in arbitrary order. This can be seen as follows: The freedom of normalization consists of adding an arbitrary polynomial in p to the self-energy (5.10.13). After Fourier transform this gives rise to a δ-potential and derivatives thereof. This potential with point support is completely undetermined. Of course, the potential picture can no longer be taken seriously for $r \to 0$. Consequently, there is not much physics in the normalization constants.

The correction in (5.10.16) to the Newtonian potential is not the whole story. There are further contributions of the same order $O(\kappa^4)$ which must be included: Since we include graviton-matter coupling, there is also another self-energy diagram with a matter-loop. In addition, one must take the vertex correction and the box diagram into account. All these diagrams can be calculated by the causal method without problem.

Finally we want to discuss vacuum graphs. It is well known that also vacuum diagrams contribute in the perturbative expansion. In the ordinary formalism they are strongly ultraviolet divergent and are usually omitted or "divided away". This is impossible in the causal theory because it would violate the causality condition (2.2.28). Therefore, we must live with the vacuum diagrams which are pretty finite and well defined in the causal theory. What is less well-known is that the vacuum contributions contain the most severe infrared divergences. To show this let us consider the second order vacuum contribution. The infrared problem requires a careful treatment of the adiabatic limit $g(x) \to 1$, where g is the test-function in the S-matrix. The second-order vacuum-to-vacuum amplitude is given by

$$\lim_{g \to 1} (\Omega, S_2(g)\Omega) = \tfrac{1}{2} \lim_{g \to 1} \int dx\, g(x) \int dp\, \hat{T}_2(p) e^{ipx} \hat{g}(-p). \quad (5.10.17)$$

Here Ω is the Fock vacuum of the free asymptotic fields. The adiabatic limit will be performed in scaling form $g(x) = g_0(\varepsilon x)$, where $g_0 \in \mathcal{S}(\mathbb{R}^4)$ is a fixed test function with $g_0(0) = 1$ and we study the limit $\varepsilon \to 0$. Since

$$\hat{g}(p) = (2\pi)^{-2} \int g_0(\varepsilon x) e^{-ipx} d^4 x = \frac{1}{\varepsilon^4} \hat{g}_0\left(\frac{p}{\varepsilon}\right), \quad (5.10.18)$$

we get

$$\lim_{g \to 1} (\Omega, S_2(g)\Omega) = \frac{(2\pi)^2}{2} \lim_{g \to 1} \frac{1}{\varepsilon^4} \int dp\, \hat{T}_2(\varepsilon p) \hat{g}_0(p) \hat{g}_0(-p). \quad (5.10.19)$$

This limit only exists if the second order vacuum diagram $\hat{T}_2(p)$ goes to 0 for $p \to 0$ at least as $\sim p^4$. Otherwise the Fock vacuum is unstable, that means the fundamental fields in the interaction $T_1(x)$ cannot be physically relevant asymptotic fields at the same time.

The factor $1/\varepsilon^4$ is proportional to the interaction volume defined by the support of g_0 in x-space. Consequently, we have found a volume divergence which must be compensated by the interaction. As in statistical mechanics the non-existence of the infinite-volume limit signals a radical rearrangement of the system, in other words its infrared instability.

Let us now apply this criterion to quantum gravity. It follows from (5.1.17) that the second-order vacuum diagrams have singular order $\omega = 6$. Consequently, in momentum space the vacuum amplitude is of the following form

$$\hat{T}_2(p) = \text{const.} \cdot p^6 \log\left(-\frac{p^2 + i0}{M^2}\right). \quad (5.10.20)$$

The normalization polynomial which can be added has the form $c_0 + c_2 p^2 + c_4 p^4 + c_6 p^6$. c_0 and c_2 must be zero, otherwise the adiabatic limit would not exist. c_4 must vanish, in order to have a unique adiabatic limit, but c_6 as well as (5.10.20) give no contribution to (5.10.19). That means quantum gravity is infrared stable. The free Fock vacuum is stable against graviton interactions and so are all graviton states. Consequently, perturbation theory is well applicable. Obviously, this is a consequence of the non-normalizability. The same argument applies to higher orders.

For comparison let us examine other field theories. In ordinary (massive) spinor QED we have (cf. FQED, Sect.4.1)

$$\hat{T}_2(p) = \text{const.} \left(\frac{p^2}{m^2}\right)^3 + \ldots \tag{5.10.21}$$

for $p \to 0$, where m is the electron mass. Hence the Fock vacuum is stable. However, if the electron were massless we would get

$$\hat{T}_2(p) = \text{const.} \cdot p^4 \log\left(-\frac{p^2 + i0}{M^2}\right) \tag{5.10.22},$$

where M is a free normalization constant as above (5.10.20). Then the Fock vacuum is unstable and massless electrons and photons cannot be real asymptotic scattering states.

The same result (5.10.22) comes out in QCD considering the second order vacuum diagram for the gluon self-interaction and the gluon-ghost coupling discussed in Chap.3. The factor p^4 is a consequence of the singular order $\omega = 4$ of the vacuum diagrams (2.7.2). Again the Fock vacuum is unstable which is a way to express colour confinement. But we stress the fact that the non-Abelian character of QCD is not exclusive for this phenomenon, it occurs already in massless QED. In all these massless normalizable theories we have "infrared slavery". That means the perturbation series must be re-summed in a clever way, or one must apply non-perturbative techniques. In this respect quantum gravity is much simpler ! Therefore, the old sentence about confinement "ultraviolet freedom gives infrared slavery" has an inversion: *Ultraviolet slavery gives infrared freedom.*

5.11 Yang-Mills fields in interaction with gravity

The gravitational interaction of matter requires more study than was made in Sect.5.9. The reason is that in quantum field theory there does not exist a conserved energy-momentum tensor in general, so that we cannot apply the result (5.9.8). Instead we use causal gauge invariance again, but this time for spin-1 and spin-2 gauge fields combined. In this case the gauge variation d_Q operates in the big Fock space generated by Yang-Mills fields A_a^μ, the gravitational field $h^{\mu\nu}$ and the corresponding ghost and anti-ghost fields.

We take the Yang-Mills fields massive which is the most interesting situation and recall the operation of d_Q on the Yang-Mills sector (1.5.36-39)

$$d_Q A_a^\mu = i\partial^\mu u_a, \quad d_Q \Phi_a = i m_a u_a \tag{5.11.1}$$

$$d_Q u_a = 0, \quad d_Q \tilde{u}_a = -i(\partial_\mu A_a^\mu + m_a \Phi_a). \tag{5.11.2}$$

We look for trilinear couplings between these fields and gravity satisfying the descent equations. As in Sect.5.2 the descent procedure starts from the antisymmetric tensor $T^{\mu\nu\varrho}$ which contains Yang-Mills and gravitational ghost fields only. However, it follows from (5.11.1) that u_a and $\partial^\mu u_a$ are co-boundaries, so that no non-trivial coupling is possible. The next descent equation then is

$$d_Q T^{\mu\nu} = 0 \tag{5.11.3}$$

where $T^{\mu\nu}$ contains two gravitational ghost fields u^μ and one Yang-Mills gauge invariant quantity (or co-cycle)

$$F_a^{\mu\nu} = \partial^\mu A_a^\nu - \partial^\nu A_a^\mu \tag{5.11.4}$$

$$\Psi_a^\mu = \partial^\mu \Phi_a - m_a A_a^\mu. \tag{5.11.5}$$

With only one such invariant no Yang-Mills scalar can be formed, therefore, we must start from

$$d_Q T^\mu = 0. \tag{5.11.6}$$

Here non-trivial co-cycles do exist:

$$T_s^\mu = c_{ab} u^\mu F_a^{\varrho\sigma} F_{b\varrho\sigma} + e_{ab} u^\varrho F_a^{\mu\nu} F_{b\nu\varrho}$$

$$+ g_{ab} u^\mu \Psi_{a\nu} \Psi_b^\nu + k_{ab} u_\nu \Psi_{a\mu} \Psi_b^\nu, \tag{5.11.8}$$

where we can impose the symmetry conditions

$$c_{ab} = c_{ba}, \quad g_{ab} = g_{ba}. \tag{5.11.8}$$

The subscript s indicates that this coupling is parity invariant. Since weak interactions do not conserve parity, we must also consider parity violating couplings:

$$T_a^\mu = c'_{ab} \varepsilon^{\mu\nu\varrho\sigma} u_\nu F_{a\varrho}^\lambda F_{b\lambda\sigma} + e'_{ab} \varepsilon^{\lambda\nu\varrho\sigma} u_\nu F_{a\lambda}^\mu F_{b\varrho\sigma} +$$

$$+ g'_{ab} \varepsilon^{\nu\lambda\varrho\sigma} u^\mu F_{a\nu\lambda} F_{b\varrho\sigma} + k'_{ab} \varepsilon^{\mu\nu\varrho\sigma} u_\nu \Psi_{a\varrho} \Psi_{b\sigma}. \tag{5.11.9}$$

Now we compute the divergence $\partial_\mu T_s^\mu$ and require it to be a co-boundary. Co-boundaries come from derivatives of u^μ

$$\partial_\mu T_s^\mu = c_{ab} \partial_\mu u^\mu F_a^{\varrho\sigma} F_{b\varrho\sigma} + e_{ab} \partial_\mu u^\varrho F_a^{\mu\nu} F_{b\nu\varrho}$$

$$+ g_{ab} \partial_\mu u^\mu \Psi_{a\nu} \Psi_b^\nu + k_{ab} \partial_\mu u_\nu \Psi_{a\mu} \Psi_b^\nu \ldots, \tag{5.11.10}$$

due to the identities

$$\partial_\mu u^\mu = id_Q h \tag{5.11.11}$$

$$\partial_\mu u_\nu + \partial_\nu u_\mu = id_Q (2h_{\mu\nu} - \eta_{\mu\nu} h). \tag{5.11.12}$$

It follows from (5.11.12) that e_{ab} and k_{ab} must also be symmetric in a, b because the antisymmetric combination cannot be compensated. The dots in (5.11.10) represent Wick monomials with no derivative on u^ν and these terms must all cancel. This gives the following relations

$$u^\mu \partial_\mu \partial^\varrho A_a^\sigma \partial_\varrho A_{b\sigma} : \quad 4c_{ab} - e_{ba} = 0$$

$$u^\varrho A_a^\nu \partial_\nu A_{b\varrho} : \quad m_b k_{ab} - m_a e_{ab} = 0$$

$$u^\varrho A_a^\nu \partial_\varrho A_{b\nu} : \quad m_a e_{ab} + 2m_b g_{ba} = 0$$

$$u^\mu \partial_\mu \partial_\nu \Phi_a \partial^\nu \Phi_b : \quad 2g_{ab} + k_{ab} = 0, \tag{5.11.13}$$

here we have also indicated the Wick monomials where the relations come from. Terms with $\partial_\mu A_a^\mu$ give co-boundaries, too, due to equation (5.11.2) for $d_Q \tilde{u}_a$. This mechanism produces ghost couplings containing both Yang-Mills and gravitational ghost fields. It is not hard to verify that all terms compensate if the indicated substitutions are made.

The above relations enable us to express all coefficients in terms of c_{ab}:

$$e_{ab} = 4c_{ab}, \quad g_{ab} = -2\frac{m_b}{m_a}c_{ba}$$

$$k_{ab} = 4\frac{m_b}{m_a}c_{ab}. \tag{5.11.14}$$

. Using the symmetry in a, b of the matrices, it follows $c_{ab} = 0$ if $m_a \neq m_b$. In a subspace of fields with equal masses we can diagonalize the symmetric matrix c_{ab} by an orthogonal transformation. Such a transformation of the fields is always possible without changing commutation relations, gauge structure etc. Therefore we arrive at the diagonal form

$$c_{ab} = c_a \delta_{ab}. \tag{5.11.15}$$

This gives the following final Q-vertex

$$T^\mu = \sum_a c_a \Big[u^\mu F_a^{\varrho\sigma} F_{a\varrho\sigma} + 4u^\varrho F_a^{\mu\nu} F_{a\nu\varrho} -$$

$$- 2u^\mu \Psi_{a\nu} \Psi_a^\nu + 4u_\nu \Psi_a^\mu \Psi_a^\nu \Big]. \tag{5.11.16}$$

The co-boundaries define the coupling by $d_Q T = i\partial_\mu T^\mu$:

$$T = \sum_a c_a \Big[4h_\mu^\varrho F_a^{\mu\nu} F_{a\varrho\nu} - h F_a^{\mu\nu} F_{a\mu\nu} + 4u_\mu \partial_\nu \tilde{u}_a F_a^{\mu\nu} -$$

$$- 4m_a u_\mu \tilde{u}_a \Psi_a^\mu - 4h_{\mu\nu} \Psi_a^\mu \Psi_a^\nu \Big]. \tag{5.11.17}$$

If one treats the parity violating coupling (5.11.9) in the same way one concludes that all coefficients must vanish, so that T (5.11.17) is the general result. In the case of massless Yang-Mills fields only the terms in the first lines in (5.11.16) and (5.11.17) appear.

The ghost coupling terms in the first line of (5.11.17) are somewhat surprising. Physical states do not contain ghost or anti-ghost modes, therefore, the above ghost couplings do not contribute to S-matrix elements between physical states. Even in loop graphs they cannot enter because there are no corresponding coupling terms with gravitational antighost and Yang-Mills ghost fields. This is in contrast to the pure gravitational ghost loops which we have found in the last section (5.10.11). Nevertheless, these couplings are necessary for gauge invariance of the theory.

The strange situation becomes clear when we write the coupling T (5.11.17) in terms of the physical part of the massive Yang-Mills field which we have introduced at the end of Sect.1.5 (1.5.48):

$$A_{a\mu}^{\text{phys}} = A_{a\mu} + \frac{1}{m_a^2}\partial^\mu \partial_\nu A_a^\nu. \qquad (5.11.18)$$

This field has the following properties:

$$d_Q A_{a\mu}^{\text{phys}} = 0, \quad \partial^\mu A_{a\mu}^{\text{phys}} = 0. \qquad (5.11.19)$$

Then one can prove by some computations the following identity

$$h_{\mu\nu}\Psi_a^\mu \Psi_a^\nu + m_a u_\mu \tilde{u}_a \Psi_a^\mu - u_\mu \partial_\nu \tilde{u}_a F_a^{\mu\nu} = m_a^2 h^{\mu\nu} A_{a\mu}^{\text{phys}} A_{a\nu}^{\text{phys}} +$$
$$+ d_Q B + \partial_\mu B^\mu, \qquad (5.11.20)$$

where the co-boundary and divergence terms do not matter in S-matrix elements between physical states. Using this identity in (5.11.17) we can express the interaction T in the standard form (5.9.8):

$$T = h^{\mu\nu} b_{\alpha\beta\mu\nu} t^{\alpha\beta} \qquad (5.11.21)$$

up to divergence and co-boundaries. The energy-momentum tensor

$$t_{\mu\nu} = \sum_a c_a \left(-F_{a\mu\varrho} F_{a\nu}^\varrho - \frac{1}{4}\eta_{\mu\nu} F_{a\varrho\sigma} F_a^{\varrho\sigma} - \right.$$
$$\left. -m_a^2 A_{a\mu}^{\text{phys}} A_{a\nu}^{\text{phys}} + \frac{m_a^2}{2}\eta_{\mu\nu} A_a^{\text{phys},\varrho} A_{a\varrho}^{\text{phys}} \right) \qquad (5.11.22)$$

appearing here contains only physical fields and it is additive in the physical degrees of freedom. In addition one can prove directly that $t_{\mu\nu}$ is conserved

$$\partial_\nu t^{\mu\nu} = 0. \qquad (5.11.23)$$

This brings us back to Sect.5.9, because we have now understood that gravity couples to the physical degrees of freedom, only. By repeating the second order calculation of Sect.5.9 one can show that the constants c_a in (5.11.22) must all be equal to $\sqrt{8\pi G}$ where G is Newton's constant. This is the universal coupling of gravity.

However, as beautiful the new expression (5.11.22) looks like, it nevertheless has a deficiency. The physical field (5.11.18) has canonical dimension 3 due to the two derivatives. The original free fields have all canonical dimension 1 by definition. Therefore, in the first form (5.11.17) the interaction has canonical dimension 5 as in pure gravity. In the new form (5.11.21) there are terms with canonical dimension 7. This may cause problems in perturbation theory. In fact, the derivatives remain in the commutation rule (1.5.47). For this reason it might be better to work with the first expression (5.11.17) and pay the price of taking along unphysical degrees of freedom. This situation is similar to the one in QED where the Coulomb gauge which avoids unphysical degrees of freedom is not practical for perturbative calculations.

5.12 Massive gravity: second order

The analysis of second order gauge invariance for the massive spin-2 theory of Sect.5.4 can be carried through along the lines of Sect.5.7 and 5.8, as far as the pure graviton and ghost sectors are concerned. The somewhat surprising result is that second order gauge invariance holds *without any additional (Higgs like) field*. This is in contrast to the spin-1 case in Sect.4.4 where without the Higgs fields gauge invariance breaks down in second order. Of course, the vector graviton field v^μ is present in addition. But it has the same mass as the graviton field $h^{\mu\nu}$ and so describes the additional degrees of freedom of the massive graviton. A Higgs-like field would have a different mass as in spin-1.

New phenomena appear when we consider massive gravity in interaction with ordinary matter which we assume to have spin 0 for simplicity. As in Sect.5.9 we take a real scalar field $\varphi(x)$ of mass M with trivial gauge variation $d_Q\varphi = 0$ and derive its coupling to massive gravity by the descent procedure. In massive gravity we have new gauge invariant quantities (co-cycles) namely the (symmetric) tensor

$$\phi_{\mu\nu} \equiv -\partial_\mu v_\nu - \partial_\nu v_\mu + \eta_{\mu\nu}\partial_\rho v^\rho + m\, h_{\mu\nu} \tag{5.12.1}$$

and its trace:

$$\phi \equiv \eta^{\mu\nu}\phi_{\mu\nu}. \tag{5.12.2}$$

These expression are immediately proved to be gauge invariant. Then the most general trilinear co-cycle involving the scalar field is given by

$$T_0 = c_1\varphi\phi_{\mu\nu}\phi^{\mu\nu} + c_2\varphi\phi^2 + c_3\varphi^2\,\phi + c_5\varphi^3. \tag{5.12.3}$$

The first nontrivial term in the descent equations is the Q-vertex:

$$T^\mu = f_1 u^\mu \varphi^2 + f_2 u^\mu \partial^\nu\varphi\partial_\nu\varphi + f_3 u^\nu \partial^\mu\varphi\partial_\nu\varphi + f_4 u^{\mu\nu}\varphi\partial_\nu\varphi + f_5 u_{\nu\rho}\varphi\partial^\mu\partial^\nu\varphi \tag{5.12.4}$$

where $u^{\mu\nu}$ is defined by (5.2.6). The term with f_4 can be dropped because it is a divergence $\partial_\nu(u^{\mu\nu}\varphi^2)$ plus a coboundary plus a term of the form f_1. We compute the divergence $\partial_\mu T^\mu$ and require it to be a co-boundary $\partial_\mu T^\mu = -id_Q t$. This fixes the coefficients in (5.12.4) with the result

$$T^\mu = c_4\left(\frac{1}{2}u^\mu\partial^\nu\varphi\partial_\nu\varphi - u^\nu\partial^\mu\varphi\partial_\nu\varphi - \frac{1}{2}M^2 u^\mu\varphi^2\right) \quad (5.12.5)$$

for some arbitrary constant f up to a divergence and co-boundary. Moreover one proves that

$$t \equiv c_4\left(h_{\mu\nu}\partial^\mu\varphi\partial^\nu\varphi - \frac{1}{2}M^2 h\varphi^2\right). \quad (5.12.6)$$

The most general first order coupling then is given by

$$T = t + d_Q B + i\partial_\mu B^\mu + T_0 \quad (5.12.7)$$

with the co-cycles T_0 according to (5.12.3).

We omit the co-boundary and divergence terms in (5.12.7) which we also call relative co-boundaries for short, and go to second order. To T and T_μ above we add the contributions from the pure graviton sector of Sect.5.4. Then the terms in the total T_μ which generate anomalies are the ones with a derivative ∂_μ (resp. ∂_α in (5.2.11)):

$$T_\mu^{an} = u^\alpha(-2\partial_\alpha h^{\varrho\nu}\partial_\mu h_{\varrho\nu} + \partial_\alpha h \partial_\mu h - 2\partial_\alpha u^\nu \partial_\mu \tilde{u}_\nu + 2\partial_\alpha \partial_\nu u^\nu \tilde{u}_\mu) -$$
$$-2\partial_\nu u^\nu h^{\alpha\varrho}\partial_\mu h_{\alpha\varrho} + \partial_\nu u^\nu h\partial_\mu h - 2\partial_\nu u_\alpha h^{\alpha\nu}\partial_\mu h + 4\partial^\nu u_\alpha \partial_\mu h^{\alpha\varrho} h_{\nu\varrho} +$$
$$+4u^\alpha \partial_\alpha v^\nu \partial_\mu v_\nu - c_4 u^\alpha \partial_\alpha \varphi \partial_\mu \varphi. \quad (5.12.8)$$

Here we have put the gravitational coupling constant $\kappa = 1$ for simplicity. According to (5.12.3) (5.12.6) the first order coupling to the scalar field φ of mass M is given by

$$T_\varphi = c_1\varphi\phi_{\mu\nu}\phi^{\mu\nu} + c_2\varphi(mh + 2\partial_\mu v^\mu)^2 + c_3\varphi^2(mh + 2\partial_\mu v^\mu) +$$
$$+c_4\left(\partial_\mu\varphi\partial_\nu\varphi h^{\mu\nu} - \frac{1}{2}M^2\varphi^2 h\right) + c_5\varphi^3. \quad (5.12.9)$$

The last term herein is the φ^3 self-coupling of the scalar field. We first consider the couplings linear in φ, i.e. with coefficients c_1, c_2 in second order.

Theorem 12.1. Second order gauge invariance implies $c_1 = 0$ and $c_2 = 0$.

Proof. To prove this result it is sufficient to find anomalies with c_1 or c_2, which cannot be compensated. For c_1 we consider the commutator

$$-8c_1 u^\lambda \partial_\lambda v^\nu [\partial_\mu v_\nu(x), \phi^{\alpha\beta}(y)]\phi_{\alpha\beta}(y)\varphi \quad (5.12.10)$$

As usual the commutator gives a causal propagator which in the chronological product becomes a Feynman propagator. Applying the derivative $\partial/\partial x^\mu$ we get a $\partial^2 D^F$ leading to the anomaly

$$A_1 = 4ic_1 u^\lambda \partial_\lambda v^\nu(x)\varphi(y)\left(2\phi_{\alpha\nu}(y)\partial_y^\alpha - \phi(y)\partial_\nu^y\right)\delta(x-y). \quad (5.12.11)$$

In the same way we consider the commutator

$$-8c_2 u^\alpha \partial_\alpha v^\nu [\partial_\mu v_\nu(x), \phi(y)]\phi(y)\varphi(y). \tag{5.12.12}$$

Here the resulting anomaly is equal to

$$A_2 = -8ic_2 u^\alpha \partial_\alpha v^\nu(x)\varphi(y)\phi(y)\partial_\nu^y \delta(x-y). \tag{5.12.13}$$

There are no other anomalies with Wick monomials $uv\phi\varphi$, $uv\phi_{\mu\nu}\varphi$, respectively. Consequently, A_1 and A_2 must cancel against each other. For the last Wick monomial we see from (5.12.11) that c_1 must be 0 and hence, c_2 must also vanish. □

The situation is non-trivial for the remaining couplings which are bilinear and trilinear in φ.

Theorem 12.2. Second order gauge invariance implies $c_4 = 2$, but c_3 and c_5 remain unrestricted. In the second-order chronological products the following finite renormalizations are necessary

$$T(x,y) = T^F(x,y) + i\delta(x-y)N(x) \qquad T_\mu(x,y) = T^F_\mu(x,y) + i\delta(x-y)N^\mu(x) \tag{5.12.14}$$

where

$$N = 2\varphi^2 \Big\{ M^2(2h^{\mu\nu}h_{\mu\nu} - h^2) + c_3\Big[m(h^2 - 2h^{\mu\nu}h_{\mu\nu}) - \frac{8}{m}(\partial_\mu v^\mu \partial_\nu v^\nu - \partial_\mu v^\nu \partial_\nu v^\mu)\Big] + \frac{12}{m}c_5 v^\mu \partial_\mu \varphi \Big\} \tag{5.12.15}$$

and

$$N^\mu = -8(u^\mu h^{\alpha\beta} - u^\beta h^{\alpha\mu})\partial_\alpha \varphi \partial_\beta \varphi + (2M^2 + 2mc_3)u^\mu h\varphi^2$$
$$+ 2c_3(2u^\mu \partial_\alpha v^\alpha - u^\alpha \partial_\alpha v^\mu)\varphi^2. \tag{5.12.16}$$

Proof. In this proof we must calculate all anomalies containing φ. We also give the commutators where the anomalies come from. From

$$(u^\alpha \partial_\alpha h + \partial_\alpha u^\alpha h - 2\partial^\nu u^\alpha h_{\alpha\nu})[\partial_\mu h(x), h(y)] \left(mc_3 - \frac{M^2}{2}c_4\right)\varphi^2$$

we get the anomaly

$$A_1 = 2i(2mc_3 - M^2c_4)(u^\alpha \partial_\alpha h + \partial_\alpha u^\alpha h - 2\partial^\nu u^\alpha h_{\alpha\nu})\varphi^2 \delta, \tag{5.12.17}$$

and

$$2(-u^\lambda \partial_\lambda h^{\alpha\nu} - \partial_\lambda u^\lambda h^{\alpha\nu} + \partial_\lambda u^\alpha h^{\lambda\nu} + \partial_\lambda u^\nu h^{\alpha\lambda})[\partial_\mu h_{\alpha\nu}(x), h(y)] \times$$
$$\times \left(mc_3 - \frac{M^2}{2}c_4\right)\varphi^2$$

leads to

$$A_2 = i(2mc_3 - M^2 c_4)(-u^\lambda \partial_\lambda h - \partial_\lambda u^\lambda h + 2\partial_\lambda u^\alpha h_{\alpha\lambda})\varphi^2 \delta, \tag{5.12.18}$$

Spin-2 gauge theory

The commutator
$$(u^\alpha \partial_\alpha h + \partial_\alpha u^\alpha h - 2\partial^\nu u^\alpha h_{\alpha\nu})[\partial_\mu h(x), h^{\beta\gamma}(y)]c_4 \partial_\beta \varphi \partial_\gamma \varphi$$
gives
$$A_3 = ic_4(u^\alpha \partial_\alpha h + \partial_\alpha u^\alpha h - 2\partial^\nu u^\alpha h_{\alpha\nu})\partial_\beta \varphi \partial^\beta \varphi \delta \qquad (5.12.19)$$
and
$$2(-u^\lambda \partial_\lambda h^{\alpha\nu} - \partial_\lambda u^\lambda h^{\alpha\nu} + \partial_\lambda u^\alpha h^{\lambda\nu} + \partial_\lambda u^\nu h^{\alpha\lambda})[\partial_\mu h_{\alpha\nu}(x), h^{\beta\gamma}(y)]c_4 \partial_\beta \varphi \partial_\gamma \varphi$$
yields
$$A_4 = -ic_4[-2u^\lambda \partial_\lambda h^{\beta\gamma} - 2\partial_\lambda u^\lambda h^{\beta\gamma} + (u^\alpha \partial_\alpha h + \partial_\lambda u^\lambda h)\eta^{\beta\gamma}$$
$$+ 2\partial_\lambda u^\beta h^{\gamma\lambda} + 2\partial_\lambda u^\gamma h^{\beta\gamma} - 2\partial_\lambda u_\alpha h^{\lambda\alpha}\eta^{\beta\gamma}]\partial_\beta \varphi \partial_\gamma \varphi \delta. \qquad (5.12.20)$$

Next the commutator
$$4u^\alpha \partial_\alpha v^\nu [\partial_\mu v_\nu(x), \partial_\beta v^\beta(y)] 2c_3 \varphi^2$$
leads to
$$A_5 = 4ic_3 u^\alpha \partial_\alpha v^\nu(x) \varphi^2(y) \partial^y_\nu \delta(x-y).$$
and finally
$$-c_4 u^\beta \partial_\beta \varphi \left[\partial_\mu \varphi(x), \varphi^2(y)\left(mc_3 h + 2c_3 \partial_\alpha v^\alpha - \frac{c_4}{2}M^2 h\right) + \right.$$
$$\left. + c_4 \partial_\alpha \varphi(y) \partial_\nu \varphi h^{\alpha\nu} + c_5 \varphi^3(y)\right] \qquad (5.12.21)$$
gives
$$A_6 = ic_4 u^\beta \partial_\beta \varphi(x) \left[2\varphi\left(mc_3 - \frac{c_4}{2}M^2\right)h + 4c_3 \varphi \partial_\nu v^\nu + \right.$$
$$\left. + 2c_4 \partial_\alpha \varphi(y) h^{\alpha\nu}(y) \partial^y_\nu + 3c_5 \varphi^2\right]\delta(x-y). \qquad (5.12.22)$$

The sum $A_1 + \ldots + A_6$ is equal to
$$B_1 = 2ic_4(u^\lambda \partial_\lambda h^{\alpha\beta} + \partial_\lambda u^\lambda h^{\alpha\beta} - \partial_\lambda u^\beta h^{\alpha\lambda} - \partial_\lambda u^\alpha h^{\beta\lambda})\partial_\alpha \varphi \partial_\beta \varphi \delta$$
$$+ 2ic_4^2 u^\beta \partial_\beta \varphi(x) h^{\mu\nu}(y) \partial_\mu \varphi(y) \partial^y_\nu \delta(x-y) \quad (T1)$$
$$+ i(2mc_3 - M^2 c_4)(u^\mu \partial_\mu h + \partial_\mu u^\mu h)\varphi^2 \delta$$
$$+ ic_4(2mc_3 - M^2 c_4) u^\beta \partial_\beta \varphi h \varphi \delta \quad (T2)$$
$$- 2i(2mc_3 - M^2 c_4)\partial_\nu u_\mu h^{\mu\nu} \varphi^2 \delta \quad (T3)$$
$$+ 4ic_3 u^\mu \partial_\mu v^\nu(x) \varphi^2(y) \partial^y_\nu \delta(x-y) + 4ic_3 c_4 u^\beta \partial_\beta \varphi \partial_\nu v^\mu \delta \quad (T4)$$
$$+ 3ic_4 c_5 u^\beta \partial_\beta \varphi \varphi^2 \delta. \quad (T5) \qquad (5.12.23)$$

Following the methods developed in Sect.5.9 we have grouped the terms according to their type of Lorentz contractions. For example, $(T1)$ has $u^\lambda h^{\alpha\beta} \varphi \varphi$ and 3 derivatives which is different from $(T3)$. Only the terms within one type $T1, \ldots T5$ can be combined to give a divergence. Due to the different coefficients c_4 and c_4^2 in T1 we must have $c_4 = 2$ in order to get a divergence. If c_4 were $\neq 2$ then the last term of $(T1)$ would remain without compensation. Since this term is not a relative co-boundary gauge invariance then would be violated.

The total anomaly is obtained by adding the contribution $x \leftrightarrow y$. For the terms with $\delta(x-y)$ this simply gives factor 2. For the terms with derivative of δ we use again the identity

$$g(x)f(y)\partial_\alpha^y \delta(x-y) + x \leftrightarrow y = (\partial_\alpha g f - g\partial_\alpha f)\delta(x-y) \quad (5.12.24)$$

Now the total anomalies of type T1 in (5.12.23) can be written in the form

$$(T1)_{\text{tot}} = 4ic_4\Big[(u^\lambda \partial_\lambda h^{\alpha\beta} + \partial_\lambda u^\lambda h^{\alpha\beta})\partial_\alpha \varphi \partial_\beta \varphi - \partial_\lambda u^\beta h^{\alpha\lambda}\partial_\alpha \varphi \partial_\beta \varphi$$

$$+u^\beta \partial_\alpha \partial_\beta \varphi \partial_\lambda \varphi h^{\alpha\lambda} - u^\beta \partial_\beta \varphi \partial_\alpha \varphi \partial_\lambda h^{\alpha\lambda} - u^\beta \partial_\beta \varphi \partial_\alpha \partial_\lambda h^{\alpha\lambda}\Big]. \quad (5.12.25)$$

This agrees with the result in massless gravity (5.9.40), and is a divergence

$$(T1)_{\text{tot}} = 4ic_4\partial_\lambda^x\Big[(u^\lambda h^{\alpha\beta} - u^\beta h^{\lambda\alpha})\partial_\alpha \varphi \partial_\beta \varphi \delta(x-y)\Big]$$

$$+x \leftrightarrow y, \quad (5.12.26)$$

where $c_4 = 2$ has been taken into account and will be assumed in the following. Type T2 is a divergence as well:

$$(T2)_{\text{tot}} = 2i(mc_3 - M^2)\partial_\mu[u^\mu h\varphi^2 \delta(x-y)] + x \leftrightarrow y. \quad (5.12.27)$$

As in the massless case (5.9.45) T3 is a co-boundary:

$$(T3)_{\text{tot}} = -2(mc_3 - M^2)d_Q[(h^2 - 2h_{\mu\nu}h^{\mu\nu})\varphi^2 \delta(x-y)]. \quad (5.12.28)$$

Using the identity (5.12.24) we write T4 as follows

$$(T4)_{\text{tot}} = 4ic_3[(\partial_\nu u^\mu \partial_\mu v^\nu + u^\mu \partial_\mu \partial_\nu v^\nu)\varphi^2 - 2u^\mu \partial_\mu v^\nu \varphi \partial_\nu \varphi +$$

$$+4u^\mu \partial_\nu v^\nu \varphi \partial_\mu \varphi]\delta(x-y). \quad (5.12.29)$$

We first split off a divergence

$$(T4)_{\text{tot}} = 4ic_3[2\partial_\mu(u^\mu \partial_\nu v^\nu \varphi^2) - \partial_\nu(u^\mu \partial_\mu v^\nu \varphi^2)$$

$$+2(\partial_\nu u^\mu \partial_\mu v^\nu - \partial_\mu u^\mu \partial_\nu v^\nu)\varphi^2]\delta(x-y). \quad (5.12.30)$$

Now the terms in the second line are a co-boundary

$$(T4)_{\text{tot}} = 2ic_3[2\partial_\mu^x(u^\mu \partial_\nu v^\nu \varphi^2 \delta) - \partial_\nu^x u^\mu \partial_\mu v^\nu \varphi^2 \delta)] + x \leftrightarrow y$$

$$-8\frac{c_3}{m}d_Q[(\partial_\nu v^\mu \partial_\mu v^\nu - \partial_\mu v^\mu \partial_\nu v^\nu)\varphi^2 \delta]. \quad (5.12.31)$$

Finally, T5 is also a co-boundary

$$T5 = -\frac{12}{m} c_5 d_Q(v^\mu \varphi^2 \partial_\mu \varphi \delta). \tag{5.12.32}$$

Adding the contribution $x \leftrightarrow y$ this gives the result of the theorem. □

The normalization terms N (5.12.15) give quartic interactions between the scalar field and the massive gravitational field. The first term proportional to κ^2 is known from the massless theory (5.9.45). The other two coupling terms with free coupling constants c_3 and c_5 are new and depend on the graviton mass m. We shall discuss them in the following section.

For later applications we also need the coupling of a complex scalar field (1.1.47) to massive gravity. A convenient way to obtain this is to realize that the complex scalar field is equivalent to two commuting real (hermitian) scalar fields φ_1, φ_2 with equal mass M according to (5.9.48)

$$\varphi_1 = \frac{1}{2}(\varphi + \varphi^+) \quad \varphi_2 = -\frac{i}{2}(\varphi - \varphi^+). \tag{5.12.33}$$

It is not hard to generalize the above construction to this situation. In the expression (5.12.8) the last term must be substituted by

$$T_\mu^{an} = \ldots - c_4(u^\alpha \partial_\alpha \varphi_1 \partial_\mu \varphi_1 - u^\alpha \partial_\alpha \varphi_2 \partial_\mu \varphi_2). \tag{5.12.34}$$

The first order coupling (5.12.9) now reads

$$T_\varphi = c_3(\varphi_1^2 + \varphi_2^2)(mh + 2\partial_\mu v^\mu) + c_4(\partial_\mu \varphi_1 \partial_\nu \varphi_1 h^{\mu\nu} - \frac{M^2}{2}\varphi_1^2 h) +$$

$$+ c_4(\partial_\mu \varphi_2 \partial_\nu \varphi_2 h^{\mu\nu} - \frac{M^2}{2}\varphi_2^2 h) + c_5(\varphi_1^3 + \varphi_2^3) + c_6(\varphi_1^2 \varphi_2 + \varphi_1 \varphi_2^2). \tag{5.12.35}$$

Here we have assumed that everything is symmetric in φ_1, φ_2. Since we shall use the complex scalar field to describe one kind of normal matter, we are interested in this symmetric case. As in theorem 12.1 it follows that the linear couplings with coefficients c_1, c_2 in (5.12.9) vanish, therefore we have left out these couplings in (5.12.35).

The second order gauge invariance can now be analysed as in the proof of theorem 12.2. Again it follows that $c_4 = 2$. The only change is in the commutator (5.12.21) where we get the new terms

$$-c_4 u^\beta \partial_\beta \varphi_1 \Big[\partial - \mu \varphi_1(x), c_5 \varphi_1(y)^3 + c_6(\varphi_1^2 \varphi_2 + \varphi_1 \varphi_2^2)\Big]$$

which instead of (5.12.32) lead to the co-boundary

$$T5 = -\frac{8}{m} d_Q \Big\{ v^\mu \partial_\mu \varphi_1 (3c_5 \varphi_1^2 + 2c_6 \varphi_1 \varphi_2 + c_6 \varphi_2^2) \delta \Big\},$$

and a similar term with φ_1, φ_2 interchanged. Then instead of (5.12.15) we arrive at the following quartic coupling

$$N = 2(\varphi_1^2 + \varphi_2^2)\Big\{ M^2(h^2 - 2h^{\mu\nu} h_{\mu\nu}) + c_3[m(h^2 - 2h^{\mu\nu} h_{\mu\nu}) +$$

$$-\frac{8}{m}(\partial_\mu v^\mu \partial_\nu v^\nu - \partial_\mu v^\nu \partial_\nu v^\mu))\Big\}$$
$$+\frac{8}{m}v^\mu \Big[\partial_\mu \varphi_1(3c_5\varphi_1^2 + 2c_6\varphi_1\varphi_2 + c_6\varphi_1^2) +$$
$$+\partial_\mu \varphi_2(3c_5\varphi_2^2 + 2c_6\varphi_1\varphi_2 + c_6\varphi_1^2)\Big]. \tag{5.12.36}$$

What remains to be done is to substitute the real scalar fields by the complex φ using (5.12.33). Then the trilinear coupling (5.12.35) is equal to

$$T_\varphi = c_3\varphi^+\varphi(mh + 2\partial_\mu v^\mu) + (\partial_\mu \varphi^+ \partial_\nu \varphi + \partial_\nu \varphi^+ \partial_\mu \varphi)h^{\mu\nu} - M^2\varphi^+\varphi h +$$
$$+\frac{1+i}{8}(c_5 - c_6)\varphi^3 + \frac{1-i}{8}(3c_5 + c_6)\varphi^2\varphi^+ +$$
$$+\frac{1+i}{8}(3c_5 + c_6)\varphi\varphi^{+2} + \frac{1-i}{8}(c_5 - c_6)\varphi^{+3}). \tag{5.12.37}$$

Similarly from (5.12.36) we get the following quartic coupling

$$N = 2\varphi^+\varphi\Big\{M^2(2h^{\mu\nu}h_{\mu\nu} - h^2) + c_3[m(h^2 - 2h^{\mu\nu}h_{\mu\nu}) +$$
$$-\frac{8}{m}(\partial_\mu v^\mu \partial_\nu v^\nu - \partial_\mu v^\nu \partial_\nu v^\mu))\Big\}$$
$$+\frac{8}{m}v^\mu\Big\{\frac{1}{8}\partial_\mu\varphi[(3c_5 - 5c_6)(1+i)\varphi^2 + (6c_5 + 2c_6)(1+i)\varphi^+\varphi) +$$
$$+(3c_5 + 3c_6)(1+i)\varphi^{+2}] + \frac{1}{8}\partial_\mu\varphi^+[(3c_5 + 3c_6)(1+i)\varphi^2 + (6c_5 + 2c_6)(1+i)\varphi^+\varphi +$$
$$+(3c_5 - 5c_6)(1+i)\varphi^{+2}]\Big\}. \tag{5.12.38}$$

5.13 Problems

5.1 If one uses the so-called de-Donder gauge condition $h^{\mu\nu}{}_{,\mu} = \frac{1}{2}h_{,\nu}$ instead of the Hilbert condition, the gauge charge must be defined as follows

$$Q = \int d^3x \, (h^{\mu\nu}{}_{,\mu} - \tfrac{1}{2}h^{,\nu}) \overleftrightarrow{\partial}_0 u_\nu. \tag{5.13.1}$$

Show that Q is nilpotent. Calculate the gauge variations of $h^{\mu\nu}, h, u^\mu, \tilde{u}^\mu$.

5.2 Expansion of the Einstein-Hilbert Lagrangian without using Goldberg variables by means of

$$g_{\mu\nu} = \eta_{\mu\nu} + \kappa h_{\mu\nu} : \tag{5.13.2}$$

a.) Find the expansion of the inverse $g^{\mu\nu}$ and of

$$\sqrt{-g} = \sqrt{-\det g_{\mu\nu}}. \tag{5.13.3}$$

b.) Express the last two terms in (5.5.5)

$$L_1 = -\frac{2}{\kappa^2}\sqrt{-g}g^{\mu\nu}\left(\Gamma^\alpha_{\alpha\beta}\Gamma^\beta_{\mu\nu} - \Gamma^\alpha_{\beta\nu}\Gamma^\beta_{\alpha\mu}\right) \qquad (5.13.4)$$

in terms of $g_{\mu\nu}$.
Result:

$$L_1 = -\frac{1}{2\kappa^2}\sqrt{-g}g_{\alpha\beta,\gamma}g_{\mu\nu,\varrho}\left(2g^{\alpha\gamma}g^{\mu\nu}g^{\beta\varrho} - g^{\alpha\beta}g^{\mu\nu}g^{\varrho\gamma} - \right.$$
$$\left. -2g^{\gamma\mu}g^{\alpha\nu}g^{\beta\varrho} + g^{\alpha\mu}g^{\beta\nu}g^{\varrho\gamma}\right). \qquad (5.13.5)$$

c.) Expand L_1 up to $O(h^3)$.
Result:

$$L_1 = -\tfrac{1}{2}h^{\mu\nu}{}_{,\varrho}h^{\mu\nu}{}_{,\varrho} + \tfrac{1}{2}h_{,\varrho}h_{,\varrho} - h^{\mu\nu}{}_{,\mu}h_{,\nu} + h^{\mu\nu}{}_{,\varrho}h^{\nu\varrho}{}_{,\mu} +$$
$$+\kappa\left(\tfrac{1}{4}h_{,\varrho}h_{,\varrho}h - \tfrac{1}{2}h^{\mu\nu}{}_{,\mu}h_{,\nu}h - \tfrac{1}{2}h^{\mu\nu}h_{,\mu}h_{,\nu} - h^{\mu\nu}h^{\mu\nu}{}_{,\varrho}h_{,\varrho} -\right.$$
$$-\tfrac{1}{4}h^{\mu\nu}{}_{,\varrho}h^{\mu\nu}{}_{,\varrho}h + \tfrac{1}{2}h^{\mu\nu}{}_{,\varrho}h^{\nu\varrho}{}_{,\mu}h + h^{\mu\nu}h^{\nu\varrho}{}_{,\varrho}h_{,\mu} + h^{\mu\nu}h^{\nu\varrho}{}_{,\mu}h_{,\varrho} +$$
$$+\tfrac{1}{2}h^{\mu\nu}h^{\varrho\sigma}{}_{,\mu}h^{\varrho\sigma}{}_{,\nu} + h^{\mu\nu}{}_{,\mu}h^{\varrho\sigma}{}_{,\nu}h^{\varrho\sigma} + h^{\mu\nu}h^{\mu\sigma}{}_{,\varrho}h^{\nu\sigma}{}_{,\varrho} - h^{\mu\nu}h^{\mu\varrho}{}_{,\sigma}h^{\nu\sigma}{}_{,\varrho} -$$
$$\left.-2h^{\mu\nu}{}_{,\sigma}h^{\mu\varrho}h^{\nu\sigma}{}_{,\varrho}\right). \qquad (5.13.6)$$

d.) Use the equation (5.5.12) to find the cubic interaction term from c.).
Result:

$$T_1^h = -\tfrac{1}{4}h_{,\varrho}h_{,\varrho}h + \tfrac{1}{2}h^{\mu\nu}{}_{,\mu}h_{,\nu}h + \tfrac{1}{2}h^{\mu\nu}h_{,\mu}h_{,\nu} + h^{\mu\nu}h^{\mu\nu}{}_{,\varrho}h_{,\varrho} +$$
$$+\tfrac{1}{4}h^{\mu\nu}{}_{,\varrho}h^{\mu\nu}{}_{,\varrho}h - \tfrac{1}{2}h^{\mu\nu}{}_{,\varrho}h^{\nu\varrho}{}_{,\mu}h - h^{\mu\nu}h^{\nu\varrho}{}_{,\varrho}h_{,\mu} - h^{\mu\nu}h^{\nu\varrho}{}_{,\mu}h_{,\varrho} -$$
$$-\tfrac{1}{2}h^{\mu\nu}h^{\varrho\sigma}{}_{,\mu}h^{\varrho\sigma}{}_{,\nu} - h^{\mu\nu}{}_{,\mu}h^{\varrho\sigma}{}_{,\nu}h^{\varrho\sigma} - h^{\mu\nu}h^{\mu\sigma}{}_{,\varrho}h^{\nu\sigma}{}_{,\varrho} + h^{\mu\nu}h^{\mu\varrho}{}_{,\sigma}h^{\nu\sigma}{}_{,\varrho} +$$
$$+2h^{\mu\nu}{}_{,\sigma}h^{\mu\varrho}h^{\nu\sigma}{}_{,\varrho}. \qquad (5.13.7)$$

e.) Show that this coupling T_1^h contains 8 terms beside divergence couplings. Consequently, this approach requires much more work than the treatment with Goldberg variables which has only three coupling terms (5.5.33) beside divergences.

5.3 Check the cancellation of the anomalies for the types G1, G2, G6-G11 (5.7.32-36).

5.4 Construct all tensors $P_{\alpha\beta\mu\nu}$ which are symmetric under the transpositions $\alpha \longleftrightarrow \beta$, $\mu \longleftrightarrow \nu$ and $(\alpha,\beta) \longleftrightarrow (\mu,\nu)$, separately, and of order 4 in p^μ. Compare the results with (5.10.7).

5.5 As in problem 5.4 construct all symmetric tensors of forth rank which are of lower orders < 4 in p^μ. Compare the results with (5.10.8-11).

5.6 Prove the identity (5.11.20).

6. Non-geometric general relativity

In the five chapters so far everything was defined perturbatively, that means as formal power series in some coupling parameter κ. This restriction can be removed by considering the classical limit. This is particularly appropriate in the spin-2 case because quantum effects cannot be observed here until now, the macrocosmos is governed by classical gravity. In Sections 5.5 and 5.9 we have seen that the classical limit of the massless spin-2 gauge theory coincides with general relativity. However in our approach the metric tensor field $g_{\mu\nu}(x)$ has no geometric meaning, it is a field on Minkowski space as all the other fields in this book. This is different from Einstein's geometrical theory. Einstein was the priest who made the marriage between geometry and gravity. This couple lived in harmony for about 60 years. In the eighties gravity took a lover called dark matter. But this was a disappointment because this dark woman did not show up by day. Since then there was an increasing crisis which, however, was ignored by most people. Now after 100 years it is time to separate the couple. We want to discuss this important point in more detail.

Einstein was led to his geometric interpretation by studying Riemannian geometry. But he was aware that this is not the only possibility. In his essay "Geometry and experience" (Collected Works, vol.7, document 52) he discusses the alternative view of H. Poincaré. In his important book "Science and Hypothesis" Poincaré has stated clearly (Chapter III, p.50, Dover publication):

The geometrical axioms are therefore neither synthetic à priori intuitions nor experimental facts. They are conventions.

Einstein explains this as follows (my translation from German):

Geometry (G) says nothing about the properties of real bodies, only geometry together with physical laws (P) can do this. We can say symbolically that only the sum (G) + (P) is controlled by experiment. So (G) can be chosen arbitrarily, some parts of (P), too; all those laws are conventions.

And he summarizes in Latin:

Sub specie aeternitatis in my opinion Poincaré is right.

A more modern critic of the geometric interpretation was S. Weinberg in his classical book "Gravitation and Cosmology" of 1971. He has written a section entitled "The geometric analogy" where he says:

...the geometric interpretation of the theory of gravitation has dwindled to a mere analogy, which lingers in our language in terms like "metric," "affine connection," and "curvature," but is not otherwise very useful.

We may add that it is not only not very useful, but dangerous because one is then forced to postulate some dark matter in order to explain the rotation curves in galaxies. If this dark matter cannot be found the theory must be revised. For this reason we shall follow completely the non-geometric approach which is suggested by gauge theory: *Geometry is Minkowski by convention* as in spin 1. Still we shall use the geometric notions metric etc. But the notion "curved space-time" will never be used; we simply say gravitational field instead.

6.1 Geodesic equation

Let us consider a massive test body moving in a gravitational field. Following Weinberg (loc. cit.) we introduce the local freely falling coordinate system attached to this test body with coordinates ξ^α. In this system the equation of motion of the body is trivially

$$\frac{d^2\xi^\alpha}{d\tau^2} = 0 \tag{6.1.1}$$

where $d\tau$ is the proper time

$$d\tau^2 = \eta_{\alpha\beta} d\xi^\alpha d\xi^\beta \tag{6.1.2}$$

and $\eta_{\alpha\beta} = \text{diag}(1, -1, -1, -1)$ is the Lorentz metric. In addition we introduce a laboratory system with coordinates x^μ and the coordinates ξ^α are functions of the x^μ. We want to express the equation of motion (6.1.1) in the laboratory frame. By the chain rule of differentiation we have

$$\frac{d^2\xi^\alpha}{d\tau^2} = \frac{d}{d\tau}\left(\frac{\partial\xi^\alpha}{\partial x^\mu}\frac{dx^\mu}{d\tau}\right) =$$

$$= \frac{\partial^2\xi^\alpha}{\partial x^\mu \partial x^\nu}\frac{dx^\mu}{d\tau}\frac{dx^\nu}{d\tau} + \frac{d\xi^\alpha}{\partial x^\mu}\frac{d^2x^\mu}{d\tau^2} = 0. \tag{6.1.3}$$

We assume that the Jacobian $\partial\xi^\alpha/\partial x^\mu$ has an inverse $\partial x^\varrho/\partial\xi^\alpha$. Applying this to (6.1.3) yields the so-called geodesic equation

$$\frac{d^2x^\varrho}{d\tau^2} + \Gamma^\varrho_{\mu\nu}\frac{dx^\mu}{d\tau}\frac{dx^\nu}{d\tau} = 0 \tag{6.1.4}$$

where

$$\Gamma^\varrho_{\mu\nu} = \frac{\partial x^\varrho}{\partial\xi^\alpha}\frac{\partial^2\xi^\alpha}{\partial x^\mu \partial x^\nu} \tag{6.1.5}$$

are the Christoffel symbols.

Geodesic equation

Similarly we can express the proper time (6.1.2) in the laboratory system. We get

$$d\tau^2 = g_{\mu\nu}dx^\mu dx^\nu \qquad (6.1.6)$$

where

$$g_{\mu\nu} = \frac{\partial \xi^\alpha}{\partial x^\mu}\frac{\partial \xi^\beta}{\partial x^\nu}\eta_{\alpha\beta} \qquad (6.1.7)$$

is the symmetric metric tensor. Since the coordinate transformation is assumed to be invertable the metric tensor considered as a 4×4 matrix has an inverse

$$g_{\mu\nu}g^{\nu\varrho} = \delta^\varrho_\mu \qquad (6.1.8)$$

given by

$$g^{\nu\varrho} = \frac{\partial x^\nu}{\partial \xi^\alpha}\frac{\partial x^\varrho}{\partial \xi^\beta}\eta^{\alpha\beta}. \qquad (6.1.9)$$

Now we want to relate the metric to the Christoffel symbols. Differentiating (6.1.7) with respect to x^ϱ we get

$$\frac{\partial g_{\mu\nu}}{\partial x^\varrho} = \frac{\partial^2 \xi^\alpha}{\partial x^\varrho \partial x^\mu}\frac{\partial \xi^\beta}{\partial x^\nu}\eta_{\alpha\beta} + \frac{\partial \xi^\alpha}{\partial x^\mu}\frac{\partial^2 \xi^\beta}{\partial x^\varrho \partial x^\nu}\eta_{\alpha\beta}. \qquad (6.1.10)$$

From the definition (6.1.5) it follows

$$\frac{\partial^2 \xi^\alpha}{\partial x^\mu \partial x^\nu} = \Gamma^\varrho_{\mu\nu}\frac{\partial \xi^\alpha}{x^\varrho}. \qquad (6.1.11)$$

Using this in (6.1.10) we have

$$\frac{\partial g_{\mu\nu}}{\partial x^\varrho} = \Gamma^\sigma_{\varrho\mu}\frac{\partial \xi^\alpha}{\partial x^\varrho}\frac{d\xi^\beta}{\partial x^\nu}\eta_{\alpha\beta} + \Gamma^\sigma_{\varrho\nu}\frac{\partial \xi^\alpha}{\partial x^\mu}\frac{d\xi^\beta}{\partial x^\varrho}\eta_{\alpha\beta}.$$

By the definition (6.1.7) this yields

$$\frac{\partial g_{\mu\nu}}{\partial x^\varrho} = \Gamma^\sigma_{\varrho\mu}g_{\sigma\nu} + \Gamma^\sigma_{\varrho\nu}g_{\sigma\mu}. \qquad (6.1.12)$$

We have succeeded in eliminating the freely falling coordinates ξ^μ. It follows from the definitions that $g_{\mu\nu}$ and Γ are symmetric in the lower indices. Then we add to (6.1.12) the same equation with μ and ϱ interchanged and subtract the same equation with ν and ϱ interchanged. This gives

$$\frac{\partial g_{\mu\nu}}{\partial x^\varrho} + \frac{\partial g_{\varrho\nu}}{\partial x^\mu} - \frac{\partial g_{\mu\varrho}}{\partial x^\nu} = \Gamma^\sigma_{\varrho\mu}g_{\sigma\nu} + \Gamma^\sigma_{\varrho\nu}g_{\sigma\mu}+$$

$$+\Gamma^\sigma_{\mu\varrho}g_{\sigma\nu} + \Gamma^\sigma_{\mu\nu}g_{\sigma\varrho} - \Gamma^\sigma_{\nu\mu}g_{\sigma\varrho} - \Gamma^\sigma_{\nu\varrho}g_{\sigma\mu}$$

$$= 2g_{\sigma\nu}\Gamma^\sigma_{\varrho\mu}.$$

Applying the inverse of $g_{\sigma\nu}$ we obtain the important relation

$$\Gamma^\sigma_{\varrho\mu} = \frac{1}{2}\left(\frac{\partial g_{\mu\nu}}{\partial x^\varrho} + \frac{\partial g_{\varrho\nu}}{\partial x^\mu} - \frac{\partial g_{\mu\varrho}}{\partial x^\nu}\right). \qquad (6.1.13)$$

It is also called Levi-Cività affine connection.

6.2 Einstein's equations and Maxwell's equations

It was the aim of this book to treat spin-1 and spin-2 gauge fields on the same footing. Then we would like to see this parallelism in the classical massless case, too. Before this we must discuss another basic question, namely, how the gauge fields which have been defined *mathematically* are defined *physically*. In an experimental science like physics the basic quantities are defined operationally i.e. by a measuring process. In electrodynamics the electric and magnetic fields are defined by measuring the motion of charged test bodies in those fields. This motion is described by the equation of motion

$$\frac{d^2 x^\mu}{d\tau^2} = \frac{e}{m} F^{\mu\nu} \frac{dx_\nu}{d\tau}, \tag{6.2.1}$$

where the Lorentz force appears on the right side. In the same way the gravitational field is defined by measuring the trajectories of massive test bodies. The corresponding equation of motion for test bodies in a gravitational field is the geodesic equation (6.1.4) derived in the last section

$$\frac{d^2 x^\mu}{d\tau^2} + \Gamma^\mu_{\alpha\beta} \frac{dx^\alpha}{d\tau} \frac{dx^\beta}{d\tau} = 0. \tag{6.2.2}$$

Consequently the Christoffel symbols are the gravitational field strengths.

The field equations for the $F^{\mu\nu}$ are the inhomogeneous Maxwell's equations

$$\partial_\nu F^{\mu\nu} = -\mu_0 j^\mu. \tag{6.2.3}$$

where the electric charge-current density is the source of the electromagnetic field. According to section 5.9 the source of the gravitational field is the energy-momentum tensor $T_{\mu\nu}$ and after section 5.5 the field equations for the gravitational field are Einstein's equations

$$R_{\mu\nu} - \frac{1}{2} g_{\mu\nu} R = \kappa T_{\mu\nu}. \tag{6.2.4}$$

The Ricci tensor is given by (5.5.5)

$$R_{\mu\nu} = \partial_\alpha \Gamma^\alpha_{\mu\nu} - \partial_\nu \Gamma^\alpha_{\mu\alpha} + \Gamma^\alpha_{\alpha\beta} \Gamma^\beta_{\mu\nu} - \Gamma^\alpha_{\nu\beta} \Gamma^\beta_{\alpha\mu} \tag{6.2.5}$$

and by (5.5.4) the scalar curvature is equal to

$$R = g^{\mu\nu} R_{\mu\nu} \tag{6.2.6}$$

A view to (6.2.5) shows that Einstein's equation (6.2.4) indeed is determining the gravitational field strength $\Gamma^\alpha_{\mu\nu}$ from its source. Both (6.2.3) and (6.2.4) are first order partial differential equations. However (6.2.3) are only four equations for the 6 components of $F^{\mu\nu}$. The gap is filled by introducing the vector potential

$$F^{\mu\nu} = \partial^\mu A^\nu - \partial^\nu A^\mu, \tag{6.2.7}$$

which is a consequence of the homogeneous Maxwell's equations. Similarly (6.2.4) are only 10 equations for the 40 components of Γ. The gap is filled by introducing the metric according to (6.1.13)

$$\Gamma^\alpha_{\beta\gamma} = \frac{1}{2}g^{\alpha\mu}(\partial_\gamma g_{\beta\mu} + \partial_\beta g_{\mu\gamma} - \partial_\mu g_{\beta\gamma}). \tag{6.2.8}$$

The 10 components of the metric tensor must be considered as the gravitational potentials. The expression (6.2.8) of the field strength (the Levi-Cività affine connection) can be regarded as the substitute for homogeneous field equations for gravity. However the correspondence between spin-1 and spin-2 is not completely perfect. Einstein's equation (6.2.4) also contains the potential in the trace term beside the field strength. This trace term is needed to guarantee the conservation of the energy-momentum tensor. On the other hand in the spin-1 case (6.2.3) current conservation is true without any modification of the left side.

Now we must taken into account that the potentials A^μ and $g_{\mu\nu}$ are gauge potentials. That implies non-uniqueness, the potentials can be changed by gauge transformations without affecting the physics. In particular this means that *the metric $g_{\mu\nu(x)}$ is not observable*. In the usual treatment one fixes the gauge by some condition as for example the Lorentz condition

$$\partial_\mu A^\mu = 0 \tag{6.2.9}$$

in electrodynamics or the Hilbert gauge (1.7.3) in gravity

$$\partial^\mu g_{\mu\nu}(x) = 0.$$

Such gauge conditions are usefull to simplify the field equations.

Simplicity is not always a good argument, so we shall follow a different method. We choose a sufficiently general ansatz for $g_{\mu\nu}$, it should take the symmetry properties of the problem into account, but no gauge condition is required. Then we compute the corresponding gravitational field strength $\Gamma^\alpha_{\mu\nu}$ (6.1.13) and substitute this into Einstein's equation. After integrating the equations we obtain the free functions in the metric and we can compute the observable quantities. Then we express the metric tensor by the observables as far as possible. In this way the gauge ambiguity is removed, the gauge is fixed by physical requirements. The other possibility is to choose the gauge on unphysical grounds, for example by some geometric convention or to simplify the solution of the differential equations. This standard approach is dangerous because one might miss some important physics. Our program of fixing the gauge by observables is a sort of inverse procedure compared with standard general relativity where one first calculates a metric by solving Einstein's equations in some special gauge and then determines the observables. Clearly, in standard general relativity one cannot be sure that one finds all physically relevant solutions. On the other hand it seems as if our non-standard method has less predictive power because some observables are needed to fix the gauge, and those cannot be predicted. However it is certainly better to make less predictions than wrong ones.

Finally we must discuss another important subject, namely the role of coordinate transformations. All basic equations are tensor equations, so that arbitrary coordinate transformations are possible. But some important observables as the field strength $\Gamma^\alpha_{\mu\nu}$ itself and the rotation velocity in the next section are non-tensors. As a consequence they may change under coordinate transformations non-trivially, that means their observed values depend on the frame of reference. In astrophysics there is always a preferred coordinate system, for example the laboratory system attached to the astronomers telescope or the cosmic rest system in cosmology. We then must calculate the observables in this preferred system and some coordinate transformation may obscure the physics.

6.3 Spherically symmetric gravitational field and the circular velocity

Standard geometric general relativity describes the solar system very well. But on the scale of galaxies which is 10^8 times bigger one observes circular velocities of stars and gas much too big. These velocities are found from redshift measurements according to the formula for the special relativistic Doppler effect

$$\frac{\nu_{\text{obs}}}{\nu} = (1 + V_r)^{-1}(1 - \boldsymbol{V}^2)^{-1/2}. \tag{6.3.1}$$

Here ν is the frequency of a spectral line as known from atomic physics and ν_{obs} actually measured with the telescope; $\boldsymbol{V}$ is the velocity of the light source and V_r the component in the direction from observer to light source. We emphasize that $\boldsymbol{V}$ is a 3-vector but not a 4-tensor. Therefore, it changes under coordinate transformations non-trivially, so it must be calculated in the rest system of the observer.

We consider a star moving in a static spherically symmetric gravitational field with the metric

$$ds^2 = g_{\mu\nu}dx^\mu dx^\nu = e^a dt^2 - e^b dr^2 - r^2 e^c(d\vartheta^2 + \sin^2\vartheta d\phi^2), \tag{6.3.2}$$

where $a(r), b(r), c(r)$ are functions of r only. The coordinates $x^\mu = (t, r, \vartheta, \phi)$ are measured in the laboratory system which is attached to the astronomers telescope. In standard geometric general relativity one says that r is the "circumference radius" and so puts $c = 0$. In our non-geometric interpretation of the $g_{\mu\nu}$ in (6.3.2) as a gravitational potentials there is no reason to do so. On the other hand a coordinate transformation $\bar{r} = \bar{r}(r)$ which removes $\exp c(r)$ is not allowed because the new radial coordinate would be unphysical. In fact, on large scales (galactic or bigger) there is only one preferred distance r which can be measured. For galaxies this is the luminosity distance d_L, and

Spherically symmetric gravitational field and the circular velocity

angular measurements then give the radial distance r from the center of the galaxy. Some other coordinate $\bar{r}(r)$ has no physical meaning and we shall not use it. We return to this important point at the end of the next section.

We now calculate the gravitational field corresponding to this metric which is given by the Christoffel symbols

$$\Gamma^0_{10} = \frac{a'}{2}, \quad \Gamma^1_{00} = \frac{a'}{2} e^{a-b}$$

$$\Gamma^1_{11} = \frac{b'}{2}, \quad \Gamma^1_{22} = -\frac{r^2 c' + 2r}{2} e^{c-b}$$

$$\Gamma^1_{33} = -\frac{r^2 c' + 2r}{2} e^{c-b}$$

$$\Gamma^2_{12} = \frac{c'}{2} + \frac{1}{r}, \quad \Gamma^2_{33} = -\sin\vartheta \cos\vartheta$$

$$\Gamma^3_{13} = \frac{c'}{2} + \frac{1}{r}, \quad \Gamma^3_{23} = \cot\vartheta. \tag{6.3.3}$$

Here the prime means d/dr, all other Christoffels vanish.

To simplify the following discussion we assume that the astronomer on earth has corrected his measurements for the motion of the earth with respect to the center of the galaxy, so that we can choose the center of the galaxy as origin of the laboratory coordinate system. Now the star moves on a geodesic (6.1.4)

$$\frac{d^2 x^\mu}{d\tau^2} + \Gamma^\mu_{\alpha\beta} \frac{dx^\alpha}{d\tau} \frac{dx^\beta}{d\tau} = 0. \tag{6.3.4}$$

We consider the motion in the equatorial plane $\vartheta = \pi/2$. Then we have to solve the following three differential equations:

$$\frac{d^2 t}{d\tau^2} + a' \frac{dt}{d\tau} \frac{dr}{d\tau} = 0 \tag{6.3.5}$$

$$\frac{d^2 r}{d\tau^2} + \frac{a'}{2} e^{a-b} \left(\frac{dt}{d\tau}\right)^2 + \frac{b'}{2} \left(\frac{dr}{d\tau}\right)^2 - r e^{c-b} \left(\frac{r}{2} c' + 1\right) \left(\frac{d\phi}{d\tau}\right)^2 = 0 \tag{6.3.6}$$

$$\frac{d^2 \phi}{d\tau^2} + \left(\frac{2}{r} + c'\right) \frac{dr}{d\tau} \frac{d\phi}{d\tau} = 0. \tag{6.3.7}$$

We want to find integrating factors for these three equations. Indeed multiplying (6.3.5) by $\exp a$ we get

$$\frac{d}{d\tau}\left(e^a \frac{dt}{d\tau}\right) = 0$$

so that

$$e^a \frac{dt}{d\tau} = \text{const.} = A$$

and

$$\frac{dt}{d\tau} = A e^{-a}. \tag{6.3.8}$$

Equation (6.3.7) is multiplied by r^2 which gives

$$\frac{d}{d\tau}\left(r^2\frac{d\phi}{d\tau}\right)+c'\frac{dr}{d\tau}r^2\frac{d\phi}{d\tau}=0.$$

Dividing this by $r^2 d\phi/d\tau$ leads to

$$\frac{d}{d\tau}\log\left(r^2\frac{d\phi}{d\tau}\right)+\frac{dc(r)}{d\tau}=0.$$

After integration we obtain

$$\log\left(r^2\frac{d\phi}{d\tau}\right)=-c+\text{const.}$$

so that finally

$$\frac{d\phi}{d\tau}=\frac{J}{r^2}e^{-c}. \tag{6.3.9}$$

Here the integration constant is chosen in such a way that J reminds of the angular momentum in the standard theory. Finally we substitute (6.3.8) and (6.3.9) into (6.3.6). The resulting equation can be written in the form

$$\frac{d^2r}{d\tau^2}+\frac{b'}{2}\left(\frac{dr}{d\tau}\right)^2+\frac{A^2}{2}a'e^{-a-b}-\frac{J^2}{r^3}e^{-b-c}\left(\frac{r}{2}c'+1\right)=0. \tag{6.3.10}$$

Here multiplication by

$$2e^b\frac{dr}{d\tau}$$

yields the integrable equation

$$\frac{d}{d\tau}\left[e^b\left(\frac{dr}{d\tau}\right)^2\right]+A^2a'\frac{dr}{d\tau}e^{-a}-\frac{J^2}{r^3}e^{-c}\frac{dr}{d\tau}(rc'+2)=0. \tag{6.3.11}$$

After integration we have

$$e^b\left(\frac{dr}{d\tau}\right)^2-A^2e^{-a}+\frac{J^2}{r^2}e^{-c}=\text{const.}=B. \tag{6.3.12}$$

. The 3-velocity appearing in (6.3.1) which is measured by the astronomers is equal to

$$\mathbf{V}=\left(\frac{dx^1}{dt},\frac{dx^2}{dt},\frac{dx^3}{dt}\right). \tag{6.3.13}$$

Since we consider motion in the equatorial plane $\vartheta=\pi/2$, only the first and third components are different from zero. To eliminate the affine parameter τ in favor of the measured time t we multiply by appropriate powers of

$$\frac{d\tau}{dt}=\frac{e^a}{A}. \tag{6.3.14}$$

Then from (6.3.12) we get

Spherically symmetric gravitational field and the circular velocity

$$e^b \left(\frac{dr}{dt}\right)^2 - e^a - \frac{J^2}{A^2} \frac{e^{2a-c}}{r^2} + \frac{B}{A^2} e^{2a}. \qquad (6.3.15)$$

In the following we are interested in the square

$$V^2 = -g_{11}\left(\frac{dr}{dt}\right)^2 - g_{33}\left(\frac{d\phi}{dt}\right)^2 =$$

$$= e^b \left(\frac{dr}{dt}\right)^2 + \frac{J^2}{A^2 r^2} e^{2a-c}. \qquad (6.3.16)$$

Inserting (6.3.15) the term with J^2 drops out and we end up with the simple result

$$V^2 = e^a + \frac{B}{A^2} e^{2a}. \qquad (6.3.17)$$

The result (6.3.17) is not yet the desired rotation velocity because the integration constants A and B must still be determined. To do so we specialize everything for circular motion $r = $ const.. For $dr/dt = 0$ in (6.3.15) we get the equation

$$\frac{J^2}{A^2} \frac{e^{-c}}{r^2} - e^{-a} - \frac{B}{A^2} = 0. \qquad (6.3.18)$$

A second equation is obtained by differentiating this equation with respect to r which is the stability condition for the circular path:

$$-2 \frac{J^2}{A^2} \frac{e^{-c}}{r^3} - \frac{J^2}{A^2} \frac{c' e^{-c}}{r^2} + a' e^{-a} = 0. \qquad (6.3.19)$$

This gives the following values for the integration constants

$$\frac{J^2}{A^2} = \frac{a' r^3}{rc' + 2} e^{c-a} \qquad (6.3.20)$$

$$\frac{B}{A^2} = \frac{ra'}{rc' + 2} e^{-a} - e^{-a}. \qquad (6.3.21)$$

Here r now stands for the constant radius of the circular orbit. Now we are able to compute the circular velocity squared from (6.3.17)

$$V_c^2 \equiv w = \frac{ra'}{rc' + 2} e^a. \qquad (6.3.22)$$

For a check we specialize the result (6.3.22) for the Schwarzschild metric where we have

$$a = \log\left(1 - \frac{r_s}{r}\right), \quad c = 0 \qquad (6.3.23)$$

and r_s is the Schwarzschild radius

$$r_s = 2MG \qquad (6.3.24)$$

with M being the central point mass and G Newton's constant. Then V_c^2 becomes

$$V_c^2 = \frac{r_s}{2r} = \frac{MG}{r}. \qquad (6.3.25)$$

This exactly coincides with Newton's theory. This circular velocity is right on the scale of the solar system. But on the scale of galaxies it is obviously not, $V_c(r)$ becomes constant for large r instead of decreasing like $r^{-1/2}$. If one keeps to the Schwarzschild metric one must postulate some dark matter everywhere in the outer part of the galaxies. If such dark matter cannot be found then one should take the non-standard solution with $c \neq 0$ into consideration. Without dark matter not only the Schwarzschild solution but also Newton's theory breaks down on large scales.

It was our program to determine the metric from the observable (6.3.22). To carry this through we must now solve Einstein's equation.

6.4 Solution of the vacuum equation

From the Christoffel symbols (6.3.3) we can calculate the Ricci tensor (6.2.5). The non-vanishing components are the diagonal elements

$$R_{tt} = \frac{1}{2}e^{a-b}(a'' + \frac{1}{2}a'^2 - \frac{1}{2}a'b' + a'c' + \frac{2}{r}a') \qquad (6.4.1)$$

$$R_{rr} = -\frac{1}{2}(a'' + 2c'') + \frac{b'}{4}(a' + 2c' + \frac{4}{r}) - \frac{a'^2}{4} - \frac{c'^2}{2} - \frac{2}{r}c' \qquad (6.4.2)$$

$$R_{\vartheta\vartheta} = e^{c-b}[-1 - \frac{r^2}{2}c'' - r(2c' + \frac{a'-b'}{2}) - \frac{r^2}{4}c'(a' - b' + 2c')] + 1 \qquad (6.4.3)$$

$$R_{\phi\phi} = \sin^2\vartheta R_{\vartheta\vartheta}, \qquad (6.4.4)$$

the prime always denotes d/dr. Let

$$G_{\mu\nu} = R_{\mu\nu} - \frac{1}{2}g_{\mu\nu}R \qquad (6.4.5)$$

be the Einstein tensor then Einstein's equations without matter can be reduced to the following three differential equations

$$G_{tt} = e^{a-b}\left[-c'' - \frac{3}{4}c'^2 + \frac{1}{2}b'c' + \frac{1}{r}(b' - 3c')\right] + \frac{1}{r^2}(e^{a-c} - e^{a-b}) = 0 \qquad (6.4.6)$$

$$G_{rr} = \frac{1}{2}a'c' + \frac{1}{r}(a' + c') + \frac{c'^2}{4} + \frac{1}{r^2}\left(1 - e^{b-c}\right) = 0 \qquad (6.4.7)$$

$$G_{\vartheta\vartheta} = \frac{r^2}{2}e^{c-b}\left[a'' + c'' - \frac{1}{r}(b' - a' - 2c') + \frac{1}{2}(a'^2 - a'b' + a'c' - b'c' + c'^2)\right] = 0. \qquad (6.4.8)$$

It is not hard to see that there are only two independent field equations. Indeed, using (6.4.7) b can be expressed by a and c. Eliminating b in (6.4.6) and (6.4.8) there results one second order differential equation for a and c:

Solution of the vacuum equation

$$c'' = \frac{a''}{a'}\left(c' + \frac{2}{r}\right) + \frac{4}{r^2} + a'c' + \frac{c'^2}{2} + \frac{2}{r}(a' + c'). \qquad (6.4.9)$$

Introducing the new metric function

$$f(r) = c(r) + 2\log\frac{r}{r_c} \qquad (6.4.10)$$

where r_c has been included for dimensional reasons, equation (6.4.9) assumes the simple form

$$\frac{f''}{f'} - \frac{a''}{a'} = a' + \frac{f'}{2}. \qquad (6.4.11)$$

This can immediately by integrated

$$\log\frac{f'}{a'} = a + \frac{f}{2} + \text{const.} \qquad (6.4.12)$$

On the other hand the circular velocity squared (6.3.22) now becomes

$$w(r) = \frac{a'}{f'}e^a. \qquad (6.4.13)$$

Using this in (6.4.12) leads to

$$f = -2\log w$$

and

$$c = -2\log\frac{rw}{r_c} \qquad (6.4.14)$$

where (6.4.10) has been used. This gives us the metric function

$$e^c = -g_{\vartheta\vartheta}r^{-2} = \frac{r_c^2}{r^2 w^2}. \qquad (6.4.15)$$

To get g_{tt} we return to (6.4.12) which can be written as

$$K_a a' e^a = f' e^{-f/2}. \qquad (6.4.16)$$

Here K_a is the integration constant in (6.4.12). From (6.3.22) we find

$$a'e^a = w\left(c' + \frac{2}{r}\right) = \frac{d}{dr}e^a. \qquad (6.4.17)$$

Combining this with (6.4.14)

$$c' = -2\frac{w'}{w} - \frac{2}{r}$$

we arrive at

$$\frac{d}{dr}e^a = -2w'.$$

This gives

$$g_{tt} = e^a = -2w + K_a. \qquad (6.4.18)$$

Finally $g_{\vartheta\vartheta}$ or $\exp b$ follows from (6.4.7). Solving for $\exp b$ we have

$$e^b = a'e^c\left(\frac{r^2}{2}c' + r\right) + c'e^c\left(\frac{r^2}{4}c' + r\right) + e^c. \qquad (6.4.19)$$

Substituting (6.4.18) and (6.4.14) we find

$$e^b = r_c^2 \frac{w'^2}{w^3}\left(\frac{1}{w} - \frac{1}{w - K_a/2}\right). \qquad (6.4.20)$$

We choose the integration constants $K_a = 1$ and $r_c = r_s/2$ where r_s is the Schwarzschild radius (6.3.24). Then we get

$$e^c = \frac{r_s^2}{4r^2 w^2} \qquad (6.4.21)$$

$$e^a = -2w + 1 \qquad (6.4.22)$$

$$e^b = \frac{r_s^2}{4} \frac{w'^2}{w^4(1 - 2w)}. \qquad (6.4.23)$$

This reduces to the Schwarzschild solution ((6.4.25) below) for $w = r_s/2r$.

Now we discuss again the subtle point of coordinate transformations. In other books the line element (6.3.2) is transformed to the so-called standard form by redefining the radial coordinate as follows

$$\bar{r} = re^{c/2} = \frac{r_s}{2w} \qquad (6.4.24)$$

according to (6.4.21). Then our metric (6.3.2) assumes the Schwarzschild form

$$ds^2 = \left(1 - \frac{r_s}{\bar{r}}\right)d\bar{t}^2 - \left(1 - \frac{r_s}{\bar{r}}\right)^{-1}d\bar{r}^2 - \bar{r}^2(d\vartheta^2 + \sin^2\vartheta d\phi^2). \qquad (6.4.25)$$

Mathematically the class of nonstandard solutions (6.4.21-23) has collapsed to the Schwarzschild solution. What does this mean physically? As was repeatedly emphasized we reject to interpret the metric physically. Instead we consider the observable $w(r) = V_c^2(r)$. From (6.4.24) we obtain

$$w(r) = \frac{r_s}{2\bar{r}} \equiv \bar{w}(\bar{r}). \qquad (6.4.26)$$

This is just the Schwarzschild expression in the new coordinate $\bar{r}$. Now it is clear what has been done: The new radius $\bar{r}$ has been chosen in such a way that the measured $w(r)$ becomes equal to the Schwarzschild expression $\bar{w}(\bar{r})$. Such a transformation is trivially possible, but it has no physical significance. We see that *solutions that are equivalent under diffeomorphisms can be physically in-equivalent*. The reason is that the physical observables transform non-trivially under coordinate transformations. One may ask the question: What is the right physical radius, r or $\bar{r}$? The astronomer must give the answer. If he would work with $\bar{r}$ then for every measured rotation curve, i.e. for every galaxy, he must define a new radial coordinate $\bar{r}$. This is not what he does. He always applies the same measuring procedure (for example measuring the apparent luminosity) to all galaxies, and this gives our radius r. After all in reality, the astronomer adds, the rotation curves in galaxies are not Schwarzschild. Summing up there is essentially only one distance definition which is related to the best measurable quantities. *If one makes arbitrary coordinate transformations one may loose the contact with physics.*

As far as the vacuum equations are concerned we are not able to predict the circular velocity; it must be given. What seems to be a weakness is a strength. Obviously the rotation curve depends on the details of the galaxy structure. A universal rotation curve outside the ordinary matter content seems not to exist. Then the physically relevant vacuum solutions must be flexable enough to match all possible inner solutions.

In addition there arises the question how the precision tests of general relativity in the solar system fit into the non-geometric theory. The answer is simple: For small distances compared to galactic ones the metric function $c(r)$ (6.4.21) is very close to $c = 0$, so that a deviation from the Schwarzschild metric cannot be measured. But on the galactic scale the difference is manifest. It comes from a different behavior of the gravitational field and not from some hypothetical dark matter.

6.5 Cosmology in the cosmic rest frame

So far we have investigated static time-independent gravitational fields. The most interesting time-dependent gravitational field is the expanding Universe, of course, which we are now going to study. We have emphasized the importance of choosing the right coordinate system, so we first ask: What is the best coordinate frame for the Universe ? There is a preferred frame which was discovered in the 1990-ties, it is the frame at rest with respect to the cosmic background radiation (CMB). The radio-astronomers have found that there is a dipole anisotropy in the CMB. This anisotropy vanishes if one transforms to a system moving with a velocity of about 300 km/sec in a direction near the Virgo cluster. This is the cosmic rest frame which can be established by any observer in the Universe. It is a global system for the whole Universe with spherical coordinates (t, r, ϑ, ϕ) with a preferred origin $t = 0$, $r = 0$ where the Big Bang has taken place. A natural and meaningful question now occurs: Where is $r = 0$ with respect to our place in the Milky Way ? We shall try to answer this question in section 6.12.

In standard cosmology one works with co-moving coordinates. Then there is no preferred $r = 0$ and the above question is meaningless. The co-moving coordinates are used because there exists a simple exact solution of Einstein's equation, the homogeneous and isotropic Friedmann, Lemaître, Robertson,

Walker (FLRW) metric where homogeneity requires these coordinates. This so-called standard model of cosmology is also in difficulty if dark matter does not exist, because the matter density comes out too big. This is one reason to study an alternative model. Another reason is theoretical: Homogeneity is an unnatural assumption if we work in the cosmic rest frame. Indeed if the Big Bang starts somewhere at $r = 0$ and the gravitational field spreads out with light speed, then at places with $r > 0$ there is still no field for some finite time. Consequently the gravitational field is certainly inhomogeneous for early times. Therefore we will study inhomogeneous but isotropic cosmology.

We start from the following metric in the cosmic rest frame

$$ds^2 = dt^2 + 2b(t,r) dt\, dr - a^2(t,r)[dr^2 + r^2(d\vartheta^2 + \sin^2\vartheta d\phi^2)]. \quad (6.5.1)$$

The reason for the off-diagonal component is the following. In the expanding Universe all matter is in radial motion. As a consequence the energy-momentum tensor $T_{\mu\nu}$ has a T_{01} component and, hence, the metric as well. The coordinates are assumed to be dimensionless, that means the measured values have been divided by suitable units. The components of the metric tensor are

$$g_{00} = 1, \quad g_{01} = b(t,r), \quad g_{11} = -a^2(t,r)$$
$$g_{22} = -r^2 a^2(t,r), \quad g_{33} = -r^2 a^2(t,r) \sin^2\vartheta, \quad (6.5.2)$$

and zero otherwise. The components of the inverse metric are equal to

$$g^{00} = \frac{a^2}{D}, \quad g^{01} = \frac{b}{D}, \quad g^{11} = -\frac{1}{D} \quad (6.5.3)$$

$$g^{22} = -\frac{1}{a^2 r^2}, \quad g^{33} = -\frac{1}{a^2 r^2 \sin^2\vartheta} \quad (6.5.4)$$

where

$$D = a^2 + b^2 \quad (6.5.5)$$

is the determinant of the 2×2 matrix of the t, r components. The corresponding non-vanishing Christoffel symbols are given by

$$\Gamma^0_{00} = \frac{b\dot{b}}{D}, \quad \Gamma^0_{01} = -\frac{ab}{D}\dot{a}, \quad \Gamma^0_{11} = \frac{1}{D}(a^3\dot{a} - aba' + a^2 b') \quad (6.5.6)$$

$$\Gamma^0_{22} = \frac{r}{D}(ra^3\dot{a} + ba^2 + rbaa'), \quad \Gamma^0_{33} = \frac{r \sin^2\vartheta}{D}(ra^3\dot{a} + ba^2 + rbaa')$$

$$\Gamma^1_{00} = -\frac{\dot{b}}{D}, \quad \Gamma^1_{01} = \frac{a\dot{a}}{D}, \quad \Gamma^1_{11} = \frac{1}{D}(ab\dot{a} + aa' + bb')$$

$$\Gamma^1_{22} = \frac{r}{D}(rb\dot{a} - a^2 - raa'), \quad \Gamma^1_{33} = \frac{r}{D}(rb\dot{a} - a^2 - raa') \sin^2\vartheta, \quad (6.5.7)$$

$$\Gamma^2_{02} = \frac{\dot{a}}{a}, \quad \Gamma^2_{12} = \frac{a'}{a} + \frac{1}{r}, \quad \Gamma^2_{33} = -\sin\vartheta \cos\vartheta \quad (6.5.8)$$

Cosmology in the cosmic rest frame

$$\Gamma^3_{03} = \frac{\dot{a}}{a}, \quad \Gamma^3_{13} = \frac{a'}{a} + \frac{1}{r}, \quad \Gamma^3_{23} = \frac{\cos\vartheta}{\sin\vartheta}. \tag{6.5.9}$$

Here the dot means $\partial/\partial t$ and the prime $\partial/\partial r$.

To write down Einstein's equations we have to calculate the Ricci tensor

$$R_{\mu\nu} = \partial_\alpha \Gamma^\alpha_{\mu\nu} - \partial_\nu \Gamma^\alpha_{\mu\alpha} + \Gamma^\beta_{\mu\nu}\Gamma^\alpha_{\alpha\beta} - \Gamma^\alpha_{\nu\beta}\Gamma^\beta_{\alpha\mu}$$

using the above Christoffel symbols. We obtain

$$R_{00} = -\frac{\ddot{a}}{D}\left(3a + 2\frac{b^2}{a}\right) - \frac{b^2\dot{a}^2}{D^2} + \frac{\dot{a}\dot{b}}{D^2}\left(3ab + 2\frac{b^3}{a}\right) + \frac{\dot{b}}{D^2}(aa' + bb') +$$
$$-\frac{\dot{b}'}{D} - 2\frac{a'\dot{b}}{aD} - \frac{2}{r}\frac{\dot{b}}{D} \tag{6.5.10}$$

$$R_{01} = -\ddot{a}\frac{ab}{D} - \dot{a}^2\frac{b}{D^2}(2a^2 + 3b^2) + \frac{ab^2}{D^2}\dot{a}\dot{b} - \frac{2b^2\dot{a}}{raD} + \frac{b^2}{D^2}b\dot{b}' - \frac{b}{D}\dot{b}' +$$
$$+ \frac{ab}{D^2}a'\dot{b} + 2\frac{\dot{a}}{D}a' - 2\frac{\dot{a}'}{a}. \tag{6.5.11}$$

$$R_{11} = \frac{a^3}{D}\ddot{a} + \frac{a^4\dot{a}^2}{D^2}\left(2 + 3\frac{b^2}{a^2}\right) - \frac{a^3b}{D^2}\dot{a}\dot{b} + \frac{2}{rD}(ab\dot{a} + aa' + bb') + \frac{a^2}{D}b'+$$
$$+2\frac{a}{D}\dot{a}b' - \frac{a^3}{D^2}a'\dot{b} - \frac{a^2b}{D^2}bb' - 2\frac{a''}{a} + \frac{2}{D}a'^2 + 2\frac{b}{aD}a'b' - \frac{4}{r}\frac{a'}{a} \tag{6.5.12}$$

$$R_{22} = r^2\left[\frac{a^3}{D}\ddot{a} + \frac{a^4\dot{a}^2}{D^2}\left(2 + 3\frac{b^2}{a^2}\right) - \frac{a^3b}{D^2}\dot{a}\dot{b}\right] + \frac{b^2}{D} + r\frac{ab}{D^2}\dot{a}(3a^2 + 4b^2) +$$
$$+ r\frac{a^4}{D^2}\dot{b} + \frac{r^2a^2}{D^2}b'\left(a\dot{a} + \frac{b}{r}\right) + \frac{r^2}{D}\left(2ba\dot{a}' - \frac{4}{r}aa' - aa''\right) +$$
$$+ \frac{r^2}{D^2}\left(a^3a'\dot{b} + 2b^3\dot{a}a' - b^2a'^2 + \frac{a^3}{r}a' + aba'b'\right) \tag{6.5.13}$$

$$R_{33} = R_{22}\sin^2\vartheta. \tag{6.5.14}$$

This gives the following scalar curvature

$$R = g^{\mu\nu}R_{\mu\nu} = -6\frac{a\ddot{a}}{D} - \frac{\dot{a}^2}{D^2}(6a^2 + 12b^2) +$$
$$+6\dot{a}\dot{b}\frac{ab}{D^2} - \frac{4}{rD^2}\left(a^2\dot{b} + 2ab\dot{a} + 3\frac{b^3}{a}\dot{a}\right) - 2\frac{b^2}{r^2a^2D} + 2\frac{b}{D^2}b\dot{b}' - 2\frac{\dot{b}'}{D} -$$
$$-4\frac{b'\dot{a}}{D^2}a - 4\frac{b}{rD^2}b' - 2a\frac{a'\dot{b}}{D^2} - 8\frac{b}{aD}\dot{a}' + 4\frac{b}{D^2}\dot{a}a'\left(1 - \frac{b^2}{a^2}\right) + 4\frac{a''}{aD} +$$
$$+2\frac{a'^2}{D^2}\left(\frac{b^2}{a^2} - 1\right) - 4\frac{b}{aD^2}a'b' + 4\frac{a'}{rD^2}\left(2a + 3\frac{b^2}{a}\right). \tag{6.5.15}$$

Next we calculate the Einstein tensor:

$$G_{00} = R_{00} - \frac{g_{00}}{2}R = \frac{\dot{a}^2}{D^2}(3a^2 + 5b^2) - 2\frac{\dot{a}\dot{b}^2}{aD} + 2\frac{\dot{a}\dot{b}}{D^2}\frac{b^3}{a} -$$

$$-2\frac{\dot{b}b^2}{rD^2}+2\frac{\dot{a}b}{rD^2}(2a+3\frac{b^2}{a})+2\frac{a\dot{a}}{D^2}b'-2\frac{b^2}{a}\frac{a'\dot{b}}{D^2}+4\frac{b}{aD}\dot{a}'+\frac{2b}{D^2}\dot{a}a'\left(\frac{b^2}{a^2}-1\right)$$

$$+2\frac{a''}{aD}+\frac{a'^2}{D^2}\left(1-\frac{b^2}{a^2}\right)+2\frac{b}{aD^2}a'b'-\frac{2a'}{rD^2}\left(2a+3\frac{b^2}{a}\right)+\frac{b^2}{r^2a^2D}+2\frac{bb'}{rD^2}. \quad (6.5.16)$$

$$G_{01}=R_{01}-\frac{g_{01}}{2}R=b\left[2\ddot{a}\frac{a}{D}+\frac{\dot{a}^2}{D^2}(a^2+3b^2)-2\dot{a}\dot{b}\frac{ab}{D^2}\right]+$$

$$+\frac{2b^2}{raD^2}\dot{a}(a^2+2b^2)+2\frac{a^2b}{rD^2}\dot{b}+2\frac{ab}{D^2}\dot{a}b'+$$

$$+2ab\frac{a'\dot{b}}{D^2}+\frac{2\dot{a}'}{aD}(b^2-a^2)+2\frac{\dot{a}a'}{D^2}\left(\frac{b^4}{a^2}-b^2\right)-2\frac{b}{aD}a''+$$

$$+\frac{b}{D^2}a'^2\left(1-\frac{b^2}{a^2}\right)+\frac{2b^2}{aD^2}a'b'-2\frac{a'b}{rD^2}\left(2a+3\frac{b^2}{a}\right)+2\frac{b^2}{rD^2}b'+\frac{b^3}{r^2a^2D}. \quad (6.5.17)$$

$$G_{11}=a^2\left[-2\frac{\ddot{a}a}{D}-\frac{\dot{a}^2}{D^2}\left(a^2+3b^2\right)+2\dot{a}\dot{b}\frac{ab}{D^2}\right]-\frac{\dot{a}}{rD^2}(2a^3b+4ab^3)-$$

$$-\frac{2a^4}{rD^2}\dot{b}+2\frac{ab^2}{D^2}\dot{a}b'-\frac{b^2}{r^2D}+2\frac{b^3}{rD^2}b'+\frac{2}{a}a''+\frac{a'^2}{D^2}(a^2+3b^2)+$$

$$+2\frac{b^3}{aD^2}a'b'+\frac{2a'}{raD^2}(a^4-2b^4)-2\frac{a'a^3\dot{b}}{D^2}-4\frac{ab}{D}\dot{a}'+2\frac{b}{D^2}\dot{a}a'(a^2-b^2). \quad (6.5.18)$$

$$G_{22}=r^2a^2\left[-2\frac{a}{D}\ddot{a}-\frac{\dot{a}^2}{D^2}\left(a^2+3b^2\right)+2\dot{a}\dot{b}\frac{ab}{D^2}\right]+r^2\left[-\dot{a}\frac{ab}{rD^2}(a^2+2b^2)-\right.$$

$$-\frac{a^4}{rD^2}\dot{b}-\frac{a^2}{D^2}b'(a\dot{a}+\frac{b}{r})+\frac{a^2b}{D^2}\dot{b}b'-\frac{a^2}{D}\dot{b}'+2\frac{\dot{a}a'}{D^2}a^2b-2\frac{ab}{D}\dot{a}'+$$

$$+\frac{aa'}{rD^2}(a^2+2b^2)-\frac{a^2}{D^2}a'^2+\frac{a}{D}a''-\frac{ab}{D^2}a'b'\Big]. \quad (6.5.19)$$

We notice that the leading terms in G_{01}, G_{11} and G_{22} agree up to some factors. This will play an important role in the following. Our aim now is to solve Einstein's equations

$$G_{\mu\nu}=8\pi GT_{\mu\nu}. \quad (6.5.20)$$

6.6 Failure of homogeneous cosmology

One might think that for large distance from the point $r = 0$ where the Big Bang has occurred the r-dependence disappears. Therefore we consider the special case of a homogeneous cosmology putting $a' = 0 = b'$ and $r \to \infty$. Then we get from (6.5.17)

$$G_{01} = b\left(2\ddot{a}\frac{a}{D} + \frac{\dot{a}^2}{D^2}(a^2 + 3b^2) - 2\frac{ab}{D^2}\dot{a}\dot{b}\right) \tag{6.6.1}$$

and (6.5.19) gives essentially the same result after division by r^2:

$$\frac{G_{22}}{a^2 r^2} = -2\ddot{a}\frac{a}{D} - \frac{\dot{a}^2}{D^2}(a^2 + 3b^2) + 2\frac{ab}{D^2}\dot{a}\dot{b}. \tag{6.6.2}$$

By (6.5.20) this implies a strange relation between two components of the energy-momentum tensor

$$T_{01} = -\frac{b}{ar^2}T_{22}. \tag{6.6.3}$$

In $T_{\mu\nu}$ we only consider ordinary matter and radiation. We first set up the radiation tensor $T^r_{\mu\nu}$. Since the metric (6.5.2) has off-diagonal elements the energy-momentum tensor of radiation in the corresponding gravitational field must have off-diagonal elements as well. Therefore we assume the mixed tensor to be of the form

$$T^{r0}_0 = \varrho_r, \quad T^{r0}_1 = q = T^{r1}_0, \quad T^{r1}_1 = T^{r2}_2 = T^{r3}_3 = -p_r. \tag{6.6.4}$$

and zero otherwise. The value of q follows from the basic property that $T^r_{\mu\nu} = T^r_{\nu\mu}$ must be symmetric: since

$$T^r_{01} = g_{01}T^{r1}_1 + g_{00}T^{r0}_1 = -bp_r + q \tag{6.6.5}$$

$$T^r_{10} = g_{10}T^{r0}_0 + g_{11}T^{r1}_1 = b\varrho_r - a^2 q, \tag{6.6.6}$$

we get

$$q = \frac{b}{a^2 + 1}(\varrho_r + p_r). \tag{6.6.7}$$

The remaining components are

$$T^r_{00} = g_{00}T^{r0}_0 + g_{01}T^{r1}_0 = \varrho_r + bq$$

$$T^r_{11} = a^2 p_r + bq \tag{6.6.8}$$

$$T^r_{22} = r^2 a^2 p_r, \quad T^r_{33} = r^2 a^2 p_r \sin^2 \vartheta.$$

In our model of the universe the ordinary matter with proper energy density ϱ_m and pressure p_m is in radial motion with velocity v_m. In previous attempts we have assumed that v_m can be calculated by solving the geodesic equations. This was not correct for the following reason. The geodesic equation describes the motion of a test body which does not disturb the gravitational field. On the other hand the moving normal matter with density ϱ_m, pressure p_m and 4-velocity

$$u^\mu = (u^0, u^0 v_m, 0, 0). \tag{6.6.9}$$

strongly influences the gravitational field. So its motion is governed by hydrodynamic equations in the gravitational field following from the conservation of the energy-momentum tensor

$$T_m^{\mu\nu} = -p_m g^{\mu\nu} + (p_m + \varrho_m) u^\mu u^\nu. \tag{6.6.10}$$

which reads

$$\nabla_\mu T_m^{\nu\mu} = \frac{\partial T_m^{\nu\mu}}{\partial x^\mu} + \Gamma^\nu_{\mu\lambda} T_m^{\lambda\mu} + \Gamma^\mu_{\mu\lambda} T_m^{\nu\lambda} = 0. \tag{6.6.11}$$

It is well known that due to the Bianchi identities Einstein's equations imply the conservation of the total energy-momentum tensor

$$\nabla_\mu (T_r^{\nu\mu} + T_m^{\nu\mu}) = 0. \tag{6.6.12}$$

So if we can neglect radiation or explicitly guarantee

$$\nabla_\mu T_r^{\nu\mu} = 0 \tag{6.6.13}$$

then the hydrodynamic equations are contained in Einstein's equations, therefore, they need not be imposed separately. This leads to a strategy which we will use in the following: We use Einstein's equations not only to determine $a(t,r)$ and $b(t,r)$ but also to find the most important part of the energy-momentum tensor. The latter is a simple task because no partial differential equation must be solved.

Let us now return to the strange relation (6.6.3). We consider the late Universe where the matter is non-relativistic so that we put $p_m = 0$. Then the matter tensor becomes

$$T_m^{\mu\nu} = \varrho_m u^\mu u^\nu. \tag{6.6.14}$$

Here u^0 follows from the normalization

$$g_{\mu\nu} u^\mu u^\nu = (u^0)^2 + 2b(u^0)^2 v_m - a^2 (u^0)^2 v_m^2 = 1. \tag{6.6.15}$$

We obtain

$$u^0 = (1 + 2b v_m - a^2 v_m^2)^{-1/2}. \tag{6.6.16}$$

We also need the covariant components

$$u_0 = g_{00} u^0 + g_{01} u^1 = (1 + b v_m)(1 + 2b v_m - a^2 v_m^2)^{-1/2} \tag{6.6.17}$$

$$u_1 = g_{10} u^0 + g_{11} u^1 = (b - a^2 v_m)(1 + 2b v_m - a^2 v_m^2)^{-1/2}. \tag{6.6.18}$$

This gives

$$u_0 u_1 = (1+bv_m)(b-a^2 v_m)(1+2bv_m - a^2 v_m^2)^{-1} \equiv wb \qquad (6.6.19)$$

where

$$w = \frac{1+(b-a^2/b)v_m - a^2 v_m^2}{1+2bv_m - a^2 v_m^2}$$

is of the order 1. Now the strange relation (6.6.3) becomes

$$w\varrho_m = -p_r = -\frac{\varrho_r}{3} \qquad (6.6.20)$$

where we have assumed the normal equation of state for the radiation. In the present Universe with $\varrho_m \gg \varrho_r$ this relation is clearly violated. The only way to escape the strange relation (6.6.3) is that the expression (6.6.2) which appears also in (6.6.1) vanishes in leading order. This then leads to the inhomogeneous universe which is considered in the following section.

6.7 An inhomogeneous universe

Because of the very many terms in the Einstein tensor it seems to be rather hopeless to solve Einstein's equations exactly. This is in sharp contrast to the simple situation in the homogeneous and isotropic universe. But a perturbative treatment is possible. In the late Universe which we should understand first, $a(t,r)$ and $b(t,r)$ are big ($\gg 1$). Then we put

$$a(t,r) = a_0 + a_1 + \ldots$$
$$b(t,r) = b_0 + b_1 + \ldots \qquad (6.7.1)$$

where a_0, b_0 satisfy (6.6.2)

$$-2\frac{a_0^3}{D_0}\ddot{a}_0 - \dot{a}_0^2 \frac{a_0^2}{D_0^2}(a_0^2 + 3b_0^2) + 2\frac{a_0^3 b_0}{D_0^2}\dot{a}_0 \dot{b}_0 = 0 \qquad (6.7.2)$$

with $D_0 = a_0^2 + b_0^2$. Then the leading terms in G_{01}, G_{11} and G_{22} vanish. Multiplying by $D_0^2/(a_0^3 \dot{a}_0)$ we get a linear equation for $y = b_0^2$:

$$\dot{y} - y\left(2\frac{\ddot{a}_0}{\dot{a}_0} + 3\frac{\dot{a}_0}{a_0}\right) - 2\frac{\ddot{a}_0}{\dot{a}_0}a_0^2 - a_0 \dot{a}_0 = 0. \qquad (6.7.3)$$

A special solution is $y_1 = -a_0^2$. Since the solution of the homogeneous equation is $C\dot{a}_0^2 a_0^3$, the general solution of (6.7.3) is equal to

$$y = C\dot{a}_0^2 a_0^3 - a_0^2. \qquad (6.7.4)$$

This must be positive for arbitrary t, r. The simplest way to chieve this is to choose a_0 independent of r and satisfying

$$\dot{a}_0^2 a_0^3 = \lambda a_0^2 \qquad (6.7.5)$$

with $\lambda > 1/C$ and constant. The solution of this equation is

$$a_0(t) = \alpha t^{2/3} \tag{6.7.6}$$

where α is a new constant of integration. This choice of a_0 defines our perturbative scheme, α is the perturbative parameter in (6.7.1). It is interesting to note that the $t^{2/3}$-law is the same as in the matter dominated universe in the standard FRW cosmology, but we have obtained this law from the vacuum equations; furthermore this law gets modified in the next order. We shall see below that the energy-momentum tensor does not contribute to the leading order equation (6.7.2). By taking $C(r)$ in (6.7.4) now r-dependent we find the following form of $b_0(t,r)$

$$b_0(t,r) = f(r)a_0(t). \tag{6.7.7}$$

To determine $f(r)$ we consider

$$G_{11} - \frac{G_{22}}{r^2} = \frac{a^2}{D}b' + \frac{\dot{a}b'}{D^2}(2ab^2 + a^3) - \frac{\dot{b}b'}{D^2}a^2b$$
$$-\dot{b}\frac{a^4}{rD^2} - \frac{\dot{a}}{rD^2}(a^3b + 2ab^3) - 2\frac{ab}{D}\dot{a}' - 2\frac{\dot{a}a'}{D^2}b^3+$$
$$+\frac{b'b}{rD^2}(2b^2 + a^2) - \frac{b^2}{r^2D} + \frac{a''}{aD}(a^2 + 2b^2) + \frac{a'^2}{D^2}(a^2 + 3b^2)+$$
$$+\frac{a'b'}{D^2}\left(2\frac{b^3}{a} + ab\right) + \frac{a'}{rD^2}\left(a^3 - 4\frac{b^4}{a} - 2ab^2\right) - \frac{a'\dot{b}}{D^2}a^3 = O(\alpha). \tag{6.7.8}$$

After multiplying with D^2/a^2 we restrict to the leading order $O(\alpha^3)$:

$$\dot{b}'_0 D_0 + \dot{a}_0 b'_0\left(2\frac{b_0^2}{a_0} + a_0\right) - b'_0 \dot{b}_0 b_0 - \frac{1}{r}\left[\dot{b}_0 a_0^2 + \dot{a}_0\left(a_0 b_0 + 2\frac{b_0^3}{a_0}\right)\right] = 0. \tag{6.7.9}$$

We have used the fact that $a'_0 = 0$ and again, as we shall see, there is no contribution from $T_{\mu\nu}$. Inserting (6.7.7) we get a simple equation for $f(r)$

$$f'(r) - \frac{1}{r}f(r) = 0 \tag{6.7.10}$$

with the solution $f(r) = Lr$. Hence

$$b_0(t,r) = Lra_0(t) = Lr\alpha t^{2/3} \tag{6.7.11}$$

is the leading order of $b(t,r)$ in (6.7.1) with another constant of integration L of dimension of a reciprocal length.

The Einstein's equations for G_{01}, G_{11} and G_{22} are now fulfilled in leading order. There remains G_{00} to be investigated. Since $G_{00} = O(\alpha^0)$ according to (6.5.16) here T_{00} does contribute. Using the leading order results we obtain

$$8\pi G T_{00} = \frac{\dot{a}_0^2}{D_0^2}(3a_0^2 + 5b_0^2) - 2\frac{\ddot{a}_0 b_0^2}{a_0 D_0} + 2\frac{\dot{a}_0 \dot{b}_0}{D_0^2}\frac{b_0^2}{a_0} =$$

$$= \frac{\dot{a}_0^2}{D_0^2} a_0^2 (3 + 5L^2 r^2 + 2L^4 r^4) - 2\frac{\ddot{a}_0}{D_0} a_0 L^2 r^2. \tag{6.7.12}$$

Substituting (6.7.6) we get the very simple result

$$T_{00} = \frac{1}{6\pi G t^2}, \tag{6.7.13}$$

that means in lowest order the inhomogeneity drops out here. This result for T_{00} is the same as in the Einstein - de Sitter universe. However, our energy density T_{00} contains a 4-velocity $u_0 > 1$ so that (6.7.13) differs considerably from the matter density ϱ_m. It is well known that taking for t the present age T of the Universe, the matter density in the Einstein - de Sitter model comes out much too big. To remove the defect one usually introduces a cosmological constant Λ and in addition some dark matter. We shall not do so but simply go to :

6.8 Next to leading order

This is also called first order correction to the leading zeroth order because it involves a_1 and b_1 in (6.7.1). Now we have to be careful about the contributions of $T_{\mu\nu}$. To get an idea of the radiation contribution we investigate energy conservation

$$\nabla_\mu T_r^{0\mu} = \frac{\partial T_r^{0\mu}}{\partial x^\mu} + \Gamma^0_{\mu\nu} T_r^{\nu\mu} + \Gamma^\mu_{\mu\nu} T_r^{0\nu} = 0. \tag{6.8.1}$$

We cannot set up energy conservation for the matter content because we do not know the radial velocity $u^1 = u^0 v_m$. As discussed above (6.6.12) this is no harm. Using the Γ's of Sect.6.4 we have

$$\partial_t T_r^{00} + \partial_r T_r^{01} + \frac{b\dot{b}}{D} T_r^{00} - 2\frac{ab}{D} \dot{a} T_r^{01} + \frac{1}{D}(a^3 \dot{a} - aba' + a^2 b') T_r^{11} +$$

$$+ \frac{r}{D}(ra^3 \dot{a} + ba^2 + raba') T_r^{22} + \frac{r}{D} \sin^2 \vartheta (ra^3 \dot{a} + ba^2 + raba') T_r^{33} +$$

$$+ \left(\frac{b\dot{b}}{D} + \frac{a\dot{a}}{D} + 2\frac{\dot{a}}{a}\right) T_r^{00} + \left(\frac{aa' + bb'}{D} + 2\frac{a'}{a} + \frac{2}{r}\right) T_r^{01} = 0. \tag{6.8.2}$$

Now we substitute the leading order results. From (6.6.8) we get

$$T_r^{00} = \frac{\varrho_r}{1 + L^2 r^2} + O(\alpha^{-2})$$

$$T_r^{01} = Lr \frac{\varrho_r}{a(1 + L^2 r^2)} + O(\alpha^{-3}) \tag{6.8.3}$$

$$T_r^{11} = \frac{p_r}{a^2(1 + L^2 r^2)} + \frac{L^2 r^2}{a^2 + 1} \frac{\varrho_r + p_r}{1 + L^2 r^2}.$$

Then in (6.8.2) the factor $1/(1+L^2r^2)$ cancels and we arrive at

$$\partial_t \varrho_r + 3\frac{\dot{a}}{a}(p_r + \varrho_r) + O(\alpha^{-1}) = 0. \tag{6.8.4}$$

This is the same equation as in standard FRW-cosmology. With the usual equation of state $p_r = \varrho_r/3$ one has

$$\varrho_r(t) = \frac{\varrho_0}{a^4(t)} \tag{6.8.5}$$

where ϱ_0 is constant.

We start the first order calculation with

$$\frac{G_{22}}{a^2r^2} = -\frac{1}{D}(2a\ddot{a} + \dot{a}^2) + \frac{2}{D^2}(ab\dot{a}\dot{b} - b^2\dot{a}^2) -$$

$$-\frac{\dot{a}b}{raD^2}(a^2 + 2b^2) - \dot{b}\frac{a^2}{rD^2} - \frac{b'}{D^2}\left(a\dot{a} + \frac{b}{r}\right) + \dot{b}b'\frac{b}{D^2} - \frac{\dot{b}'}{D} +$$

$$+2b\frac{\dot{a}a'}{D^2} - 2\frac{b}{aD}\dot{a}' + O(\alpha^{-2}). \tag{6.8.6}$$

We insert (6.7.1) and collect the first order contributions $O(\alpha^{-1})$:

$$\left(\frac{G_{22}}{a^2r^2}\right)_1 = -\frac{1}{D_0}(2a_1\ddot{a}_0 + 2a_0\ddot{a}_1 + 2\dot{a}_0\dot{a}_1) - \frac{2}{D_0^2}(2b_0b_1\dot{a}_0^2 + 2b_0^2\dot{a}_0\dot{a}_1) +$$

$$+\frac{8}{D_0^2 a_0 \beta}(a_1 + Lrb_1)(b_0^2\dot{a}_0^2 - \dot{a}_0 b_0 a_0 b_0) + \frac{2}{D_0^2}(\dot{a}_1 b_0 a_0 b_0 + b_1 \dot{a}_0 a_0 b_0 +$$

$$+a_1\dot{a}_0\dot{b}_0 b_0 + b_1\dot{a}_0\dot{b}_0 a_0) - \frac{\dot{a}_0 b_0}{ra_0 D_0^2}(a_0^2 + 2b_0^2) - \dot{b}_0\frac{a_0^2}{rD_0^2} - L\frac{a_0^2}{D_0^2}\dot{a}_0 +$$

$$+\dot{b}_0 L a_0 \frac{b_0}{D_0^2} - L\frac{\dot{a}_0}{D_0} + O(\alpha^{-2}) = 8\pi G p_r. \tag{6.8.7}$$

We assume pressureless matter so that only radiation pressure contributes on the right-hand side. In (6.8.7) we have used the expansion

$$D = D_0\left(1 + \frac{2}{a_0}\frac{a_1 + Lrb_1}{\beta}\right) \tag{6.8.8}$$

and the abbreviation

$$\beta = 1 + L^2r^2. \tag{6.8.9}$$

The right-hand side in (6.8.7) must be of order α^{-1}. Therefore we write (6.8.5) in the form

$$p_r = \frac{\varrho_r}{3} = \frac{p_0}{a_0 t^2} + O(\alpha^{-2}), \tag{6.8.10}$$

where $p_0 = \varrho_0/3a^3$.

Substituting the lowest order results into (6.8.7) we arrive at

$$\ddot{a}_1 + \frac{\dot{a}_1}{t}\left(\frac{4}{3} - \frac{2}{3\beta}\right) - \frac{a_1}{t^2}\left(\frac{2}{3} - \frac{4}{9\beta}\right) - \frac{\dot{b}_1}{t}\frac{2}{3}\frac{Lr}{\beta} +$$
$$+ \frac{b_1}{t^2}\frac{4Lr}{9\beta} + \frac{2L}{3t}\left(\frac{1}{\beta} + 1\right) + 8\pi G \frac{\beta p_0}{2t^2} = 0. \qquad (6.8.11)$$

To solve the time-dependence we set

$$a_1(t,r) = tg_1(r) + h_1(r) \qquad (6.8.12)$$
$$b_1(t,r) = tg_2(r) + h_2(r). \qquad (6.8.13)$$

Now the terms proportional to t^{-1} give

$$-(3L^2 r^2 + 2)g_1 + Lrg_2 = 3L(\beta + 1)$$

or

$$g_2 = (3Lr + \frac{2}{Lr})g_1 + \frac{3}{r}(L^2 r^2 + 2) \equiv \gamma_1 g_1 + \gamma_2. \qquad (6.8.14)$$

The terms proportional to t^{-2} lead to

$$h_1\left(\frac{4}{9\beta} - \frac{2}{3}\right) + h_2\frac{4Lr}{9\beta} + 4\pi G\beta p_0 = 0$$

or

$$h_1 = \frac{2}{3L^2 r^2 + 1}(Lrh_2 + 9\pi G\beta^2 p_0). \qquad (6.8.15)$$

Next we consider

$$\frac{1}{b}G_{01} + \frac{G_{22}}{a^2 r^2} = \frac{\dot{a}b}{rD^2}\left(a + 2\frac{b^2}{a}\right) + \dot{b}\frac{a^2}{rD^2} + b'\frac{a\dot{a}}{D^2} +$$
$$+ 2\frac{a\dot{b}}{D^2}a' - 2\frac{\dot{a}}{D}\left(\frac{a}{b} + \frac{b}{a}\right) - \frac{\dot{b}'}{D} + \frac{b}{D^2}\dot{b}b' +$$
$$+ 2\frac{\dot{a}a'}{D^2}\frac{b^3}{a^2} - \frac{a''}{aD} - \frac{a'^2}{D^2}\frac{b^2}{a^2} + \frac{b}{D^2}\frac{a'}{a}b' -$$
$$- \frac{a'}{rD^2}\left(3a + 4\frac{b^2}{a}\right) + \frac{bb'}{rD^2} + \frac{b^2}{ra^2 D} + O(\alpha^{-3}) = 8\pi G\frac{T_{01}^m}{b}. \qquad (6.8.16)$$

This combination has been chosen in such a way that radiation does not contribute. In lowest order this gives

$$\frac{2b_0}{ra_0 D_0^2}\dot{a}_0(a_0^2 + 2b_0^2) + 2\dot{b}_0\frac{a_0^2}{rD_0^2} + 2\frac{a_0}{D_0^2}\dot{a}_0 b_0' = 8\pi G\frac{T_{01}^m}{b}. \qquad (6.8.17)$$

Using (6.7.6) and (6.7.7) we finally obtain

$$8\pi G T_{01}^m = \frac{4}{3}\frac{L^2 r}{\beta t} \qquad (6.8.18)$$

which is of the order $O(\alpha^0)$.

We treat $G_{11} - G_{22}/r$ (6.7.8) in the same way, here we need T_{11}^m. By (6.6.14) this can now be calculated according to

$$T_{11}^m = \frac{(T_{01}^m)^2}{T_{00}^m} \tag{6.8.19}$$

which yields

$$8\pi G T_{11}^m = \frac{4}{3}\frac{L^4 r^2}{\beta^2}. \tag{6.8.20}$$

Substituting the leading order expressions we finally obtain

$$\left(G_{11} - \frac{G_{22}}{r^2}\right)_1 = \frac{1}{\beta}\dot{b}_1' - \frac{2}{\beta}Lr\dot{a}_1' + \frac{2}{3t\beta}b_1' - \frac{4L^3 r^3}{3t\beta^2}a_1' - \frac{1}{\beta r}\dot{b}_1 -$$

$$- \frac{2}{3\beta t r}b_1 + \frac{L^4 r^2}{\beta^2} = 8\pi G T_{11}^m. \tag{6.8.21}$$

Inserting (6.8.20), substituting (6.8.12-13) and separating the t-dependence we obtain the following ODE for the g's

$$\frac{5}{3}g_2' - \frac{5}{3r}g_2 - g_1'\left(2Lr + \frac{4L^3 r^3}{3\beta}\right) = \frac{L^2}{3}\left(1 - \frac{1}{\beta}\right), \tag{6.8.22}$$

and in addition we get a homogeneous equation for the h's:

$$h_2' - \frac{h_2}{r} - 2\frac{L^3 r^3}{\beta}h_1' = 0. \tag{6.8.23}$$

Now we are able to calculate the metric functions in first order. We eliminate g_2 in (6.8.22) by means of (6.8.14) and obtain the following linear equation for $g_1(r)$ alone

$$g_1'\left(\frac{5}{3}Lr + \frac{10}{3Lr} + \frac{4Lr}{3\beta}\right) - \frac{20}{3Lr^2}g_1 = L^2\left(\frac{1}{3} - \frac{1}{3\beta} + \frac{20}{L^2 r^2}\right). \tag{6.8.24}$$

This equation can be solved by quadratures which is carried out below. Similarly we substitute (6.8.15) into (6.8.23) and get an equation for h_2 alone:

$$h_2' + \frac{h_2}{r}\frac{3L^6 r^6 - 19L^4 r^4 - 7L^2 r^2 - 1}{\beta(3L^2 r^2 + 1)^2} - \frac{q_0 L^5 r^4}{(3L^2 r^2 + 1)^2}(3L^2 r^2 - 1) = 0 \tag{6.8.25}$$

where

$$q_0 = 72\pi G p_0. \tag{6.8.26}$$

This equation, too, can be solved by quadratures as folllows.
Equation (6.8.25) is of the form

$$h_2' + f(r)h_2 = g(r). \tag{6.8.27}$$

It is well known that the solution of this linear equation is given by

$$h_2(r) = e^{-F}\left(A + \int^r g(r')e^{F} dr'\right) \tag{6.8.28}$$

with

$$F(r) = \int^r f(r') dr' \tag{6.8.29}$$

and A is a constant of integration. According to (6.8.29) we must calculate the integral

$$F(r) = \frac{1}{18}\int dx \frac{3x^3 - 19x^2 - 7x - 1}{x(x+1)(x+1/3)}$$

where we have used the substitution

$$x = L^2 r^2, \quad \frac{dr}{r} = \frac{dx}{2x}. \tag{6.8.30}$$

After decomposition into partial fractions we can integrate:

$$F(r) = \frac{1}{18}\int\left(-\frac{9}{x} + \frac{36}{x+1} - \frac{24}{x+1/3} + \frac{1}{(x+1/3)^2}\right) dx$$

$$= -\frac{1}{2}\log x + 2\log(x+1) - \frac{4}{3}\log(x+1/3) - \frac{1}{18(x+1/3)}. \tag{6.8.31}$$

This gives

$$e^F = \frac{(x+1)^2}{\sqrt{x}(x+1/3)^{4/3}} \exp\left(-\frac{1}{18(x+1/3)}\right). \tag{6.8.32}$$

The remaining integral in (6.8.28) cannot be expressed in terms of elementary functions. Therefore we perform an expansion for $x \gg 1$. This is not bad because we shall see in the next section that in the present Universe we have $x = L^2 R^2 \approx 46.3$. One finds

$$e^F = x^{1/6}\left(1 + \frac{3}{2x} + O(x^{-2})\right). \tag{6.8.33}$$

The right side $g(r)$ in (6.8.27) is equal to

$$g(r) = q_0 L^5 r^4 \frac{3L^2 r^2 - 1}{(3L^2 r^2 + 1)^2}. \tag{6.8.34}$$

After expansion for large x and multiplying by (6.8.32) we can integrate

$$\int^r g e^F = \frac{q_0}{16} x^{8/3}\left(1 + \frac{4}{5x} + O(x^{-2})\right). \tag{6.8.35}$$

This finally gives

$$h_2(r) = \frac{q_0}{16} x^{17/6}\left(1 - \frac{7}{10x} + O(x^{-2})\right) + H_2 x^{1/6}\left(1 - \frac{3}{2x} + O(x^{-2})\right) \tag{6.8.36}$$

where H_2 is a constant of integration.

The equation (6.8.24) is solved in exactly the same way. We only give the results in the expanded form:

$$F = \frac{2}{x}\left(1 - \frac{7}{5x} + O(x^{-2})\right) \qquad (6.8.37)$$

$$\int^r g e^F = \frac{L}{10}\left(\log x - \frac{291}{5x} + O(x^{-2})\right) \qquad (6.8.38)$$

$$g_1(r) = \left(1 - \frac{2}{x} + O(x^{-2})\right)\left[G_1 + \frac{L}{10}\left(\log x - \frac{291}{5x} + O(x^{-2})\right)\right]. \qquad (6.8.39)$$

The leading terms in the present Universe are

$$Tg_1(R) = G_1 T + \frac{LT}{10}\left(\log(L^2 R^2) - \frac{291}{5L^2 R^2}\right). \qquad (6.8.40)$$

This must be dimensionless. Since L has dimension of an inverse length there is a factor c (light speed) in the second term when physical units are used. To compare this first order contribution with the zeroth order a_0 we need the fundamental constant L. To get this we must investigate the redshift - distance relation, this will be done in Sect.6.12.

6.9 Calculation of the energy-momentum tensor

After the metric functions have been calculated the remaining two Einstein's equations determine the first order contributions to the energy-momentum tensor. We recall our results for the energy-momentum tensor of normal matter in lowest order (6.7.13), (6.8.18),(6.8.20)

$$T^m_{00} = \varrho_m (u_0)^2 = \frac{1}{6\pi G t^2} + O(\alpha^{-1}) \qquad (6.9.1)$$

$$T^m_{01} = \varrho_m u_0 u_1 = \frac{1}{6\pi G t}\frac{L^2 r}{\beta} + O(\alpha^{-1}) \qquad (6.9.2)$$

$$T^m_{11} = \varrho_m (u_1)^2 = \frac{1}{6\pi G}\frac{L^4 r^2}{\beta^2} + O(\alpha^{-1}). \qquad (6.9.3)$$

This is of order α^0 only, therefore, the expansion in the inhomogeneous universe is mainly driven by the gravitational field in vacuum.

To determine the observable quantities we divide (6.9.3) by (6.9.2)

$$\frac{u_1}{u_0} = \frac{L^2 r t}{\beta}. \qquad (6.9.4)$$

The 4-velocity is normalized according to

$$g^{00}(u_0)^2 + 2g^{01}u_0 u_1 + g^{11}(u_1)^2 = 1 =$$

$$-\frac{a^2}{D}(u_0)^2 + 2\frac{b}{D}u_0 u_1 - \frac{1}{D}(u_1)^2. \tag{6.9.5}$$

Dividing this by $(u_0)^2$ and substituting (6.9.4) we get

$$\frac{1}{(u_0)^2} = \frac{a^2}{D} + 2\frac{b}{D}\frac{L^2 rt}{\beta} - \frac{1}{D}\frac{L^4 r^2 t^2}{\beta^2} =$$

$$= \frac{1}{\beta} + \frac{2}{\alpha}\frac{L^3 r^2}{\beta^2} t^{1/3} + O(\alpha^{-2}). \tag{6.9.6}$$

Multiplying by T_{00}^m we obtain the matter density

$$\varrho_m = \left(\frac{1}{6\pi G t^2} + O(\alpha^{-1})\right)\left(\frac{1}{\beta} + \frac{2}{\alpha}\frac{L^3 r^2}{\beta^2} t^{1/3}\right) =$$

$$= \frac{1}{6\pi G t^2 \beta} + O(\alpha^{-1}). \tag{6.9.7}$$

This differs from the Einstein-de Sitter value by the factor $1/\beta$.

Let us also compute the 4-velocity. Since we know ϱ_m we can calculate

$$(u_1)^2 = \frac{T_{11}^m}{\varrho_m} = \frac{L^4}{\beta} r^2 t^2$$

or

$$u_1 = \frac{L^2 rt}{\sqrt{\beta}} + O(\alpha^{-1}). \tag{6.9.8}$$

The zeroth component follows from (6.9.6)

$$u_0 = \sqrt{\beta} + O(\alpha^{-1}). \tag{6.9.9}$$

Now we turn to first order. We write

$$u_0 = u_{00} + u_{01}, \quad u_1 = u_{10} + u_{11} \tag{6.9.10}$$

where u_{00} and u_{10} are given by (6.9.8-9) and we use the normalization (6.9.5) again. In first order $O(\alpha^{-1})$ we obtain

$$0 = 2\frac{a_0^2}{D_0}u_{00}u_{01} + 2\frac{a_0 a_1}{D_0}(u_{00})^2 + O(\alpha^{-2}) +$$

$$+\left(\frac{1}{D}\right)_1 [a_0^2(u_{00})^2 + 2b_0 u_{00} u_{10} - (u_{10})^2] \tag{6.9.11}$$

because the second and third terms in (6.9.5) are one or two orders smaller than the first. The square bracket in (6.9.11) is equal to D_0 by zero order normalization. Therefore we arrive at

$$\frac{2}{\beta}u_{00}u_{01} + \frac{2}{\beta a_0}a_1(u_{00})^2 - \frac{2}{D_0}(a_0 a_1 + b_0 b_1) = 0. \tag{6.9.12}$$

This allows to calculate

274   Non-geometric general relativity

$$u_{01} = \frac{1}{\sqrt{\beta}a_0}(a_1 + Lrb_1) - \sqrt{\beta}\frac{a_1}{a_0}. \qquad (6.9.13)$$

Now we turn to G_{00} in first order. From the terms up to $O(\alpha^{-1})$ in (6.5.16) we find

$$(G_{00})_1 = \frac{1}{D_0^2}\Big[2\dot{a}_0\dot{a}_1(3a_0^2 + 5b_0^2) + \dot{a}_0^2(6a_0a_1 + 10b_0b_1) +$$

$$+2\dot{a}_1\dot{b}_0\frac{b_0^3}{a_0} + 2\dot{b}_1\dot{a}_0\frac{b_0^3}{a_0} + 3\dot{b}_1\dot{a}_0b_0\frac{b_0^2}{a_0} - a_1\frac{b_0^3}{a_0^2}\dot{a}_0\dot{b}_0\Big] -$$

$$-\frac{4}{D_0^2 a_0}\frac{a_1 + Lrb_1}{\beta}\Big[\dot{a}_0^2(3a_0^2 + 5b_0^2) + 2\dot{a}_0\dot{b}_0\frac{b_0^3}{a_0}\Big] - 4b_1\frac{\ddot{a}_0 b_0}{a_0 D_0} + 2\frac{a_1}{a_0^2}\frac{\ddot{a}_0 b_0^2}{D_0} +$$

$$+4\frac{\ddot{a}_0 b_0^2}{a_0^2 D_0}\frac{a_1 + Lrb_1}{\beta} + \frac{2}{rD_0^2}\Big(-\dot{b}_0 b_0^2 + 4\dot{a}_0 a_0 b_0 + 6\dot{a}_0\frac{b_0^3}{a_0}\Big) + 2\frac{\ddot{a}_0 a_0}{D_0^2}La_0. \qquad (6.9.14)$$

Here we have used again $a_0' = 0$ and $\ddot{a}_1 = 0$. This yields

$$(G_{00})_1 = \frac{1}{a_0}\Big[\frac{g_1}{t}\Big(\frac{1}{3} + \frac{1}{\beta} - \frac{1}{3\beta^2}\Big) + \frac{4}{3}\frac{g_2}{t}\frac{L^3 r^3}{\beta^2} +$$

$$+\Big(\frac{g_1}{t} + \frac{h_1}{t^2}\Big)\Big(\frac{8}{9\beta^2} - \frac{20}{9\beta} - \frac{4}{3}\Big) + Lr\Big(\frac{g_2}{t} + \frac{h_2}{t^2}\Big)\frac{8}{9}\Big(\frac{1}{\beta^2} - \frac{1}{\beta}\Big) + \frac{20L}{3\beta t}\Big] =$$

$$= 8\pi G\Big[\frac{3p_0}{a_0 t^2}(1 + \frac{4}{3}L^2 r^2) + \beta\varrho_{m1} + 2\sqrt{\beta}\frac{u_{01}}{6\pi G t^2 \beta}\Big]. \qquad (6.9.15)$$

We substitute u_{01} (6.9.13) and then the only unknown is the first order matter density ϱ_{m1}. Collecting the many terms we obtain

$$8\pi G\beta a_0 \varrho_{m1} = \frac{g_1}{t}\Big(\frac{5}{3} - \frac{35}{9\beta} + \frac{5}{9\beta^2}\Big) - \frac{g_2}{t}\frac{Lr}{9}\Big(\frac{20}{\beta} + \frac{4}{\beta^2}\Big) +$$

$$+\frac{h_1}{t^2}\Big(\frac{4}{3} - \frac{44}{9\beta} + \frac{8}{9\beta^2}\Big) + \frac{h_2}{t^2}\frac{Lr}{9}\Big(\frac{8}{\beta^2} - \frac{32}{9\beta}\Big) + \frac{20L}{3\beta t} - 8\pi G\frac{3p_0}{t^2}(1 + \frac{4}{3}L^2 r^2). \qquad (6.9.16)$$

There remains to calculate u_{11} which follows from the last Einstein's equation for G_{01}. We return to (6.8.16) and compute the first order. We use

$$\Big(\frac{1}{D}\Big)_1 = -\frac{2}{D_0^2}(a_0 a_1 + b_0 b_1) \qquad (6.9.17)$$

and

$$\Big(\frac{1}{D^2}\Big)_1 = -\frac{4}{D_0^3}(a_0 a_1 + b_0 b_1).$$

Inserting the zero order results and collecting the terms we find

$$\Big(\frac{1}{b}G_{01} + \frac{G_{22}}{a^2 r^2}\Big)_1 = \frac{1}{a_0^2}\Big\{-\frac{2}{Lr}\dot{a}_1' - \frac{\dot{b}_1'}{\beta} + \frac{a_1'}{t}\Big(\frac{4}{3}\frac{Lr}{\beta} + \frac{Lr}{\beta^2} + \frac{4}{3\beta^2 Lr}\Big) +$$

$$+\frac{b_1'}{t}\frac{2}{3\beta}+\frac{2L}{\beta^2}\dot{a}_1+\frac{\dot{b}_1}{r\beta}-\frac{a_1}{t}\left(\frac{20}{3}\frac{L}{\beta^2}+\frac{4}{3}\frac{L^3r^2}{\beta^2}\right)+$$

$$+\frac{b_1}{t}\left(\frac{2}{3r\beta^2}+\frac{10}{3}\frac{L^2r}{\beta^2}-\frac{8L}{\beta^2}\right)+L^2\left(\frac{1}{\beta^2}+\frac{1}{\beta}\right)\bigg\}=8\pi G\left(\frac{\varrho_m}{b}u_0u_1\right)_1. \quad (6.9.18)$$

Here the right-hand side is equal to

$$8\pi G\left[\varrho_{m1}\frac{u_{00}u_{10}}{b_0}-\varrho_{m0}\frac{b_1}{b_0^2}u_{00}u_{10}+\right.$$

$$\left.+\frac{\varrho_{m0}}{b_0}(u_{01}u_{10}+u_{00}u_{11})\right]. \quad (6.9.19)$$

Since everything except u_{11} is known here we can determine it. We see that it is of order α^{-1}. As a consequence all field equations are satisfied in first order.

We now want to put some numbers in for our present Universe. The most interesting quantity is the density of ordinary matter (6.9.7). The factor $\beta=L^2r^2+1$ in (6.9.7) enables us to fit any value of the matter density. Let us assume a "realistic" density of normal matter

$$\varrho_m=0.01\times\varrho_{\text{crit}}=\frac{1}{6\pi G\beta T^2} \quad (6.9.20)$$

with a critical density

$$\varrho_{\text{crit}}=1.878\times10^{-29}h^2\text{g/cm}^3 \quad (6.9.21)$$

and a Hubble constant $h=0.7$ in the usual unit km/(s Mpc). Taking an age $T=14\times10^9$ years of the Universe, this corresponds to a rather small value

$$\beta=44.3 \quad \text{or} \quad LR=6.6 \quad (6.9.21)$$

where R is the distance of the Milky Way from the origin $r=0$ where the Big Bang has taken place. To determine this distance R we must use some other observable to fix the integration constant L. This will be done below where we work out the redshift - distance relation for the inhomogeneous universe. But the small value of LR suggest already that we live not far away from $r=0$. The appearance of β in (6.8.7) can be traced back to the non-diagonal element $b(t,r)$ in the metric and to the radial motion of the matter. If one uses co-moving coordinates this motion is transformed away and then the matter density is a big problem. But we must be aware that (6.9.21) only is a lowest order orientation because first order may strongly alter the picture.

Next we consider the radial velocity v_m of the galaxies. We restrict to lowest order only. To determine v_m from $u^1=u^0v_m$ we need the components with upper indices. We find

$$u^0=g^{00}u_0+g^{01}u_1=\frac{1}{\sqrt{\beta}}+\frac{L^3r^2t}{a_0\beta^{3/2}}=$$

$$= \frac{1}{\sqrt{\beta}} + O(\alpha^{-1}). \tag{6.9.22}$$

and
$$u^1 = g^{11}u_1 + g^{10}u_0 = \frac{Lr}{a_0\sqrt{\beta}} - \frac{L^2rt}{a_0^2\beta^{3/2}} =$$

$$= \frac{Lr}{\alpha\sqrt{\beta}} t^{-2/3} + O(\alpha^{-2}). \tag{6.9.23}$$

At present time $t = T$, v_m is of the same order of magnitude as the local velocity of the Galaxy due to gravitational attraction from nearby galaxy clusters and therefore cannot be measured easily. To have simple numbers let us assume a radial velocity of 300 km/sec, so that $v_m = 0.001$ because the light speed is $c = 1$. Then

$$v_m = \frac{LR}{a(T)} \tag{6.9.24}$$

and using (6.9.21) we get

$$a(T) = \frac{LR}{v_m} = 6.6 \times 10^3 \gg 1. \tag{6.9.25}$$

This value of the spatial scale function is the relevant quantity in our perturbative scheme. Since $1/a(T) \ll 1$ this scheme is consistent in the late Universe. One should remember that we have discussed the lowest order results only, we expect considerable changes in the next order.

The radiation constant p_0 in (6.8.10) is directly related to the energy density of CMB, or to the temperature $T = 2.725$ K due to the Stephan-Boltzmann law. So the only constants of integration which are not known at present are L and R separately. Of course the redshift - distance relation will give further interesting information.

6.10 Null geodesics

The propagation of radiation in the cosmic gravitational field is described by the geodesic equation

$$\frac{dk^\mu}{dp} + \Gamma^\mu_{\alpha\beta}k^\alpha k^\beta = 0 \tag{6.10.1}$$

where

$$k^\mu = \frac{dx^\mu}{dp}, \quad k_\mu k^\mu = 0 \tag{6.10.2}$$

is the wave vector and p is the affine parameter along the ray. First we determine the radial null geodesics. From

$$ds^2 = dt^2 + 2bdtdr - a^2dr^2 = 0 \tag{6.10.3}$$

we get the quadratic equation

$$\left(\frac{dr}{dt}\right)^2 - 2\frac{b}{a^2}\frac{dr}{dt} - \frac{1}{a^2} = 0 \tag{6.10.4}$$

with the solution

$$\frac{dr}{dt} = \frac{1}{a^2}(b - \sqrt{D}) = \frac{1}{a}(Lr - \sqrt{\beta}) = \frac{k^1}{k^0}. \tag{6.10.5}$$

Here we have chosen the minus sign for the square root because we consider radiation propagating from large r towards $r = 0$. To find k^0 and k^1 separately we insert

$$k^1 = \frac{1}{a}(Lr - \sqrt{\beta})k^0 \tag{6.10.5}$$

into the geodesic equation

$$\frac{dk^0}{dp} + \frac{L^2 r^2}{\beta}\frac{\dot{a}}{a}(k^0)^2 - 2\frac{Lr}{\beta}\dot{a}k^0 k^1 + \frac{a\dot{a}}{\beta}(k^1)^2 = 0. \tag{6.10.6}$$

Here and in the following the Christoffel symbols are taken in leading order in α. The resulting equation is divided by k^0 and then easily integrated which yields

$$\log k^0 = -\log a + \text{const..} \tag{6.10.7}$$

This gives the desired radial null geodesics

$$k^0 = \frac{K}{a} \tag{6.10.8}$$

$$k^1 = \frac{K}{a^2}(Lr - \sqrt{\beta}) \tag{6.10.9}$$

where

$$\beta = L^2 r^2 + 1.$$

We now want to determine the null geodesics which start from a point on the z-axis under a certain angle and lie in a plane $\phi = $const. Then in addition to k^0, k^1 also k^2 is different from zero and the geodesic equations in leading order read as follows

$$\frac{dk^0}{dp} + \frac{L^2 r^2}{\beta}\frac{\dot{a}}{a}(k^0)^2 - 2\frac{Lr}{\beta}\dot{a}k^0 k^1 + \frac{a\dot{a}}{\beta}(k^1)^2 +$$

$$+ \frac{r^2}{\beta}a\dot{a}(k^2)^2 = 0 \tag{6.10.10}$$

$$\frac{dk^1}{dp} - \frac{Lr}{\beta}\frac{\dot{a}}{a^2}(k^0)^2 + 2\frac{\dot{a}}{\beta a}k^0 k^1 + \frac{Lr}{\beta}\dot{a}(k^1)^2 +$$

$$+ \frac{Lr^3}{\beta}\dot{a}(k^2)^2 = 0 \tag{6.10.11}$$

$$\frac{dk^2}{dp} + 2\frac{\dot{a}}{a}k^0 k^2 + \frac{2}{r}k^1 k^2 = 0 \qquad (6.10.12)$$

Equation (6.10.12) can immediately be integrated. We divide by k^2 and use
$$k^0 = \frac{dt}{dp}, \quad k^1 = \frac{dr}{dp}.$$

This leads to
$$\frac{d}{dp}\log k^2 + 2\frac{d}{dp}(\log a + \log r) = 0$$

and
$$k^2 = \frac{C}{(ar)^2} = \frac{d\vartheta}{dp} \qquad (6.10.13)$$

where C is another constant of integration which fixes the angle under which the trajectory starts at the z-axis. To solve the two remaining equations (6.10.10-11) we use the ansatz
$$k^0 = \frac{K}{a}, \quad k^1 = K\frac{Lr - \sqrt{\beta}}{a^2} + f_1(t,r). \qquad (6.10.14)$$

Indeed, it turns out that in leading order only k^1 (6.10.9) must be modified. From the normalization
$$g_{\mu\nu}k^\mu k^\nu = 0 \qquad (6.10.15)$$

we obtain a quadratic equation for f_1:
$$f_1^2 - 2\frac{K}{a^2}\sqrt{\beta}f_1 + \frac{C^2}{a^4 r^2} = 0 \qquad (6.10.16)$$

with the solutions
$$f_1 = \frac{K}{a^2}\left(\sqrt{\beta} \pm \sqrt{\beta - r_0^2/r^2}\right). \qquad (6.10.17)$$

Here we have introduced the integration constant
$$r_0 = \frac{C}{K} \qquad (6.10.18)$$

instead of C. Then the radial component is equal to
$$k^1 = \frac{K}{a^2}\left(Lr - \sqrt{\beta - \frac{r_0^2}{r^2}}\right). \qquad (6.10.19)$$

where we choose the minus sign to get a light path from large to small r. It is easy to verify that (6.10.14) and (6.10.19) satisfy the geodesic equation (6.10.10). They also satisfy (6.10.11) in leading order $O(a^{-3})$. Higher orders do not interest us because we have neglected them in the Christoffel symbols already.

We consider a trajectory starting on the z-axis at a point $r = r_1$ at time t_1 so that $\vartheta(t_1) = 0$. This is the place of the radiating galaxy. To find the light path we divide (6.10.19) by k^0

$$\frac{k^1}{k^0} = \frac{dr}{dt} = \frac{1}{a}\left(Lr - \sqrt{\beta - \frac{r_0^2}{2r^2}}\right) \quad (6.10.20)$$

and separate the variables:

$$\frac{dt}{a(t)} = \frac{dr}{Lr - \sqrt{L^2r^2 + 1 - r^2/r_0^2}}. \quad (6.10.21)$$

. In the r-integral

$$I(r) = \int \frac{r\,dr}{Lr^2 - \sqrt{L^2r^4 + r^2 - r_0^2}} \quad (6.10.22)$$

we use the substitution

$$\sqrt{L^2r^4 + r^2 - r_0^2} = s + Lr^2, \quad r^2 = \frac{s^2 + r_0^2}{1 - 2Ls}. \quad (6.10.23)$$

The resulting rational integral yields

$$I(r) = -Lr_0^2 \log|s| + \frac{1 + 4L^2 r_0^2}{8L^2(s - s_0)} + \left(\frac{1}{4L} + r_0^2\right)\log|s - s_0| \quad (6.10.24)$$

where

$$s_0 = \frac{1}{2L} \quad (6.10.25)$$

and $s(r)$ is given by (6.10.23).

We integrate (6.10.21) from an emission time t_1 to the present time T:

$$\int_{t_1}^{T} \frac{dt}{\alpha t^{2/3}} = \frac{3}{\alpha}(T^{1/3} - t_1^{1/3}) = I(R) - I(r_1). \quad (6.10.26)$$

Here r_1 and R are the radial coordinates of the emitter and observer, respectively, $r_1 > R$. Solving for $T = t$ and writing $R = r$ we obtain the first equation $t = t(r)$ of the light path

$$t = \left[t_1^{1/3} - \frac{\alpha}{3}\left(I(r) - I(r_1)\right)\right]^3. \quad (6.10.27)$$

For large times $t \to \infty$ we must have a diverging piece in $I(r)$. For $r_0 > 1/2L$ we find $s > 1/2L = s_0$ for all r so that there is no diverging piece in this case. Hence we must choose

$$r_0 < \frac{1}{2L} = s_0. \quad (6.10.28)$$

Then the first logarithm $\log|s|$ in (6.10.24) diverges for $r \to r_0$ because $s \to 0$. That means the trajectory approaches a limit circle $r = r_0$.

There remains the equation for the polar angle ϑ to be integrated. As in (6.10.20) it is best to determine $\vartheta(r)$ as a function of r instead of a function of t. So we divide (6.10.13) by (6.10.19) and obtain

$$\frac{k^2}{k^1} = \frac{d\vartheta}{dr} = \frac{r_0}{Lr^3 - r\sqrt{L^2r^4 + r^2 - r_0^2}} \equiv r_0 \frac{dJ(r)}{dr}. \qquad (6.10.29)$$

Using the same substitution (6.10.23) as above we get

$$J(r) = -\frac{1}{2L}\left[\frac{L}{s_0}\log|s| + \frac{L(s_0^2 - r_0^2) - s_0}{s_0(s_0^2 + r_0^2)}\log|s - s_0| +\right.$$

$$\left.+\frac{1 - 2Ls_0}{2(s_0^2 + r_0^2)}\log(s^2 + r_0^2) + \frac{s_0 + 2Lr_0^2}{r_0(s_0^2 + r_0^2)}\arctan\frac{s}{r_0}\right]. \qquad (6.10.30)$$

Since the initial value $\vartheta(r_1) = 0$ the polar angle is given as

$$\vartheta(r) = r_0[J(r) - J(r_1)]. \qquad (6.10.31)$$

Here for large times where $r \to r_0$, again the first logarithm gives $J(r) \to \infty$. This means that the trajectory spirals infinitely many times along the limit circle $r = r_0$ when it approaches it. A similar behavior is known from the critical null-geodesic of a Schwarzschild black hole.

6.11 The redshift

The electromagnetic radiation in the expanding universe is described by Maxwell's equations in the gravitational field. The homogeneous equations are

$$\nabla_\lambda F_{\mu\nu} + \nabla_\mu F_{\nu\lambda} + \nabla_\nu F_{\lambda\mu} = 0 \qquad (6.11.1)$$

where ∇ stands for the covariant derivative with respect to our inhomogeneous metric (6.5.1-5). In addition the field tensor $F^{\mu\nu}$ satisfies the inhomogeneous Maxwell's equation

$$\nabla_\mu F^{\mu\nu} = 0 \qquad (6.11.2)$$

away from the radiating galaxy. As in Minkowski vacuum, equation (6.11.1) implies the existence of a vector potential

$$F_{\mu\nu} = \nabla_\mu A_\nu - \nabla_\nu A_\mu \qquad (6.11.3)$$

and similarly for the upper indices. Here we can impose the "Lorentz" gauge condition

$$\nabla^\mu A_\mu = 0. \qquad (6.11.4)$$

As in classical electrodynamics we write (6.11.2) in terms of the vector potential

$$\nabla_\mu \nabla^\mu A^\nu - \nabla_\mu \nabla^\nu A^\mu = 0. \qquad (6.11.5)$$

Now we want to commute the derivatives in the second term and use (6.11.4). Since the covariant derivatives do not commute we obtain an additional term with the Ricci tensor

The redshift

$$\nabla_\mu \nabla^\mu A^\nu - R^\nu_\alpha A^\alpha = 0. \qquad (6.11.6)$$

In cosmology it is sufficient to solve this equation in the eikonal approximation by considering a solution of the form

$$A^\nu(x) = f^\nu(x) \sin[\varphi(x)].$$

Here f^ν is a slowly varying amplitude, φ is the phase which varies from 0 to 2π over the wave length of the radiation. Since the latter is extremely small compared to cosmic distances, the derivatives of φ are big so that we shall consider only the first two orders of φ in (6.11.6). The leading order $O(\varphi^2)$ gives

$$\nabla^\mu \varphi \nabla_\mu \varphi = 0 \qquad (6.11.7)$$

and in order $O(\varphi)$ we obtain

$$2\nabla^\mu f^\nu \nabla_\mu \varphi + f^\nu \nabla_\mu \nabla^\mu \varphi = 0. \qquad (6.11.8)$$

As in optics one introduces the wave vector

$$k^\mu = \nabla^\mu \varphi = \partial^\mu \varphi, \qquad (6.11.9)$$

the second equality follows because φ is a scalar field. Then the two equations to be solved read

$$k^\mu k_\mu = 0 \qquad (6.11.10)$$

$$k^\mu \nabla_\mu f^\nu = -\frac{1}{2} f^\nu \nabla_\mu k^\mu. \qquad (6.11.11)$$

From equation (6.11.10) we have

$$k^\mu \nabla_\nu k_\mu = 0, \qquad (6.11.12)$$

and since (6.11.9) implies

$$\nabla_\nu k_\mu = \nabla_\mu k_\nu$$

we arrive at

$$k^\mu \nabla_\mu k^\nu = k^\mu \left(\frac{\partial k^\nu}{\partial x^\mu} + \Gamma^\nu_{\mu\lambda} k^\lambda \right) = 0. \qquad (6.11.13)$$

Let now $x^\mu(p)$ be the curve

$$\frac{dx^\mu}{dp} = k^\mu, \qquad (6.11.14)$$

then we have

$$\frac{\partial k^\nu}{\partial x^\mu} = \frac{dk^\nu}{dp} \frac{dp}{dx^\mu} = \frac{dk^\nu}{dp} \frac{1}{k^\mu}$$

and (6.11.13) becomes

$$\frac{dk^\nu}{dp} + \Gamma^\nu_{\mu\lambda} k^\mu k^\lambda = 0, \qquad (6.$$

This is the geodesic equation (6.10.1) which we have solved in the last

282   Non-geometric general relativity

For an observer moving with the star with 4-velocity $dx^\mu/d\tau = u^\mu$ in the cosmic gravitational field the frequency of a spectral line is equal to

$$\omega = \frac{d\varphi}{d\tau} = \partial_\mu \varphi \frac{dx^\mu}{d\tau} = k_\mu u^\mu \qquad (6.11.16)$$

where τ is the proper time of the moving star. Comparing this with the frequency measured by an observer on earth we obtain the redshift z:

$$\frac{\omega_{em}}{\omega_{obs}} = 1 + z = \frac{(u_\mu k^\mu)_{em}}{(u_\mu k^\mu)_{obs}}. \qquad (6.11.17)$$

The radial velocity of the galaxy we know from equations (6.9.8-9):

$$u_0 = \sqrt{\beta}, \quad u_1 = \frac{L^2 rt}{\sqrt{\beta}}.$$

Then, since by (6.10.19)

$$\omega \equiv u_\mu k^\mu = K \frac{\sqrt{\beta}}{a(t)}\left[1 + \frac{L^2 rt}{\beta a(t)}\left(Lr - \sqrt{\beta - \frac{r_0^2}{r^2}}\right)\right] \qquad (6.11.18)$$

the redshift becomes

$$1 + z = \frac{\sqrt{\beta}}{\sqrt{\beta_R}} \frac{a(T)}{a(t)}\left[1 + \frac{L^2 rt}{\beta a(t)}\left(Lr - \sqrt{\beta - \frac{r_0^2}{r^2}}\right)\right] \times$$

$$\times \left[1 + \frac{L^2 RT}{\beta_R a(T)}\left(LR - \sqrt{\beta_R - \frac{r_0^2}{R^2}}\right)\right]^{-1}, \qquad (6.11.19)$$

where T is the present age of the Universe and $R > 0$ the radial coordinate of the observer. The emission time t follows from the coordinate distance $r = r_1$ of the galaxy by means of (6.10.27). For later use we introduce the expression

$$K_r = 1 + \frac{L^2 rt}{\beta a(t)}\left(Lr - \sqrt{\beta - \frac{r_0^2}{r^2}}\right). \qquad (6.11.20)$$

Then the redshift is equal to

$$1 + z = \frac{\sqrt{\beta}}{\sqrt{\beta_R}} \frac{a(T)}{a(t)} \frac{K_r}{K_R} \qquad (6.11.21)$$

where

$$\beta_R = L^2 R^2 + 1 \qquad (6.11.22)$$

according to (6.8.9).

6.12 Area and luminosity distances

In this section we must specify the radial coordinate R of the observer. To simplify the discussion we choose

$$R \approx r_0 \qquad (6.12.1)$$

close to the limit circle. This greatly simplifies the following calculations because we have

$$\left.\frac{dr}{dt}\right|_R \approx 0 \qquad (6.12.2)$$

$$K_R = 1 \qquad (6.12.3)$$

so that the redshift (6.11.21) becomes

$$1 + z = \frac{\sqrt{\beta}}{\sqrt{\beta_R}} \frac{a(T)}{a(t)} K_r. \qquad (6.12.4)$$

Theoretically the simplest radial distance is the area or angular-diameter distance

$$D_A = a(t) r(t) \qquad (6.12.5)$$

where $r(t)$ is the radial coordinate of the galaxy at the time of emission t which can be found by inverting (6.10.31). Assuming reciprocity D_A is related to the luminosity distance D_L which is measured by the astronomers by

$$D_L = (1+z)^2 D_A = a(t) r(t) (1+z)^2. \qquad (6.12.6)$$

Considering t as a parameter the equations (6.12.4) and (6.12.6) determine the distance-redshift relation $D_L = D_L(z)$ in parametric form.

We want to calculate the derivative

$$\left.\frac{dD_L}{dz}\right|_{z=0} = \frac{c}{H_0} \qquad (6.12.7)$$

which gives the Hubble constant H_0 at present time T (we have put the velocity of light c equal to 1). From (6.12.6) we obtain

$$\left.\frac{dD_L}{dz}\right|_{z=0} = \dot{a}(T) \left.\frac{dt}{dz}\right|_T R + 2a(T) R \qquad (6.12.8)$$

where (6.12.2) has been taken into account. Again using (6.12.2) we get from (6.12.4)

$$\left.\frac{dz}{dt}\right|_T = \left.\frac{\partial z}{\partial t}\right. = -\left.\frac{\dot{a}}{a}\right|_T + \partial_t K_r|_T. \qquad (6.12.9)$$

Since

$$\partial_t K_r|_T = 0$$

because $r(T) = R$ is near the limit circle, we get in leading order $O(\alpha)$

$$\left.\frac{dz}{dt}\right|_T = -\frac{2}{3T}. \qquad (6.12.10)$$

Using

$$\dot{a}(T) = \frac{2}{3}\frac{a}{T} \qquad (6.12.11)$$

we finally find

$$\left.\frac{dD_L}{dz}\right|_{z=0} = a(T)R. \qquad (6.12.12)$$

Now it is time to put some simple numbers in. Let

$$\varrho_{crit} = 1.878 \times 10^{-29} h^2 g/cm^3 \qquad (6.12.13)$$

be the critical density and we assume a Hubble constant

$$h = 0.7, \quad H_0 = 70\,\text{km}\,\text{s}^{-1}\text{Mpc}^{-1}. \qquad (6.12.14)$$

We use the empirical fact that the corresponding Hubble time

$$\frac{1}{H_0} = 14 \times 10^9 \text{years} = T \qquad (6.12.15)$$

coincides with the age T of the Universe. A "realistic" density of normal matter at present time T is

$$\varrho_m = 0.01 \times \varrho_{crit} = \frac{1}{6\pi G \beta_R T^2} \qquad (6.12.16)$$

where the last equality is the lowest order result eq.(6.9.7). As before this allows to determine

$$\beta_R = L^2 R^2 + 1 = 44.3 \qquad (6.12.17)$$

which gives

$$LR = 6.6. \qquad (6.12.18)$$

Next we assume a radial velocity of our Galaxy of 300 km/sec that means

$$v_m = 0.001 = \frac{LR}{a(T)} \qquad (6.12.19)$$

where the last equality comes from eq.(6.9.24). Using (6.12.18) this allows to determine

$$a(T) = 6.6 \times 10^3. \qquad (6.12.20)$$

Now we are able to calculate R from (6.12.12) using (6.12.7) and the Hubble time (6.12.15)

$$R = \frac{T}{a(T)} = 2.1 \times 10^6 \text{ly} = 640\,\text{kpc}. \qquad (6.12.21)$$

This is the distance between our Galaxy and the origin $r = 0$ where the Big Bang has taken place. It is even smaller than the distance of the Andromeda galaxy which is 2.5×10^6 light years away. From the dipole anisotropy of the CMB one has derived a net velocity of the local group of galaxies of 627 ± 22 km/sec in a direction between the Hydra and Centaurus clusters of galaxies (see the bibliographical notes). We expect that the origin $r = 0$ lies in the opposite direction. The velocity about 600 km/sec seems to be rather high. But this velocity is the sum of the cosmic radial velocity and a local velocity due to the attraction of the Hydra and Centaurus clusters. Since the latter has not been estimated we have made the simple assumption that 300 km/sec is the cosmic velocity (6.12.19). If we calculate the second derivative $d^2 D_L/dz^2$ in the same way we get zero. This shows that a lowest order calculation is not good enough to determine the acceleration parameter q_0.

The surprisingly small distance (6.12.21) must be taken with care, because the use of the area distance is not quite in the spirit of our non-geometric view of general relativity. A better treatment would be the direct calculation of the luminosity distance from the energy-momentum tensor of radiation. This will be done elsewhere (see arXiv).

6.13 The Riemann and Weyl tensors

As a reference for further investigations of the inhomogeneous universe we calculate the Riemann and Weyl tensors for our inhomogeneous metric (6.5.2-5) in this section.

The components of the Riemann tensor have simpler expressions than the Ricci tensor. They can be obtained from the formula

$$R_{\alpha\beta\mu\nu} = \frac{1}{2}(\partial_\alpha \partial_\nu g_{\beta\mu} - \partial_\beta \partial_\nu g_{\alpha\mu} + \partial_\mu \partial_\beta g_{\alpha\nu} - \partial_\mu \partial_\alpha g_{\beta\nu}) +$$
$$+ g_{\varrho\sigma}(\Gamma^\varrho_{\alpha\nu}\Gamma^\sigma_{\beta\mu} - \Gamma^\varrho_{\alpha\mu}\Gamma^\sigma_{\beta\nu}) \quad (6.13.1)$$

where the Christoffel symbols of Sect.6.5 have to be used. Up to permutation of the indices only the following components are different from zero:

$$R_{0101} = a\ddot{a} + \frac{1}{D}(\dot{a}^2 b^2 - \dot{a}\dot{b}ab - a'\dot{b}a - \dot{b}b'b) + \dot{b}'. \quad (6.13.2)$$

The terms are always ordered according to their importance in our perturbation theory.

$$R_{0202} = r^2\left[a\ddot{a} + \frac{1}{D}(-\dot{a}\dot{b}ab + a'\dot{b}a + \frac{a^2}{r}\dot{b})\right] \quad (6.13.3)$$

$$R_{0303} = R_{0202}\sin^2\vartheta$$

$$R_{1220} = r^2 a\left[\frac{1}{D}(-\dot{a}^2 ab + \dot{a}a'a - \frac{b^2}{r}\dot{a}) - a'\right] \quad (6.13.4)$$

$$R_{1212} = r^2 a \left\{ \frac{1}{D}\left[-\dot{a}^2 a^3 - \dot{a}b'a^2 - \frac{\dot{a}}{r}a^2 b + \frac{a'}{r}(a^2 + 2b^2) - a'^2 a - \right.\right.$$
$$\left.\left. - a'b'b - \frac{b'}{r}ab \right] + a'' \right\} \tag{6.13.5}$$

$$R_{1330} = R_{1220} \sin^2 \vartheta$$

$$R_{2323} = \frac{a^2 r^2}{D} \sin^2 \vartheta \left[-r^2 \dot{a}^2 a^2 - 2r\dot{a}ab - 2r^2 \dot{a}a'b + r^2 a'^2 + 2ra'a - b^2 \right]. \tag{6.13.6}$$

The Weyl tensor is obtained from the Riemann tensor by subtracting all possible contractions:

$$C_{\alpha\beta\mu\nu} = R_{\alpha\beta\mu\nu} - \frac{1}{2}\left(g_{\alpha\mu} R_{\beta\nu} - g_{\alpha\nu} R_{\beta\mu} + g_{\beta\nu} R_{\alpha\mu} - g_{\beta\mu} R_{\alpha\nu} + \right.$$
$$\left. + \frac{1}{3} R g_{\alpha\nu} g_{\beta\mu} - \frac{1}{3} R g_{\alpha\mu} g_{\beta\nu} \right) \tag{6.13.7}$$

where the Ricci tensor and R are given in Sect.6.5. As in Sect.6.5 this gives more complicated expressions:

$$C_{0101} = \frac{b'}{3} - \frac{2b}{3a}\dot{a}' + \frac{1}{3D}\left[-2a a' \dot{b} - b\dot{b}b' - a\dot{a}b' - \frac{b}{a}a'b' + \dot{a}a'\left(4b + 2\frac{b^3}{a^2}\right) + \right.$$
$$\left. + \frac{\dot{a}}{r}ab - \frac{\dot{b}}{r}a^2 \right] + \frac{a''}{3a} - \frac{a'^2}{3D}\left(\frac{b^2}{a^2} + 2\right) - \frac{1}{3rD}(a'a + b'b) + \frac{b^2}{3r^2 a^2} \tag{6.13.8}$$

$$C_{0202} = \frac{r^2}{6D}\left[2\dot{a}'ab - a''a - \frac{b^2}{r^2} + \frac{1}{D}\left(-a'\dot{b}a^3 + \dot{b}b'a^2 b + \dot{a}b'a^3 - 2\dot{a}a'(b^3 + 2a^2 b) + \right.\right.$$
$$\left.\left. + \frac{\dot{b}}{r}a^4 - \frac{\dot{a}}{r}a^3 b + a'^2(2a^2 + b^2) + a'b'ab + \frac{b'}{r}a^2 b + \frac{a'}{r}a^3 \right) \right] \tag{6.13.9}$$

$$C_{0303} = C_{0202} \sin^2 \vartheta$$

$$C_{1220} = \frac{r^2}{6D}\left[-2\dot{a}'ab^2 + \dot{b}'ab^2 + \frac{1}{D}\left(\dot{a}a'(4a^2 b^2 + 2b^4) - \dot{a}b'a^3 b - 2a'\dot{b}a^3 b - \right.\right.$$
$$\left.\left. - \dot{b}b'a^2 b^2 + \frac{\dot{a}}{r}a^3 b^2 - \frac{\dot{b}}{r}a^4 b \right) + a''ab + \frac{b^3}{r^2} - a'^2(b^3 + 2a^2 b) - \right.$$
$$\left. - a'b'ab^2 - \frac{a'}{r}a^3 b - \frac{b'}{r}a^2 b^2 \right] \tag{6.13.10}$$

$$C_{1330} = C_{1220} \sin^2 \vartheta$$

$$C_{2323} = \frac{r^4 a^2}{3D}\sin^2\vartheta \left[2\dot{a}'ab - \dot{b}'a^2 + \frac{1}{D}\left(2a'\dot{b}a^3 + \dot{a}b'a^3 - 2\dot{a}a'(2a^2 b + b^3) + \right.\right.$$
$$\left.\left. + \frac{\dot{b}}{r}a^4 - \frac{\dot{a}}{r}a^3 b + \dot{b}b'a^2 b + a'b'ab + a'^2(2a^2 + b^2) + \frac{a'}{r}a^3 + \right.\right.$$
$$\left.\left. + \frac{b'}{r}a^2 b \right) - a''a - b^2 \right]. \tag{6.13.11}$$

This tensor satisfies the conditions

$$g^{\alpha\mu} C_{\mu\beta\alpha\nu} = 0$$

for all β, ν.

Bibliographical Notes

We first list a few textbooks on gauge theories in chronological order:
J.C.Taylor, *Gauge theories of weak interactions, Cambridge 1976.*
C.Itzykson, J.-B. Zuber, *Quantum Field Theory, McGraw-Hill Inc. 1980.*
T.P.Cheng, L.F.Li, *Gauge theory of elementary particle physics, Oxford University Press 1984.*
M.Kaku, *Quantum field theory, Oxford University Press 1993.*
K.Huang, *Quantum field theory: from operators to path integrals, Wiley-Interscience 1998.*
S.Weinberg, *The quantum theory of fields, vol.I and II, Cambridge University Press 1996, vol.III 2000.*
G.Scharf, *Finite quantum electrodynamics - the causal approach, Dover Publications, INC, Mineola, NY 2014.* This book is referred to as FQED in the text.

All these books except the last one follow the traditional approach based on classical gauge theory which is then quantized by Feynman path integrals. Compared with our direct approach to the quantum theory this leads to considerable differences in concepts and methods: For example, in the traditional approach the classical Lagrangian must be given; the ghost fields play a minor and rather formal role in the Faddeev - Popov method; the mechanism of spontaneous symmetry breaking, i.e. the Higgs potential, is introduced by hand to get massive gauge fields and leptons. We have seen that all these features are consequences of perturbative gauge invariance. In comparison with other textbooks this one is an anti-Feynman book: no Feynman path integrals and no Feynman rules have appeared.

We now want to collect some important references related to the various chapters in the book. No attempt of completeness is made, the reader will find further references in the cited papers and the above books.

Chapter 1

Ordinary free fields are treated in any text on quantum field theory. This is not the case for the fermionic ghost fields which are typical for quantum gauge theories. We have followed our paper M.Dütsch, T.Hurth, G.Scharf, *Nuovo Cim. 108A (1994) 737*. The nilpotent gauge charge Q was introduced by T.Kugo, I.Ojima (*Progr. Theoret. Phys. Suppl. 66 (1979) 1*. Earlier work is due to G.Curci and R.Ferrari, *Nuov. Cimento 30 A, 155 (1975), 35 A, 273 (1976)*. The rigorous discussion of Wick polynomials in free fields was given by A.S.Wightman, L.Gårding, *Arkiv f. Fysik 28 (1964) 129*.

Chapter 2

The causal construction of time-ordered products goes back to E.C.G.Stückelberg (see E.C.G.Stückelberg, T.A.Green, *Helv. Phys. Acta 24 (1951) 153*) and N.N.Bogoliubov, D.V.Shirkov, *Introduction to the theory of quantized fields*, Wiley-Interscience 1959. It was further developed by H.Epstein and V.Glaser (*Ann. Inst. Poincaré A 19 (1973) 211*). R.Stora (*Differential algebras in Lagrangian field theory, ETH lectures 1993*) developed an alternative method based on extension of distributions instead of splitting. The problem of distribution splitting was already solved by the mathematician B.Malgrange (*Seminaires Schwartz 21 (1960)*). Quasi-asymptotics of distributions were introduced by V.S. Vladimirov, Y.N. Drozzinov, B.I. Zavialov, *Tauberian Theorems for Generalized Functions*, Kluwer Acad. Publ. 1988 and lead to the singular order ω in a natural way. Another simpler definition of the so-called scaling degree has been given by O.Steinmann, *Perturbation expansions in axiomatic field theory*, Lect. Notes in Physics 11, Springer-Verlag 1971, but there are cases were this definition cannot be applied.

The S-matrix $S(g)$ can also be used as starting point for the construction of interacting fields by functional differentiation with respect to some component in g. This goes back to Bogoliubov and Shirkov (*loc. cit.*) The method was recently extended to a construction of the local algebras of observables by M. Dütsch and K. Fredenhagen (*Comm. Math. Phys. 203, 71 (1999), hep-th/0001129*). In the last paper these authors also derive a quantum action principle first introduced by J.H. Lowenstein (*Comm. Math. Phys. 24, 1 (1971)*); another derivation was given by G. Pinter (*Annalen Phys. 10, 33 (2001), hep-th/9911063*), see also F.-M.Boas (*hep-th/0001014*). These results make contact with the method of algebraic renormalization (see O. Piguet, S.P. Sorella, *Algebraic Renormalization, Springer-Verlag (1995)*). The older BPHZ renormalization scheme (see W. Zimmermann, *Comm. Math. Phys. 15, 208 (1969)* and references therein) is closer to the causal method which was already demonstrated by Epstein and Glaser (*loc. cit*). The massless case in the BPHZ scheme has been analyzed by J.H. Lowenstein (*Comm. Math. Phys. 47, 53 (1976)*).– The whole theory was also developed on curved space-time by R.Brunetti and K.Fredenhagen (*Comm.Math. Phys. 208, 623 (2000), hep-th/9709011*). In this situation where the whole construction has to be carried out in x-space, the extension method of R.Stora mentioned above is appropriate.

Chapter 3

In the usual approach to non-Abelian gauge theories path integral quantization methods are used. The functional integral must be restricted to sections transversal to the gauge orbits by choosing a gauge fixing. This gives rise to the Faddeev-Popov determinant (L.D.Faddeev, V.N.Popov, *Phys. Lett. B25 (1967) 29*) which can be attributed to the ghost fields. The total Lagrangian is then invariant under the BRST transformation (C.Becchi, A.Rouet, R.Stora, *Comm. Math. Phys. 42 (1975) 127, Ann. Phys. 98 (1976) 287*, I.V.Tyutin, *Lebedev Institute preprint N39 (1975)*). Although the BRST transformation has some similarity to perturbative gauge invariance, there are the following essential distinctions: The BRST variation of the *total* Lagrangian is *zero*, but the gauge variation of the interaction T_1, only, is not zero but a *divergence*; the BRST transformation operates on the interacting fields whereas the gauge variation transforms free fields.

Perturbative gauge invariance was introduced in our paper M.Dütsch, T.Hurth, F.Krahe, G.Scharf, *Nuovo Cim. 106 A (1993) 1029*. The analysis of self-coupling of gauge fields is contained in A.Aste, G.Scharf, *Int. J. Mod. Phys. A14 (1999) 3421*, it has been investigated independently by R.Stora (private communication) and by D.R. Grigore, *Romanian J. Phys. 44, 853 (1999)*. Gauge invariance to all orders and unitarity are worked out in M.Dütsch, T.Hurth, G.Scharf, *Nuovo Cim. 108A (1994) 737*. The Slavnov-Taylor identities were first investigated by J.C.Taylor, *Nucl.Phys. B 33 (1971) 436*, and A.A.Slavnov, *Theor.Math.Phys. 10 (1972) 99*. It is interesting that Taylor has already introduced the Q-vertex without mentioning it. For the relation between Slavnov-Taylor and Cg-identities see M.Dütsch, *Int.J. Mod. Phys. A 12 (1997) 3205*. The discussion of the λ-gauges and gauge independence is taken from A.Aste, G.Scharf,M.Dütsch, *J. Phys. A31 (1998) 1563*. The problematic nature of the proof by functional integral methods has been demonstrated by X.S. Chen, W.M. Sun, F. Wang (*J.Phys. G 25 (1999) 2021*).

Chapter 4

Massive gauge fields are usually generated by the Higgs mechanism (P.W. Higgs, *Phys. Rev. 145 (1966) 1156*, F.Englert, R.Brout, *Phys. Rev. Lett. 13 (1964) 321*). One starts with the massless gauge fields, couples them to scalar fields and introduces a gauge-invariant potential which has a form that gives rise to spontaneous symmetry breaking. In this process the gauge boson acquire a mass.

We start directly with the massive, asymptotic gauge fields. The modification of the nilpotent gauge charge Q requires the introduction of scalar fields. The couplings are determined by perturbative gauge invariance. Second order gauge invariance requires an additional physical scalar (Higgs) field. The Higgs potential comes out from third order gauge invariance. This mechanism was demonstrated for the simple Abelian Higgs model in A.Aste, G.Scharf, M.Dütsch, *J. Phys. A30 (1997) 5785*. The electroweak theory is treated in M.Dütsch, G.Scharf, *Ann. Physik 8 (1999) 359, 389*. The complete couplings, including all ghost couplings, are derived there; it is not easy to find these in the literature. The discussion of general massive gauge theories is taken from G.Scharf, *Nuovo Cim. 112A (1999) 619*.; see also the papers of D.R. Grigore, *Journ. Phys. A33, 8443 (2000), A34, 5429 (2001) and A35, 1655 (2002)*. Axial anomalies have been discovered by S.Adler (*Phys. Rev. 177 (1969) 2426*) and J.S.Bell, R.Jackiw (*Nuovo Cim. 60A (1969) 47*). Its relevance for possible models of particle physics was pointed out by D.J.Gross, R.Jackiw (*Phys. Rev. D6 (1972) 477*).

It is possible to substitute perturbative gauge invariance by other assumptions in the construction of massive gauge theories. C.H.Llewellyn Smith (in *Proc. 5th Hawaii Topical Conf. Part. Phys.*, ed. P.N.Dobson et al., Univ. of Hawaii Press, Honolulu 1993), L.S.Bell (*Nucl. Phys. B 60, 427 (1973)*) and J.M.Conwall et al. (*Phys. Rev. D 10, 1145 (1974)*) have used bounds on the high-energy behavior of cross sections which follow from unitarity. In this way the introduction of the Z-boson and the Higgs boson was suggested. M.Dütsch and B.Schroer [*Journ. Phys. A33, 4317 (2000)*] worked with the requirement that the S-matrix induces a well-defined unitary operator on the physical Hilbert space. They show that this assumption determines the theory to the same extent as perturbative gauge invariance. But this method cannot be used for massless gauge fields. The cohomological aspect of gauge invariance was worked out in great detail by D.R. Grigore in *Rom. J. Phys. 55, 386 (2010), arXiv:0711.3986 (2007)*

Chapter 5

The early attempts to quantize general relativity are mainly due to B.S.De Witt (see *Phys. Rev. 162 (1967) 1195, 1239* and *Gen Rel. Grav. 1 (1970) 181*). The non-geometrical, field-theoretic route to gravitation was followed by several authors: S.N.Gupta (*Rev. Mod. Phys. 29 (1957) 334*), R.P.Feynman (*Acta Phys. Polon. 24 (1963) 697*), V.I.Ogivetsky, I.V.Polubarinov (*Ann. Phys. 35 (1965) 167*). The corresponding quantum theory was investigated by D.G.Boulware, S.Deser (*Ann. Phys. 89 (1975) 193*). Progress was blocked by the non-renormalizability, because in the traditional approach with ultraviolet regularization it is not clear how to obtain cutoff-independent finite results. This problem is relieved in the causal theory where we have only to fix finite normalization constants.

Quantization of the gravitational field has been treated by many authors. The most recent papers are D.R.Grigore, *Class. Quant. Grav. 19 (2000)*

319 and N.Grillo, *hep-th/9911118*. The result that the second order coupling determines the whole classical Einstein theory uniquely can be found in W.Wyss, *Helv. Phys. Acta 38 (1965) 469*. The convenience of Goldberg variables (J.N.Goldberg, *Phys. Rev. 111 (1958) 315*) is known to relativists (see L.D.Landau, E.M.Lifshitz, *Classical field theory, Berlin 1971, Sect.101*). The alternative simple expansion $g^{\mu\nu} = \eta^{\mu\nu} + \kappa h^{\mu\nu}$ corresponds to the de Donder gauge instead of the Hilbert gauge. It has also been used by several authors (e.g. M.H.Goroff, A.Sagnotti, *Nucl. Phys. B 266 (1986) 709*). The choice of the ghost coupling for quantum gravity at the end of Sect.5.4 is not unique, one can add divergences and co-boundaries. We have chosen the coupling suggested by T.Kugo, I.Ojima (*Nucl. Phys. B144 (1978) 234*) which has been often used in the literature. Second order gauge invariance of quantum gravity was first studied by I.Schorn (*Class. Quantum Grav. 14 (1997) 653, 671*). One-loop radiative corrections have been investigated by D.M.Capper, G.Leibbrandt, M.Medrano (*Phys. Rev. D8 (1973) 4320*) and G. 't Hooft, M.Veltman (*Ann. Poincaré Phys. Theor. A20 (1974) 69*). We have followed N.Grillo (*hep-th/9912114*). Two-loop quantum gravity was considered by Goroff, Sagnotti (loc. cit.) and by A.E. van de Ven (*Nucl. Phys. B 378 (1992) 309*). Quantum corrections to the Newtonian potential were calculated by J.F. Donoghue, (*Phys. Rev. D50 (1994) 3875*) and H.W.Hamber and S.Liu, (*Phys. Lett. B 357 (1995) 51*). For recent work on higher order quantum gravity see M.Reuter and H.Weyer, *Phys.Rev. D79: 105005 (2009)*, and references therein.

Massive gravity was developed by D.R.Grigore and G.Scharf, *Gen.Rel. Gravit. 37(6), 1075 (2005)*. The discussion of physical states is taken from J.B.Berchtold and G.Scharf, *Gen.Rel.Gravit. 39(9), 1489 (2007)*. The elegant method of descent or ladder equations for massive gravity was used in D.R.Grigore and G.Scharf, *Class.Quant.Grav. 25, 225008 (2008)*. An important earlier work on the descent equations is F.Brandt, N.Dragon, M.Kreuzer, *Nucl.Phys. B 340, 187 (1990)*, but they have used a different (vielbein) formulation. The implications to modify general relativity are discussed in G.Scharf, *Gen.Rel.Gravit. 42, 471 (2010)* and in D.R.Grigore and G.Scharf, *Gen. Rel. Grav. 43, 1323 (2011), arXiv:0912.1112 (2009)*. There exist many attempts to modify general relativity; the most recent are: P.D.Mannheim, *ApJ 391 (1992) 429, astro-ph/0505266v2 (2005)* and J.D.Bekenstein, *Contem.Phys. 47 (6) (2006) 387*. These approaches have classical field theory as the very basis and are motivated by the open dark matter problem. Our modification of general relativity is not contained in the vector-tensor theories previously investigated (see e.g. C.M.Will, *Theory and experiment in gravitational physics, Cambridge 1993*).

Chapter 6

This chapter is the result of various attempts to understand the dark matter problem. Since many years a standard idea is that some new particles predicted by supersymmetry are the constituents of the dark matter. Therefore we (the author and D.R. Grigore) have studied supersymmetric gauge theories for some years. In our strict understanding of quantum gauge theories as presented in this book we finally have found a no-go result (*No-go result for supersymmetric gauge theories in the causal approach*, Annalen der Physik (2008) 17, 864): A SUSY quantum gauge theory in this strict sense does not exist. Of course this result was not appreciated by other people which have a different understanding of gauge theory. Who has the right rules can only be decided by experiment. Until today the strongly hoped for SUSY particles did not appear.

Thereafter a revision of gravity theory itself seems to be the only option. A natural candidate for a modification of general relativity is the massless limit of the massive spin-2 gauge theory discussed in Chapter 5.4 (see also the bibliographical notes above).When investigating static spherically symmetric solutions of this theory I realized that the same kind of solutions with arbitrary rotation curves exists in ordinary general relativity, too. They have not been considered before because of the geometrical prejudice from the days of Einstein. Curiously enough the mild revision of general relativity was the result of a continues retreat from the revolutionary supersymmetric quantum gauge theory to a little unconventional general relativity.

Concerning bibliography a serious problem must be mentioned: the collaps of the peer-reviewed system. When one submits a paper with really new ideas to a first class journal one gets it back with unsuitable comments after about 10 days. Obviously due to the very many manuscripts the peers have given up and left the review job to their butlers. It seems as if those butlers try to prevent the publication of nonstandard research. I dare mentioning some famous journals where this might be the case: Physical Review D, Classical and Quantum Gravity, General Relativity and Gravitation. The way out I would propose is economical: The author pays a publication fee, but this money completely goes to the reviewer who then is obliged to discuss with the author on the appropriate scientific level.

The modification of general relativity mentioned above is discussed in G.Scharf,*Dark matter in galaxies according to the tensor-four-scalars theory*, Phys.Rev. D 84, 084045 (2011). All further publications about non-geometric general relativity are available in the arXiv and in the Research Gate. The beginning was the paper *Against geometry - non-standard general relativity*, arXiv 1208.3749 of 2012 and its later improvements. After hundred years of the geometric dogma the non-geometric approach to gravity theory is hard to except. Here the more pedagogic introduction in Chap.6 may be helpful. The most important experimental search for dark matter particles is done by several independent groups with different techniques. The main references are:

Xenon100 collaboration, Phys.Rev.Lett.109, 181301 (2012), arXiv 1207.5988 and LUX collaboration, Phys.Rev.Lett.112, 091303 (2014), arXiv 1310.8214. The China Dark Matter Experiment (CDEX) is described in the Phys.Rev. D 90, 091701 (2014), arXiv:1404.4946. The indirect probe of dark matter is by means of gravitational lensing. There is no doubt that the lensing signals can be understood on the basis of non-standard solutions of Einstein's equations without assuming dark matter, in a similar way as the rotation curve in Sect.6.4.

Inhomogeneous cosmology is less controversial. There exist various models in the literature. A rather complete overview can be obtained from the book by A. Krasiski, *Inhomogeneous Cosmological Models*, Cambridge 1997 and from the recent book by G.F.R. Ellis, R. Maartens and M.A.H. MacCallum, *Relativistic Cosmology*, Cambridge 2012. However the cosmic rest frame, although mentioned by many authors seems not to have been taken as the basis of cosmology. Experimentally the dipole anisotropy was measured by: (at least here the long list of authors must be given completely because otherwise there is too much honour to Dr. et al.) Kogut A., Lineweaver C., Smoot G.F., Bennett C.I., Banday A., Boggess N.W., Cheng E.S., De Amici G., Fixsen D.J., Hinshaw G., Jackson P.D., Janssen M., Keegstra P., Loewenstein K., Lubin P., Mather J.C., Tenorio L., Weiss R., Wilkinson D.T., and Wright E.L., Astrophys. J. 419, 1 1993. In sections 6.5-12 we have mainly followed our papers *Inhomogeneous cosmology in the cosmic rest frame without dark stuff*, arXiv 1312.2695 and *Redshift-distance relation in inhomogeneous cosmology*, arXiv 1503.05878.

Subject Index

The numbers in this index are equation numbers: the first number gives the chapter, the second the section and the third the relevant equation. This gives a more precise localization of the relevant places. After some change in the manuscript the index need not be changed, too.

Abelian Higgs model (4.1.17), (4.1.38)
Absorption operator (1.1.18), (1.1.42), (1.2.3), (1.3.6), (1.5.10), (1.10.8)
Adiabatic limit (2.2.1), (3.1.25)
 - switching (2.1.21), (2.2.1)
Adjoint (1.1.5), (1.1.43), (1.2.15)
 - Dirac field (1.8.12)
 - representation (3.5.5), (3.11.11)
 - spinor (1.8.12)
Advanced function (1.1.15), (2.2.53)
 - distribution (1.1.15), (2.2.53)
Advanced part (2.2.65)
Angular diameter distance (6.12.6)
Annihilation operators (1.1.18), (1.1.42), (1.2.3), (1.3.6), (1.5.10)
Anomaly (3.4.6), (5.7.1)
 - axial (4.9.6)
Anticommutation relations (1.8.15)
Antisymmetric tensor (5.2.2)
Area distance (6.12.5)
Asymptotic completeness (2.1.18)
Axial anomaly (4.9.6)

Big Bang (6.5.1)
Bilinear form (1.10.9)
Bose field (1.1.8)

Cabibbo-Kobayashi-Maskawa (CKM) matrix (4.8.45), (4.10.7)
Canoniacal commutation relation (1.1.6)
 - dimension (5.11.33)
Causal connection (2.1.27)
 - distribution (2.2.54), (2.3.1)
 - gauge invariance(3.1.15-18), (3.7.1), (5.1.12)
 - support (1.1.14)

Causality (2.1.27), (2.1.30), (2.2.23)
Central splitting solution (2.4.16), (2.6.37)
Cg-identities (3.7.6), (3.7.9), (3.7.40)
Charge (1.4.9)
 - conjugate spinor (1.8.18)
 - conjugation (1.8.18)
Charged scalar field (1.1.47)
Chiral fermions (4.7.2)
Christoffel symbols (5.5.6), (6.1.5), (6.2.2)
Chronological products (2.1.19), (2.2.67), (5.7.8)
Circular velocity (6.3.22)
CKM matrix (4.8.45), (4.10.7)
Co-boundary (1.4.32), (3.2.28), (3.3.22), (5.4.10)
Cohomology theory (1.4.32)
Commutation functions (1.1.13)
 - relations (1.1.7), (1.1.24), (1.7.11)
Completeness relations (2.5.24)
Confinement (5.10.22)
Conjugation (1.3.25), (1.5.9), (1.6.7), (3.8.7), (3.9.29)
Conserved current (3.1.14)
Contraction of tensors (5.2.1)
 - of field operators (1.9.9)
Cosmic rest frame (6.5.1)
Cosmological constant (5.6.1), (5.6.14)
Covariance (1.3.36)
Covariant (2.6.45)
 - derivative (4.4.71)
 - projection operators (1.8.14)
 - splitting solution (2.6.45)
CP-violation (4.10.17-18)
Creation operators (1.1.18), (1.1.38), (1.2.3), (1.3.6), (1.5.10)
Cross section (2.1.41), (2.5.35)
Current-current interaction (4.10.1)

Decay rate (4.10.12)
 - width (4.10.15)
De Donder gauge (5.14.1)
Descent equation (5.2.2)
Differentiation of tensors (5.2.1)
Differential cross section (2.1.41), (2.5.38)
Dimension canonical (5.11.33)
Dipole distribution (3.11.4)
Dirac adjoint (1.8.12)
 - equation (1.8.1), (4.8.2)
 - field (1.8.3)
 - spinor (1.8.3)
Dispersion integral (2.4.14), (2.4.16)
 - with subtractions (2.4.13)
Distribution operator-valued (1.1.4)
Distribution splitting (2.2.65), (2.3.30), (2.6.33), (2.6.55)
 - non-trivial (2.3.39), (2.4.16)
Distances (6.12.6)
Divergence (3.3.2), (5.2.13)

Subject Index

Divergences, infrared (3.3.1)
- ultraviolet (2.4.16)
- coupling (3.3.4), (5.4.3), (B.1)

Dyson series (1.1.15)

Eikonal approximation (6.11.7)
Einstein-de Sitter universe (6.7.13)
Einstein's equation (6.2.4)
Einstein tensor (6.4.5), (6.5.16)
Einstein-Hilbert Lagrangian (5.5.3), (5.11.2), (5.6.1)
Electromagnetic field (1.3.1), (1.3.45)
- classical (1.3.45)
- quantization (1.3.14)

Electric charge (4.7.7), (4.8.24)
Emission operator (1.1.18), (1.1.38), (1.2.3), (1.3.6), (1.5.10)
Energy (1.1.3)
- momentum conservation (2.5.31), (5.9.24), (5.11.23), (6.6.12)
- momentum tensor (2.5.5), (5.9.24), (5.9.33), (5.11.22), (6.6.8), (6.9.1)

Epstein-Glaser method (2.2.1)
Euler - Lagrange equation (5.5.9), (5.6.9)

Fermi constant (4.10.3)
- operators (1.2.1)
- theory (4.10.1)

Feynman diagrams (1.1.9)
- gauge (3.9.13)
- propagator (2.5.12)
- rules (2.5.12)

Fibration (1.3.46)
Fields Bose (1.1.8)
- free (1.1.1)
- ghost (1.2.1), (1.4.11), (1.6.2)
- Fermi (1.2.1), (1.6.2)
- scalar (1.1.2), (1.1.45), (1.2.1)
- spinor (1.8.1)
- tensor (1.7.2)
- vector (1.3.1), (1.6.2)

Fock representation (1.1.32)
- space (1.1.25), (1.1.31), (1.1.52), (1.2.13), (1.5.19)

Free fields (1.1.1)
Fundamental representation (3.6.20), (3.11.6)

Gamma matrices (1.8.2)
Gauge (1.3.46)
- charge (1.4.6), (1.4.12), (1.5.31), (1.5.42), (1.7.8), (3.10.1), (4.2.6), (5.3.10)
- field (1.2.1), (1.5.33), (2.6.1), (3.11.1)
- independence (3.9.1), (3.10.20)
- invariance (3.1.25)
- invariance, perturbative (3.1.0), (3.1.18), (3.7.1)
- potentials (6.2.9)
- transformation, classical (1.3.46), (1.7.5)

- transformation, operator (1.4.1), (1.4.33)
- variation (1.4.29), (1.5.36), (1.7.12), (4.3.2), (5.1.8), (5.3.8)

Geodesic equation (6.1.4), (6.2.2), (6.3.4)
Ghost fields (1.1.1), (1.2.1), (1.5.33), (2.6.1), (3.11.1)
Goldberg variables (5.5.15), (5.9.26)
Gravitational force (2.5.6)
Graviton (1.7.49), (5.3.44)

Hamiltonian (1.3.21)
Helicity (1.7.46)
Higgs boson (4.0.1)
- couplings (4.3.24), (4.6.10)
- field (4.0.1), (4.1.1)
- mechanism (4.0.1), (4.5.16)
- potential (4.1.34), (4.2.7), (4.5.15), (4.6.11)

Hilbert gauge (1.7.3), (5.5.30), (5.6.11)
Hilbert space (1.3.11)
Hubble constant (6.12.7)

Indefinite metric (1.3.28)
Infrared divergence (5.10.19)
Inhomogeneous Lorentz transformation (1.3.35)
Inverse S-matrix (2.1.22), (2.2.4), (3.8.13)
Irreducible (1.4.5)

Jacobi identity (3.4.17), (3.5.3), (4.4.6)
Jordan-Pauli distribution (1.1.13), (1.10.1), (2.3.21)

Killing form (3.5.10)
Klein - Gordon equation (1.1.1), (1.5.1), (1.5.32)
Krein operator (1.3.27), (1.5.21)

Leptonic coupling (4.7.20)
Levi-Cività connection (6.1.13), (6.2.8)
Lie algebra (3.5.1)
- simple, semisimple (3.5.10)

Lifting (1.3.36)
Light cone (2.2.48)
- path (6.11.31)

Local distributions (2.3.46), (3.4.6)
Loop graphs (2.6.7), (5.10.3)
- expansion (2.7.12)

Lorentz condition (1.3.2), (1.3.30), (1.5.2), (6.2.9), (6.11.4)
- covariance (1.3.7), (1.3.13), (1.3.37), (2.2.19)
- force (6.2.1)
- group (1.1.32), (1.3.35)
- group, proper (1.1.32)
- transformations (1.1.32)
- transformations, special (0.1.13), (1.1.19), (1.4.49)

- transformations, inhomogeneous (1.1.33)
- tensor (1.7.2)
Luminosity distance (6.12.6)

Matter coupling (3.6.1)
Majorana field (1.8.20)
Mass density (6.9.7), (6.9.20), (6.12.16)
Mass scale (5.10.22)
Massive gravity (5.4.9)
- vector field (1.5.1)
- tensor field (5.3.1)
Maxwell's equations (6.2.3), (6.11.1)
Metric, Minkowski (1.1.1)
Metric tensor (6.1.7), (6.2.9), (6.5.2)
Minkowski metric (1.1.1)
- scalar product (1.1.33)

Newton's constant (2.5.45), (5.5.3)
- potential (2.5.44), (5.10.16)
Nilpotent (1.4.9), (1.4.31), (1.5.31)
Non - abelian gauge theories (3.2.1)
Non - normalisable (2.7.1), (5.1.17)
Norm (1.3.11)
Normal order (1.9.1), (1.9.4), (3.1.7), (3.2.4), (3.4.7)
Normalisability (2.7.1)
Normalisation conditions (3.1.18)
- point (2.4.12), (2.6.41)
- term (3.4.28)
Null geodesic (6.10.2), (6.10.8), (6.10.21)

One - particle operator (1.1.25)
- sector (1.1.26)
Operator - valued distribution (1.1.4)
Operator character (1.10.8)
- products (1.9.1)
Order, singular (2.3.16)

Parity violation (4.7.20)
Particle number operator (1.3.22)
Pauli - Jordan distribution (1.1.13)
Perturbation series (2.1.15), (2.2.1)
- first order (3.2.1)
Perturbative gauge invariance (3.1.0), (3.1.18)
Photon (1.3.1), (1.3.45)
Photon, transversality (1.3.33)
Physical subspace (1.3.33), (1.4.21), (1.4.26), (1.5.44), (1.7.39),
- (3.8.14), (3.9.43), (3.10.12), (5.3.44)
- equivalence (3.3.1)
- part of vector field (1.5.45), (5.11.18)
Poincaré covariance (1.3.37), (1.5.23)

- group (1.1.35)
- lemma (5.2.2)
- transformation (1.5.26)

Polarizations (1.5.3)
Polarization vectors (1.5.3), (1.5.45)
- tensors (1.7.45)

Power counting function (2.3.16)
Probability (2.5.31), (4.10.9)
Proliferation of graviton couplings (5.5.23), (5.7.47)
Projections (1.8.7)
Propagator (2.5.12), (5.10.12)
Proper Lorentz group (1.1.33)
- transformations (1.1.32)
Proper time (6.1.2-6)
Pseudo - unitarity (1.3.38), (1.5.24), (3.8.13)

Quantization (1.1.6)
Quantum chromodynamics (QCD) (3.6.8)
Quantum electrodynamics (3.1.12)
- massive (4.1.1)
Quasi - asymptotics (2.3.6)
Q - vertex (3.1.16), (3.2.57), (3.6.9), (4.3.17), (5.2.11), (5.11.16), (5.8.18)

Radiation field (1.3.1)
- gauge (1.3.45)
Radiative corrections (5.10.1)
Redshift (6.3.1), (6.11.17)
Reductive Lie algebra (3.5.1)
Regularisation (2.4.12)
Regularly varying functions (2.3.20)
Renormalisable theories (2.7.1)
Renormalization (3.7.32)
Representation (1.1.25)
Rest system (4.10.4)
Retarded distribution (1.1.15), (2.2.52), (2.4.14), (2.4.19)
- function (2.1.33)
Retarded part (2.2.52)
Ricci tensor (5.5.5), (6.2.5)
Rotation curve (6.4.26)

Salam - Weinberg model (4.6.7)
Scalar field (1.1.1), (1.1.45)
- field, charged (1.1.47)
- product (1.3.28)
Scaling limit (2.3.6)
Scaling transformation (2.3.14)
Scattering amplitude (2.1.40), (2.5.39)
- cross section (2.1.41), (2.5.35)
- matrix (S-matrix) (2.1.7), (2.2.1)
- operator (2.1.7)
- theory (2.1.7)

Schrödinger equation (2.1.9)
Schwarzschild metric (6.3.23), (6.4.25)
Second quantization (1.1.6)
Self - conjugate (1.3.27), (1.6.6)
 - coupling (3.2.1), (4.1.17)
Self - energy, gravitational (5.10.3)
Singular order (2.3.16), (2.7.1), (5.1.17)
Singularities, infrared (5.10.22)
Slash (1.8.9)
Slavnov - Taylor identities (3.7.40)
S-matrix (2.1.7), (2.1.18)
 - element (2.5.16)
Spin (1.7.2)
Spin - 2 gauge theory (1.7.2), (5.1.14)
Spinor field (1.8.3)
 - representation (1.7.1)
Spin - statistics theorem (1.2.14)
Splitting of distributions (2.3.30), (2.3.44)
 - trivial (2.3.30)
Stability (5.10.20)
Structure constants (2.6.3), (3.2.28)
SU(N) Yang-Mills theories (3.11.6)
Subtractions (2.4.13)
Super-normalizable theories (5.1.19)
Switching function (2.1.21), (2.2.1)

Tensor field (1.7.2), (5.3.1)
 - product (1.1.29), (1.1.52)
Time evolution (2.1.1)
 - ordered exponential (2.1.20)
 - ordered product (2.1.19), (2.2.30), (2.2.67), (2.2.71), (5.7.8)
 - ordering (2.1.30)
Time - like (2.2.48)
 - reversal (4.10.8)
T - product, n-point function (2.2.30), (2.2.67), (2.2.71)
Transition probability (2.1.8), (2.5.23)
Translation invariance (2.2.16), (2.3.5)
Transversal polarisations (1.5.45), (3.9.42)
Transverse modes (3.9.43)
Tree graph (2.5.12)
Triangular anomalies (4.9.6)

Ultraviolet divergences (2.4.16)
Unitarity (1.3.54), (1.5.30), (2.2.14)
 - physical (3.8.28)
 - pseudo (4.7.20)
Unitary group (3.5.11)
 - propagator (2.1.4)
 - representation (1.5.24)
 - time evolution (2.1.1), (2.1.4)
Universe, inhomogeneous (6.7.1)

Vacuum (1.1.25)
 - expectation value (1.1.46), (1.9.33)
 - graphs (5.10.17)
 - polarization (3.11.5)
 - stability (5.10.20)
Vector field, massless (1.3.1)
 - field, massive (1.5.1), (1.5.45), (5.11.18)
 - graviton field (5.3.7)
 - potential (6.2.7), (6.11.3)
Velocity, radial (6.9.8), (6.9.24), (6.12.19)
Vertex (1.9.9)

Ward - Takahashi identities (3.1.22)
Wave equation (1.3.1)
 - operator (1.1.1)
 - vector (6.10.2), (6.11.9)
W - boson (4.6.5), (4.10.4)
Weak mixing angle (4.6.6)
Weinberg - Salam model (4.6.7)
Wick monomial (1.9.20), (3.1.7)
Wick ordering, product (1.9.1), (1.9.4), (3.1.7), (3.2.4), (3.4.7)
Wick's theorem (1.9.8), (1.9.34), (2.8.1)
Width (4.10.12)

Yang - Mills coupling (2.6.1), (3.2.52), (5.11.1)

Z - boson (4.10.16)

A CATALOG OF SELECTED
DOVER BOOKS
IN SCIENCE AND MATHEMATICS

Physics

THEORETICAL NUCLEAR PHYSICS, John M. Blatt and Victor F. Weisskopf. An uncommonly clear and cogent investigation and correlation of key aspects of theoretical nuclear physics by leading experts: the nucleus, nuclear forces, nuclear spectroscopy, two-, three- and four-body problems, nuclear reactions, beta-decay and nuclear shell structure. 896pp. 5 3/8 x 8 1/2. 0-486-66827-4

QUANTUM THEORY, David Bohm. This advanced undergraduate-level text presents the quantum theory in terms of qualitative and imaginative concepts, followed by specific applications worked out in mathematical detail. 655pp. 5 3/8 x 8 1/2. 0-486-65969-0

ATOMIC PHYSICS AND HUMAN KNOWLEDGE, Niels Bohr. Articles and speeches by the Nobel Prize–winning physicist, dating from 1934 to 1958, offer philosophical explorations of the relevance of atomic physics to many areas of human endeavor. 1961 edition. 112pp. 5 3/8 x 8 1/2. 0-486-47928-5

COSMOLOGY, Hermann Bondi. A co-developer of the steady-state theory explores his conception of the expanding universe. This historic book was among the first to present cosmology as a separate branch of physics. 1961 edition. 192pp. 5 3/8 x 8 1/2. 0-486-47483-6

LECTURES ON QUANTUM MECHANICS, Paul A. M. Dirac. Four concise, brilliant lectures on mathematical methods in quantum mechanics from Nobel Prize-winning quantum pioneer build on idea of visualizing quantum theory through the use of classical mechanics. 96pp. 5 3/8 x 8 1/2. 0-486-41713-1

THE PRINCIPLE OF RELATIVITY, Albert Einstein and Frances A. Davis. Eleven papers that forged the general and special theories of relativity include seven papers by Einstein, two by Lorentz, and one each by Minkowski and Weyl. 1923 edition. 240pp. 5 3/8 x 8 1/2. 0-486-60081-5

PHYSICS OF WAVES, William C. Elmore and Mark A. Heald. Ideal as a classroom text or for individual study, this unique one-volume overview of classical wave theory covers wave phenomena of acoustics, optics, electromagnetic radiations, and more. 477pp. 5 3/8 x 8 1/2. 0-486-64926-1

THERMODYNAMICS, Enrico Fermi. In this classic of modern science, the Nobel Laureate presents a clear treatment of systems, the First and Second Laws of Thermodynamics, entropy, thermodynamic potentials, and much more. Calculus required. 160pp. 5 3/8 x 8 1/2. 0-486-60361-X

QUANTUM THEORY OF MANY-PARTICLE SYSTEMS, Alexander L. Fetter and John Dirk Walecka. Self-contained treatment of nonrelativistic many-particle systems discusses both formalism and applications in terms of ground-state (zero-temperature) formalism, finite-temperature formalism, canonical transformations, and applications to physical systems. 1971 edition. 640pp. 5 3/8 x 8 1/2. 0-486-42827-3

QUANTUM MECHANICS AND PATH INTEGRALS: Emended Edition, Richard P. Feynman and Albert R. Hibbs. Emended by Daniel F. Styer. The Nobel Prize–winning physicist presents unique insights into his theory and its applications. Feynman starts with fundamentals and advances to the perturbation method, quantum electrodynamics, and statistical mechanics. 1965 edition, emended in 2005. 384pp. 6 1/8 x 9 1/4. 0-486-47722-3

Browse over 9,000 books at www.doverpublications.com

CATALOG OF DOVER BOOKS

Physics

INTRODUCTION TO MODERN OPTICS, Grant R. Fowles. A complete basic undergraduate course in modern optics for students in physics, technology, and engineering. The first half deals with classical physical optics; the second, quantum nature of light. Solutions. 336pp. 5 3/8 x 8 1/2. 0-486-65957-7

THE QUANTUM THEORY OF RADIATION: Third Edition, W. Heitler. The first comprehensive treatment of quantum physics in any language, this classic introduction to basic theory remains highly recommended and widely used, both as a text and as a reference. 1954 edition. 464pp. 5 3/8 x 8 1/2. 0-486-64558-4

QUANTUM FIELD THEORY, Claude Itzykson and Jean-Bernard Zuber. This comprehensive text begins with the standard quantization of electrodynamics and perturbative renormalization, advancing to functional methods, relativistic bound states, broken symmetries, nonabelian gauge fields, and asymptotic behavior. 1980 edition. 752pp. 6 1/2 x 9 1/4. 0-486-44568-2

FOUNDATIONS OF POTENTIAL THERY, Oliver D. Kellogg. Introduction to fundamentals of potential functions covers the force of gravity, fields of force, potentials, harmonic functions, electric images and Green's function, sequences of harmonic functions, fundamental existence theorems, and much more. 400pp. 5 3/8 x 8 1/2. 0-486-60144-7

FUNDAMENTALS OF MATHEMATICAL PHYSICS, Edgar A. Kraut. Indispensable for students of modern physics, this text provides the necessary background in mathematics to study the concepts of electromagnetic theory and quantum mechanics. 1967 edition. 480pp. 6 1/2 x 9 1/4. 0-486-45809-1

GEOMETRY AND LIGHT: The Science of Invisibility, Ulf Leonhardt and Thomas Philbin. Suitable for advanced undergraduate and graduate students of engineering, physics, and mathematics and scientific researchers of all types, this is the first authoritative text on invisibility and the science behind it. More than 100 full-color illustrations, plus exercises with solutions. 2010 edition. 288pp. 7 x 9 1/4. 0-486-47693-6

QUANTUM MECHANICS: New Approaches to Selected Topics, Harry J. Lipkin. Acclaimed as "excellent" (*Nature*) and "very original and refreshing" (*Physics Today*), these studies examine the Mössbauer effect, many-body quantum mechanics, scattering theory, Feynman diagrams, and relativistic quantum mechanics. 1973 edition. 480pp. 5 3/8 x 8 1/2. 0-486-45893-8

THEORY OF HEAT, James Clerk Maxwell. This classic sets forth the fundamentals of thermodynamics and kinetic theory simply enough to be understood by beginners, yet with enough subtlety to appeal to more advanced readers, too. 352pp. 5 3/8 x 8 1/2. 0-486-41735-2

QUANTUM MECHANICS, Albert Messiah. Subjects include formalism and its interpretation, analysis of simple systems, symmetries and invariance, methods of approximation, elements of relativistic quantum mechanics, much more. "Strongly recommended." – *American Journal of Physics.* 1152pp. 5 3/8 x 8 1/2. 0-486-40924-4

RELATIVISTIC QUANTUM FIELDS, Charles Nash. This graduate-level text contains techniques for performing calculations in quantum field theory. It focuses chiefly on the dimensional method and the renormalization group methods. Additional topics include functional integration and differentiation. 1978 edition. 240pp. 5 3/8 x 8 1/2. 0-486-47752-5

Browse over 9,000 books at www.doverpublications.com

CATALOG OF DOVER BOOKS

Physics

MATHEMATICAL TOOLS FOR PHYSICS, James Nearing. Encouraging students' development of intuition, this original work begins with a review of basic mathematics and advances to infinite series, complex algebra, differential equations, Fourier series, and more. 2010 edition. 496pp. 6 1/8 x 9 1/4. 0-486-48212-X

TREATISE ON THERMODYNAMICS, Max Planck. Great classic, still one of the best introductions to thermodynamics. Fundamentals, first and second principles of thermodynamics, applications to special states of equilibrium, more. Numerous worked examples. 1917 edition. 297pp. 5 3/8 x 8. 0-486-66371-X

AN INTRODUCTION TO RELATIVISTIC QUANTUM FIELD THEORY, Silvan S. Schweber. Complete, systematic, and self-contained, this text introduces modern quantum field theory. "Combines thorough knowledge with a high degree of didactic ability and a delightful style." – *Mathematical Reviews*. 1961 edition. 928pp. 5 3/8 x 8 1/2. 0-486-44228-4

THE ELECTROMAGNETIC FIELD, Albert Shadowitz. Comprehensive undergraduate text covers basics of electric and magnetic fields, building up to electromagnetic theory. Related topics include relativity theory. Over 900 problems, some with solutions. 1975 edition. 768pp. 5 5/8 x 8 1/4. 0-486-65660-8

THE PRINCIPLES OF STATISTICAL MECHANICS, Richard C. Tolman. Definitive treatise offers a concise exposition of classical statistical mechanics and a thorough elucidation of quantum statistical mechanics, plus applications of statistical mechanics to thermodynamic behavior. 1930 edition. 704pp. 5 5/8 x 8 1/4. 0-486-63896-0

INTRODUCTION TO THE PHYSICS OF FLUIDS AND SOLIDS, James S. Trefil. This interesting, informative survey by a well-known science author ranges from classical physics and geophysical topics, from the rings of Saturn and the rotation of the galaxy to underground nuclear tests. 1975 edition. 320pp. 5 3/8 x 8 1/2. 0-486-47437-2

STATISTICAL PHYSICS, Gregory H. Wannier. Classic text combines thermodynamics, statistical mechanics, and kinetic theory in one unified presentation. Topics include equilibrium statistics of special systems, kinetic theory, transport coefficients, and fluctuations. Problems with solutions. 1966 edition. 532pp. 5 3/8 x 8 1/2. 0-486-65401-X

SPACE, TIME, MATTER, Hermann Weyl. Excellent introduction probes deeply into Euclidean space, Riemann's space, Einstein's general relativity, gravitational waves and energy, and laws of conservation. "A classic of physics." – *British Journal for Philosophy and Science*. 330pp. 5 3/8 x 8 1/2. 0-486-60267-2

RANDOM VIBRATIONS: Theory and Practice, Paul H. Wirsching, Thomas L. Paez and Keith Ortiz. Comprehensive text and reference covers topics in probability, statistics, and random processes, plus methods for analyzing and controlling random vibrations. Suitable for graduate students and mechanical, structural, and aerospace engineers. 1995 edition. 464pp. 5 3/8 x 8 1/2. 0-486-45015-5

PHYSICS OF SHOCK WAVES AND HIGH-TEMPERATURE HYDRO DYNAMIC PHENOMENA, Ya B. Zel'dovich and Yu P. Raizer. Physical, chemical processes in gases at high temperatures are focus of outstanding text, which combines material from gas dynamics, shock-wave theory, thermodynamics and statistical physics, other fields. 284 illustrations. 1966–1967 edition. 944pp. 6 1/8 x 9 1/4. 0-486-42002-7

Browse over 9,000 books at www.doverpublications.com

CATALOG OF DOVER BOOKS

Engineering

FUNDAMENTALS OF ASTRODYNAMICS, Roger R. Bate, Donald D. Mueller, and Jerry E. White. Teaching text developed by U.S. Air Force Academy develops the basic two-body and n-body equations of motion; orbit determination; classical orbital elements, coordinate transformations; differential correction; more. 1971 edition. 455pp. 5 3/8 x 8 1/2. 0-486-60061-0

INTRODUCTION TO CONTINUUM MECHANICS FOR ENGINEERS: Revised Edition, Ray M. Bowen. This self-contained text introduces classical continuum models within a modern framework. Its numerous exercises illustrate the governing principles, linearizations, and other approximations that constitute classical continuum models. 2007 edition. 320pp. 6 1/8 x 9 1/4. 0-486-47460-7

ENGINEERING MECHANICS FOR STRUCTURES, Louis L. Bucciarelli. This text explores the mechanics of solids and statics as well as the strength of materials and elasticity theory. Its many design exercises encourage creative initiative and systems thinking. 2009 edition. 320pp. 6 1/8 x 9 1/4. 0-486-46855-0

FEEDBACK CONTROL THEORY, John C. Doyle, Bruce A. Francis and Allen R. Tannenbaum. This excellent introduction to feedback control system design offers a theoretical approach that captures the essential issues and can be applied to a wide range of practical problems. 1992 edition. 224pp. 6 1/2 x 9 1/4. 0-486-46933-6

THE FORCES OF MATTER, Michael Faraday. These lectures by a famous inventor offer an easy-to-understand introduction to the interactions of the universe's physical forces. Six essays explore gravitation, cohesion, chemical affinity, heat, magnetism, and electricity. 1993 edition. 96pp. 5 3/8 x 8 1/2. 0-486-47482-8

DYNAMICS, Lawrence E. Goodman and William H. Warner. Beginning engineering text introduces calculus of vectors, particle motion, dynamics of particle systems and plane rigid bodies, technical applications in plane motions, and more. Exercises and answers in every chapter. 619pp. 5 3/8 x 8 1/2. 0-486-42006-X

ADAPTIVE FILTERING PREDICTION AND CONTROL, Graham C. Goodwin and Kwai Sang Sin. This unified survey focuses on linear discrete-time systems and explores natural extensions to nonlinear systems. It emphasizes discrete-time systems, summarizing theoretical and practical aspects of a large class of adaptive algorithms. 1984 edition. 560pp. 6 1/2 x 9 1/4. 0-486-46932-8

INDUCTANCE CALCULATIONS, Frederick W. Grover. This authoritative reference enables the design of virtually every type of inductor. It features a single simple formula for each type of inductor, together with tables containing essential numerical factors. 1946 edition. 304pp. 5 3/8 x 8 1/2. 0-486-47440-2

THERMODYNAMICS: Foundations and Applications, Elias P. Gyftopoulos and Gian Paolo Beretta. Designed by two MIT professors, this authoritative text discusses basic concepts and applications in detail, emphasizing generality, definitions, and logical consistency. More than 300 solved problems cover realistic energy systems and processes. 800pp. 6 1/8 x 9 1/4. 0-486-43932-1

THE FINITE ELEMENT METHOD: Linear Static and Dynamic Finite Element Analysis, Thomas J. R. Hughes. Text for students without in-depth mathematical training, this text includes a comprehensive presentation and analysis of algorithms of time-dependent phenomena plus beam, plate, and shell theories. Solution guide available upon request. 672pp. 6 1/2 x 9 1/4. 0-486-41181-8

Browse over 9,000 books at www.doverpublications.com

CATALOG OF DOVER BOOKS

HELICOPTER THEORY, Wayne Johnson. Monumental engineering text covers vertical flight, forward flight, performance, mathematics of rotating systems, rotary wing dynamics and aerodynamics, aeroelasticity, stability and control, stall, noise, and more. 189 illustrations. 1980 edition. 1089pp. 5 5/8 x 8 1/4. 0-486-68230-7

MATHEMATICAL HANDBOOK FOR SCIENTISTS AND ENGINEERS: Definitions, Theorems, and Formulas for Reference and Review, Granino A. Korn and Theresa M. Korn. Convenient access to information from every area of mathematics: Fourier transforms, Z transforms, linear and nonlinear programming, calculus of variations, random-process theory, special functions, combinatorial analysis, game theory, much more. 1152pp. 5 3/8 x 8 1/2. 0-486-41147-8

A HEAT TRANSFER TEXTBOOK: Fourth Edition, John H. Lienhard V and John H. Lienhard IV. This introduction to heat and mass transfer for engineering students features worked examples and end-of-chapter exercises. Worked examples and end-of-chapter exercises appear throughout the book, along with well-drawn, illuminating figures. 768pp. 7 x 9 1/4. 0-486-47931-5

BASIC ELECTRICITY, U.S. Bureau of Naval Personnel. Originally a training course; best nontechnical coverage. Topics include batteries, circuits, conductors, AC and DC, inductance and capacitance, generators, motors, transformers, amplifiers, etc. Many questions with answers. 349 illustrations. 1969 edition. 448pp. 6 1/2 x 9 1/4.
0-486-20973-3

BASIC ELECTRONICS, U.S. Bureau of Naval Personnel. Clear, well-illustrated introduction to electronic equipment covers numerous essential topics: electron tubes, semiconductors, electronic power supplies, tuned circuits, amplifiers, receivers, ranging and navigation systems, computers, antennas, more. 560 illustrations. 567pp. 6 1/2 x 9 1/4. 0-486-21076-6

BASIC WING AND AIRFOIL THEORY, Alan Pope. This self-contained treatment by a pioneer in the study of wind effects covers flow functions, airfoil construction and pressure distribution, finite and monoplane wings, and many other subjects. 1951 edition. 320pp. 5 3/8 x 8 1/2. 0-486-47188-8

SYNTHETIC FUELS, Ronald F. Probstein and R. Edwin Hicks. This unified presentation examines the methods and processes for converting coal, oil, shale, tar sands, and various forms of biomass into liquid, gaseous, and clean solid fuels. 1982 edition. 512pp. 6 1/8 x 9 1/4. 0-486-44977-7

THEORY OF ELASTIC STABILITY, Stephen P. Timoshenko and James M. Gere. Written by world-renowned authorities on mechanics, this classic ranges from theoretical explanations of 2- and 3-D stress and strain to practical applications such as torsion, bending, and thermal stress. 1961 edition. 560pp. 5 3/8 x 8 1/2. 0-486-47207-8

PRINCIPLES OF DIGITAL COMMUNICATION AND CODING, Andrew J. Viterbi and Jim K. Omura. This classic by two digital communications experts is geared toward students of communications theory and to designers of channels, links, terminals, modems, or networks used to transmit and receive digital messages. 1979 edition. 576pp. 6 1/8 x 9 1/4. 0-486-46901-8

LINEAR SYSTEM THEORY: The State Space Approach, Lotfi A. Zadeh and Charles A. Desoer. Written by two pioneers in the field, this exploration of the state space approach focuses on problems of stability and control, plus connections between this approach and classical techniques. 1963 edition. 656pp. 6 1/8 x 9 1/4.
0-486-46663-9

Browse over 9,000 books at www.doverpublications.com

CATALOG OF DOVER BOOKS

Astronomy

CHARIOTS FOR APOLLO: The NASA History of Manned Lunar Spacecraft to 1969, Courtney G. Brooks, James M. Grimwood, and Loyd S. Swenson, Jr. This illustrated history by a trio of experts is the definitive reference on the Apollo spacecraft and lunar modules. It traces the vehicles' design, development, and operation in space. More than 100 photographs and illustrations. 576pp. 6 3/4 x 9 1/4. 0-486-46756-2

EXPLORING THE MOON THROUGH BINOCULARS AND SMALL TELESCOPES, Ernest H. Cherrington, Jr. Informative, profusely illustrated guide to locating and identifying craters, rills, seas, mountains, other lunar features. Newly revised and updated with special section of new photos. Over 100 photos and diagrams. 240pp. 8 1/4 x 11. 0-486-24491-1

WHERE NO MAN HAS GONE BEFORE: A History of NASA's Apollo Lunar Expeditions, William David Compton. Introduction by Paul Dickson. This official NASA history traces behind-the-scenes conflicts and cooperation between scientists and engineers. The first half concerns preparations for the Moon landings, and the second half documents the flights that followed Apollo 11. 1989 edition. 432pp. 7 x 10.
0-486-47888-2

APOLLO EXPEDITIONS TO THE MOON: The NASA History, Edited by Edgar M. Cortright. Official NASA publication marks the 40th anniversary of the first lunar landing and features essays by project participants recalling engineering and administrative challenges. Accessible, jargon-free accounts, highlighted by numerous illustrations. 336pp. 8 3/8 x 10 7/8. 0-486-47175-6

ON MARS: Exploration of the Red Planet, 1958-1978--The NASA History, Edward Clinton Ezell and Linda Neuman Ezell. NASA's official history chronicles the start of our explorations of our planetary neighbor. It recounts cooperation among government, industry, and academia, and it features dozens of photos from Viking cameras. 560pp. 6 3/4 x 9 1/4. 0-486-46757-0

ARISTARCHUS OF SAMOS: The Ancient Copernicus, Sir Thomas Heath. Heath's history of astronomy ranges from Homer and Hesiod to Aristarchus and includes quotes from numerous thinkers, compilers, and scholasticists from Thales and Anaximander through Pythagoras, Plato, Aristotle, and Heraclides. 34 figures. 448pp. 5 3/8 x 8 1/2.
0-486-43886-4

AN INTRODUCTION TO CELESTIAL MECHANICS, Forest Ray Moulton. Classic text still unsurpassed in presentation of fundamental principles. Covers rectilinear motion, central forces, problems of two and three bodies, much more. Includes over 200 problems, some with answers. 437pp. 5 3/8 x 8 1/2. 0-486-64687-4

BEYOND THE ATMOSPHERE: Early Years of Space Science, Homer E. Newell. This exciting survey is the work of a top NASA administrator who chronicles technological advances, the relationship of space science to general science, and the space program's social, political, and economic contexts. 528pp. 6 3/4 x 9 1/4.
0-486-47464-X

STAR LORE: Myths, Legends, and Facts, William Tyler Olcott. Captivating retellings of the origins and histories of ancient star groups include Pegasus, Ursa Major, Pleiades, signs of the zodiac, and other constellations. "Classic." – *Sky & Telescope*. 58 illustrations. 544pp. 5 3/8 x 8 1/2. 0-486-43581-4

A COMPLETE MANUAL OF AMATEUR ASTRONOMY: Tools and Techniques for Astronomical Observations, P. Clay Sherrod with Thomas L. Koed. Concise, highly readable book discusses the selection, set-up, and maintenance of a telescope; amateur studies of the sun; lunar topography and occultations; and more. 124 figures. 26 halftones. 37 tables. 335pp. 6 1/2 x 9 1/4. 0-486-42820-6

Browse over 9,000 books at www.doverpublications.com